Dielectric Materials and Applications
ISyDMA'2016

Edited by

ACHOUR Mohammed Essaid
TOUAHNI Rajaa
MESSOUSSI Rochdi
ELAATMANI Mohammed
AIT ALI Mustapha

The First International Symposium on Dielectric Materials and Applications (ISyDMA'2016) was held in Kenitra (4 May, 2016) and in Rabat (May 5-6, 2016), Morocco. ISyDMA'2016 provided an international forum for reporting the most recent developments in Advanced Dielectric Materials and Applications.

The goal of this collection of peer reviewed papers is to provide researchers and scientists from all over the world with recent developments in dielectric materials and their innovative applications. The book will be useful for materials scientists, physicists, chemists, biologists, and electrical engineers engaged in fundamental and applied research or technical investigations of such materials.

Dielectric Materials and Applications

ISyDMA'2016

Edited by

Mohammed Essaid Achour[1], Rajaa Touahni[1], Rochdi Messoussi[1], Mohammed Elaatmani[2] and Mustapha Ait Ali[2]

[1]LASTID Laboratory, Physics Department, Faculty of Sciences, Ibn Tofail University, BP. 133, 14000 Kenitra, Morocco

[2]Faculty of Science Semlalia, Cadi Ayyad Univeristy, Marrakech, Morocco

Peer review statement

All papers published in this volume of "Materials Research Proceedings" have been peer reviewed. The process of peer review was initiated and overseen by the above proceedings editors. All reviews were conducted by expert referees in accordance to Materials Research Forum LLC high standards.

Published under license by **Materials Research Forum LLC**
Millersville, PA 17551, USA

Published as part of the proceedings series
Materials Research Proceedings
Volume 1 (2016)

ISSN 2474-3941 (Print)
ISSN 2474-395X (Online)

ISBN 978-1-945291-18-0 (Print)
ISBN 978-1-945291-19-7 (eBook)

This book contains information obtained from authentic and highly regarded sources. Reasonable efforts have been made to publish reliable data and information, but the author and publisher cannot assume responsibility for the validity of all materials or the consequences of their use. The authors and publishers have attempted to trace the copyright holders of all material reproduced in this publication and apologize to copyright holders if permission to publish in this form has not been obtained. If any copyright material has not been acknowledged please write and let us know so we may rectify this in any future reprint.

Distributed worldwide by

Materials Research Forum LLC
105 Springdale Lane
Millersville, PA 17551
USA
http://www.mrforum.com

Manufactured in the United State of America
10 9 8 7 6 5 4 3 2 1

Table of Contents

Preface

The First International Symposium on Dielectric Materials and Applications (ISyDMA'2016) was held in Morocco, Kenitra city and Rabat (the capital of Morocco). The opening ceremony and first day of the conference took place May 4th, 2016 at the Faculty of Sciences, Ibn Tofaïl University, Kenitra, Morocco. The ISyDMA'2016 Symposium with keynotes & tutorials took place May 5-6, 2016, at the Mohammed VI Conference Centre, Foundation for Promoting Social Actions in Education and Training, in Rabat. ISyDMA'2016 was organized by the Moroccan Association of Advanced Materials (A2MA) in collaboration with the Laboratory Systems of Telecommunication & Decision Engineering (LASTID) of the Sciences Faculty of Kenitra, Morocco.

The aim of the ISyDMA'2016 Symposium was to provide an international forum for the discussion of current research on electrical insulation, dielectric phenomena and related topics. The conference provides an opportunity for specialists from around the world to meet and to discuss ongoing research during these three days.

The goal of the ISyDMA'2016 is to provide a platform for researchers, scientists from all over the world to exchange ideas and to hold wide ranging discussions on recent developments in dielectric materials and their innovative applications. Materials scientists, physicists, chemists, biologists, and electrical engineers engaged in fundamental and applied research or technical investigations on such materials were invited to attend the conference.

The scope of the ISyDMA'2016 meeting included the following topics:

- Dielectric properties, polarization phenomena and applications - Physics of space charge in non-conductive materials - Polymers, composites, ceramics, glasses - Biodielectrics - Nanodielectrics, metamaterials, piezoelectric, pyroelectric and ferroelectric materials.
- Dielectrics for superconducting applications - Industrial and biomedical applications - Dielectric materials for electronics and photonics - New diagnostic applications for dielectrics - New and functional dielectrics for electrical systems - Electrical conduction and breakdown in dielectrics - Surface and interfacial phenomena.

- Electrical insulation in high voltage power equipment and cables - Ageing, partial discharges and life expectancy of HV insulation - Space charge and its effects in dielectrics. Gaseous electrical breakdown and discharges.
- Impedance spectroscopy applications to electrochemical and dielectric phenomena
- High voltage insulation design using computer; based analysis - Partial discharges in insulation: detection methods and impact on ageing - Monitoring and diagnostic methods for electrical insulation degradation - Measurement techniques - Modeling and theory

We would like to extend our sincere thanks to all the sponsors for their generous assistance. We are also very grateful to the organizing committee for their efforts without which this event would not have taken place. So, it is a personal privilege to offer to the International Community some of the results of the Symposium and we acknowledge the President at Materials Research Forum LLC Thomas Wohlbier and his staff for the good job made to extend the fruitful three days of the ISyDMA'2016 Symposium to readers of this edition.

ACHOUR Mohammed Essaid
Chair of ISyDMA'1 Symposium

TOUAHNI Rajaa
Co-chair of ISyDMA'1 Symposium

MESSOUSSI Rochdi
Co-chair of ISyDMA'1 Symposium

ELAATMANI Mohammed
Reception Committee Chair
of ISyDMA'1 Symposium

AIT ALI Mustapha
Awards Committee Chair
of ISyDMA'1 Symposium

Honor Committee

EL MIDAOUI Azzeddine	President of the Ibn Tofail University, Kenitra
MIRAOUI Abdellatif	President of the Cadi Ayyad University
MERNARI Bouchaïb	President of the Sultan Moulay Slimane University
ESSAIDI Mohammed	Director of ENSIAS, President of Institute of Electrical and Electronics Engineers (IEEE), Morocco Section
ESSAMRI Azzouz	Dean of the Sciences Faculty of Kenitra
HBID Hassan	Dean of the Sciences Faculty Semlalia of Marrakech
ZEGHAL Ahmed	Dean of the Faculty of Sciences and Technics, Beni Mellal
ADDOU Mohammed	Dean of the Faculty of Sciences and Technics, Tangier

Scientific Committee

ABD EL-KADER Kamal Marei	Suez Canal University, Ismailia, Egypt
ABOU-DAKKA Mahmoud	National Research Council of Canada, INMS, Montreal , Canada
ACHOUR Mohammed Essaid	Faculty of Sciences, University Ibn Tofail, Kenitra, Morocco
ADDOU Mohammed	FST, Abelmalek Essaidi University , Tangier, Morocco
AIT ALI Mustapha	Faculty of Science Semlalia, Cadi Ayyad Univeristy , Marrakech, Morocco
ALONTSEVA Darya	East Kazakhstan State Technical University, Kazakhstan
AROF Abdul Kariem Bin Hj Mohd	University of Malaya, Kuala Lumpur, Malaysia
BAYKAL Abdülhadi	Fatih University, Faculty of Art and Sciences, Istanbul, Turkey
BENNANI Faycal	Faculty of Sciences, University Ibn Tofail, Kenitra, Morocco
BERRADA Khalid	Faculty of Science Semlalia, Cadi Ayyad Univeristy , Marrakech, Morocco
BOUTEJDAR Ahmed	Microwave and Electrical Engineering, Magdeburg University, Germany
BOUZIANE Khalid	Université Internationale de Rabat, Maroc, Morocco
BRI Seddik	EST, University Moulay Ismail, Meknes, Morocco
BROSSEAU Christian	Université de Bretagne Occidentale, France
BUKA Agnes	Wigner Research Centre of the Hungarian Academy of Sciences, Budapest, Hungary
COSTA Luís Cadillon	University of Aveiro, Portugal
ÉBER Nándor	Wigner Research Centre of the Hungarian Academy of Sciences, Budapest, Hungary
ELAATMANI Mohamed	Faculty of Science Semlalia, Cadi Ayyad Univeristy , Marrakech, Morocco
ELAISSARI Hamid	LAGEP laboratory, Claude Bernard University Lyon-1, France
ELHARFI Ahmed	Faculty of Sciences, University Ibn Tofail, Kenitra, Morocco
EL AZHARI Youssef	Centre National des Innovations Pédagogiques et d'Expérimentation (CNIPE) Rabat, Maroc
EL-KHATEEB Mohammad	University of Science and Technology, Irbid, Jordan
EL-TANTAWY Farid	Suez Canal University, Faculty of Science, Ismailia, Egypt
FAHOUME Mounir	Faculty of Sciences, University Ibn Tofail, Kenitra, Morocco
FATTOUM Arbi	Sciences Faculty of Gafsa, Tunisia
FONSECA Maria Alexandra	TEMA –Mechanical Engineering Department, University of Aveiro, Portugal
FONTANELLA John	U. S. Naval Academy in Annapolis, USA
FRECHERO Marisa Alejandra	INQUISUR -Universidad Nacional del Sur, Argentine
FUJITA Shizuo	Graduate School of Engineering, Kyoto University, Kyoto, Japan
GARCIA COLON Rodolfo	Instituto de Investigaciones Electricas Cuernavaca, Morelos, Mexico
GOUMRI-Said Souraya	Georgia Institute of Technology, Atlanta, USA
HADDAD Mustapha	Faculty of Sciences, Moulay Ismail University, Meknes, Morocco
HAMRAOUI Ahmed	CEA SACLAY, France
HENINI Mohamed	NNNC School of Physics & Astronomy University of Nottingham, UK
INGROSSO Chiara	CNR-IPCF Bari Division, Italy
JOMNI Fathi	Faculté des Sciences de Bizerte, Université de Carthage, Tunisia
KASSIBA Abdelhadi	Maine University, France
KITYK Iwan V.	Faculty of Electrical Engineering, Czestochowa, Poland
LACHGAR Abdessadek	Wake Forest University • Department of Chemistry, USA
LAHJOMRI Fouad	National School of Applied Sciences, Tangier, Morocco

LEBLANC Roger M	University of Miami, Florida, USA	
MABROUKI Mustapha	FST, Sultan Moulay Slimane University, Beni-Mellal, Morocco	
MEKHALDI Abdelouahab	Ecole Nationale Polytechnique, Algeria	
MENENDEZ Jose Luis	CINN-CSIC, Asturias, Spain	
NEBDI Hamid	Faculty of Sciences, Chouaïb Doukkali University, El Jadida, Morocco	
NOUNEH Khalid	Faculty of Sciences, University Ibn Tofail, Kenitra, Morocco	
OUERIAGLI Amane	Faculty of Science Semlalia, Cadi Ayyad Univeristy , Marrakech, Morocco	
OUTZOURHIT Abdelkader	Faculty of Science Semlalia, Cadi Ayyad Univeristy , Marrakech, Morocco	
PARK Seungyong Eugene	Nelson Mandela African Institute of Science and Technology, Arusha, Tanzania	
PINTILIE Lucian	National Institute for Scientific and Technical Creation, Bucharest, Romania	
PISSIS Polycarpos	National Technical University of Athens, Greece	
PETKOV Plamen	Department of Physics, University of Chemical Technology and Metallurgy, Sofia, Bulgaria	
PETKOVA Tamara	IEES, Bulgarian Academy of Sciences, Sofia, Bulgaria	
RAHMOUN Khadidja	Sciences Faculty, Tlemcen University, Algeria	
REMIENS Denis	IEMN – DOAE - UMR, Valenciennes University, France	
RESHAK Ali H.	Research Centre, University of West Bohemia, Pilsen, Czech Republic	
SAADAOUI Hassan	Paul Pascal Research Center (CRPP/ CNRS), Bordeaux I University, France	
SAN Sait Eren	Department of Physics, Gebze Institute of Technology, Gebze, Turkey	
STRICCOLI Marinella	CNR-IPCF Bari Division, Italy	
TACHAFINE Amina	LEMCEL, Littoral-Côte d'Opale University, France	
TEGUAR Madjid	Ecole Nationale Polytechnique, Alger, Algeria	
THALAL Abdelmalek	Faculty of Science Semlalia, Cadi Ayyad Univeristy , Marrakech, Morocco	
TRIKI Asma	Sciences Faculty, Sfax University, Tunisia	
TUNCER Enis	GE G.R.C. Niskayuna, USA	
VIGNERAS Valérie	IMS-ENSCBP-Bordeaux INP, France	
WINTERSGILL Mary	U. S. Naval Academy in Annapolis, USA	

Organizing Committee

ACHOUR Mohammed Essaid	Faculty of Sciences, University Ibn Tofail, Kenitra	Chair
AIT ALI Mustapha	FSSM , Cadi Ayyad Univeristy , Marrakech	Awards Committee Chair
BENKIRANE Hasna	Faculty of Sciences, University Ibn Tofail, Kenitra	
BERRADA Khalid	Faculty of Science Semlalia, Cadi Ayyad Univeristy , Marrakech	Co-chair
BOUJIHA Tarik	ENSA, University Ibn Tofail, Kenitra	Local Arrangements Chairs
CHOUGDALI Khalid	ENSA, University Ibn Tofail, Kenitra	
DIYADI Jaouad	Faculty of Sciences, University Ibn Tofail, Kenitra	
EBN TOUHAMI Mohamed	Faculty of Sciences, University Ibn Tofail, Kenitra	
ELAATMANI Mohamed	FSM , Cadi Ayyad Univeristy , Marrakech	Reception Committee Chair
EL BOUJLAIDI Abdelaziz	FSSM , Cadi Ayyad Univeristy , Marrakech	Posters Chair
EL HASNAOUI Mohamed	Faculty of Sciences, University Ibn Tofail, Kenitra	Publicity Chair
EL BARI Hassan	Faculty of Sciences, University Ibn Tofail, Kenitra	
ELMANSOURI Abelmajid	ISMAC, Rabat,	
ELMERABET Youssef	Faculty of Sciences, University Ibn Tofail, Kenitra	Local Arrangements Chairs
GUESSOUSS Amina	Faculty of Sciences, University Ibn Tofail, Kenitra	Registration Chair, Conference Secretary
LAHJOMRI Fouad	National School of Applied Sciences, Tangier	Program Chair
LOTFI Noureddine	Faculty of Sciences, University Ibn Tofail, Kenitra	-
MABROUKI Mustapha	FST, Sultan Moulay Slimane University, Beni-Mellal	Co-chair
MESSOUSSI Rochdi	Faculty of Sciences, University Ibn Tofail, Kenitra	Co-chair

NARJIS Abdelfatah	FSSM , Cadi Ayyad Univeristy , Marrakech	-
NOUNEH Khalid	Faculty of Sciences, University Ibn Tofail, Kenitra	Publication Chair
OUERIAGLI Amane	FSSM , Cadi Ayyad Univeristy , Marrakech	Co-chair
OUTZOURHIT Abdelkader	FSSM , Cadi Ayyad Univeristy , Marrakech	Co-chair
SRIFI Mohamed Nabil	ENSA, University Ibn Tofail, Kenitra	Proceedings Chair
TOUAHNI Rajaa	Faculty of Sciences, University Ibn Tofail, Kenitra	Co-chair
ACHOUR Safaa	Faculty of Sciences, University Ibn Tofail, Kenitra	
ADAR Mutapha	FST, Sultan Moulay Slimane University, Beni-Mellal	
ARIBOU Najoia	Faculty of Sciences, University Ibn Tofail, Kenitra	
BELLEMALLEM Salah	Faculty of Sciences, University Ibn Tofail, Kenitra	Webmasters
BOUKHEIR Sofia	FSSM , Cadi Ayyad Univeristy , Marrakech	
BOUKNAITIR Ilham	Faculty of Sciences, University Ibn Tofail, Kenitra	
CHAABI Youness	Faculty of Sciences, University Ibn Tofail, Kenitra	Webmasters
EL HAD KASSIM Saïd Ahmed Bourhane	Faculty of Sciences, University Ibn Tofail, Kenitra	
ELALLAOUI Meryem	Faculty of Sciences, University Ibn Tofail, Kenitra	
ELBOUAZZAOUI Salaheddine	Faculty of Sciences, University Ibn Tofail, Kenitra	
EL ALAOUI Imane	Faculty of Sciences, University Ibn Tofail, Kenitra	
EL JADID Sara	Faculty of Sciences, University Ibn Tofail, Kenitra	
FEDALI Achraf	FST, Sultan Moulay Slimane University, Beni-Mellal	
FELLIR Fadoua	Faculty of Sciences, University Ibn Tofail, Kenitra	
HADRI Amal	Faculty of Sciences, University Ibn Tofail, Kenitra	
KREIT Lamyaa	Faculty of Sciences, University Ibn Tofail, Kenitra	
LEKDIOUI Khadija	Faculty of Sciences, University Ibn Tofail, Kenitra	
MESTADI Walid	Faculty of Sciences, University Ibn Tofail, Kenitra	
MOQQADDEM Safaa	Faculty of Sciences, University Ibn Tofail, Kenitra	
NAJIH Youssef	FST, Sultan Moulay Slimane University, Beni-Mellal	
NIOUA Yassine	Faculty of Sciences, University Ibn Tofail, Kenitra	
SAMIR Zineb	Faculty of Sciences, University Ibn Tofail, Kenitra	
SLIMANI Khadija	Faculty of Sciences, University Ibn Tofail, Kenitra	

Dielectric Materials and Applications: ISyDMA'2016   Materials Research Forum LLC
Materials Research Proceedings 1 (2016) 1-4   doi: http://dx.doi.org/10.21741/2474-395X/1/1

Effect of particle size on dielectric and photoluminescence spectroscopy of ZnS nanoparticles

P.K. GHOSH

Department of Physics, Abhedananda Mahavidyalaya, Sainthia, Birbhum-731234, W.B, India.

*Corresponding author: E-mail: pradipghosh2002@gmail.com

Keywords: ZnS Nanoparticles, Photoluminescence, Dielectric

Abstract. Nanoparticles of ZnS in thin film form have been synthesized by radio frequency magnetron sputtering technique on glass and Si substrates at substrate temperature 300 K. X-ray diffraction and selected area electron diffraction studies confirmed the formation of nanocrystalline cubic phase of ZnS in the films. TEM micrographs of the thin films revealed the manifestation of ZnS nanoparticles with sizes lying in the range 3.00 – 5.83 nm. The room temperature photoluminescence spectra of the films showed two peaks centered around 315 nm and 450 nm. We assigned the first peak due to bandgap transitions while the latter due to sulfur vacancy in the films. The composition analysis by energy dispersive X-rays also supported the existence of sulfur deficiency in the films. The dielectric property study showed high dielectric permittivity (85-100) at a higher frequency (>5 KHz).

Introduction

The creation of nanostructure materials opens up the opportunity of observing the evolution of physical properties of the materials with sizes [1-2]. Nanosized particles of semi conducting compounds in particular display grain size dependent optoelectronic properties, due to the size quantization effects [3-4]. The materials in nanostructure form exhibit high transparency, high dielectric constant, low field electron emission, etc. Almost all these properties could be well utilized to save the energy to produce environmental friendly atmosphere [5-7]. If the solar cell with such transparent materials be made, it could be used as transparent photovoltaic which might be used as architectural windows that permits the visible light through the window and generate the electricity from UV part of it. Also high dielectric materials are utilized as source of energy, because dipole oscillation radiates energy.

Experimental

Zinc sulfide (ZnS) target was fabricated by taking a suitable aluminium holder (5 cm dia.) and compacting the ZnS polycrystalline powder (99.99 %, Aldrich) by applying suitable hydrostatic pressure (~ 100 kg / cm^2). The fabricated ZnS target was placed in the radio frequency magnetron-sputtering chamber (13.56 MHz) for the deposition of nanocrystalline thin films on various substrates such as glass and Si. The deposited films were characterized by studying mainly nanostructural, compositional, luminescence and dielectric properties. The nanostructures and diffraction patterns of the films were studied by a transmission electron microscope (TEM, Hitachi-H600). The thicknesses of the films have been measured by an ellipsometer (Nano-view SM HG 1000) and lie in the range 260 nm to 530 nm. Compositional analysis was done by energy dispersive X-ray analysis (EDX, GEOL JSM 6300 Oxford-ISIS). The photoluminescence (PL) spectra of the nanocrystalline ZnS thin films have been measured by a fluorimeter (FL 4500). The dielectric properties of the films were also studied with varying frequency by using an L-C-R meter (HP - 4284 A) at room temperature under vacuum (~10^{-3} mbar).

Dielectric Materials and Applications: ISyDMA'2016   Materials Research Forum LLC
Materials Research Proceedings **1** (2016) 1-4   doi: http://dx.doi.org/10.21741/2474-395X/1/1

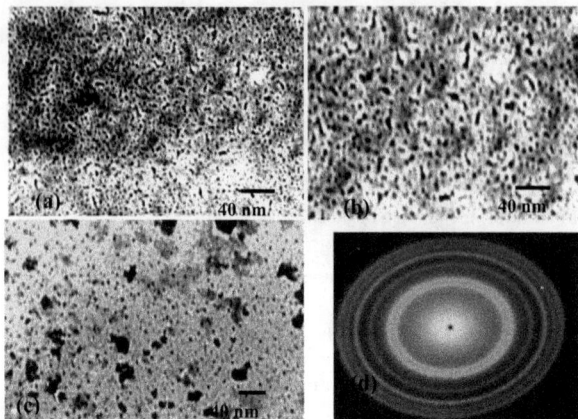

Figure 1 TEM micrographs and selected area electron diffraction pattern for (a) deposition time 10 min., (b) deposition time 20 min., (c) deposition time 30 min. and (d) SAED pattern.

Results and discussion

The nanostructures of the films were studied at room temperature by a using transmission electron microscope (TEM) as shown in Figure 1 for (a) deposition time 10 min., (b) deposition time 20 min., (c) deposition time 30 min. and (d) selected area electron diffraction (SAED) pattern. The average particles sizes were lying in the range of 3.00 – 5.83 nm and the selected area electron diffraction (SAED) pattern confirms that the materials are ZnS. Here it is observed that the particle size increases with the increase of deposition time due to increase of agglomeration probability for higher deposition time. It is also clear from the figure 1 (c) for deposition time 30 minutes, in some places there are very large particles. This is due to the fact that the huge numbers of particles are agglomerated there. The Zn/S atomic ratio in the films was determined by energy dispersive X-ray analysis (not shown here).

Dielectric property study

Figure 2 shows the variation of dielectric permittivity as a function of frequency of the applied a.c. electric field for ZnS nanoparticle thin films with different particle size (for films deposited with different deposition time) such as (a) particle size 3 nm; deposition time 10 min., (b) particle size 4.60 nm; deposition time 20 min. and (c) particle size 5.83 nm; deposition time 30 min. The dielectric permittivity for thin films of ZnS nanoparticles have been measured under vacuum (10^{-3} mbar) by using an L-C-R meter and the value of it lies in the range 85 to 100 at higher frequencies (as shown in Figure 2) which are much higher than that of bulk ZnS having dielectric constant ~9 [8]. The large value of dielectric permittivity of ZnS nanoparticle thin film is due to the fact that the nanoparticles of ZnS under the application of electric field, act as nanodipoles. As the particle size is in nanometer order, the number of particles per unit volume is very large; hence the dipole moment per unit volume increases, so the dielectric permittivity increases. Again, it is clear from the Figure 1 that the dielectric permittivity of the films decreases with increase of particle sizes. This is due to the fact that when the particle size increases, the number of nanoparticles per unit volume decreases, which implies that under application of electric field, the number of nanodipoles per unit volume decreases. Hence dielectric permittivity decreases. It has also been observed that, for each particle size of ZnS in the films, the dielectric constant is almost constant at high frequency range i.e. there is no dispersion at high frequency. At higher frequencies the nanodipoles cannot follow the rapid variations of the electric field and hence they show practically no dispersion. Bhattacharya et al [9], also obtained very high dielectric constant without dispersion in their silver nanowires formatted in a polymeric film and in the pores of a silica gel, and Ghosh et al. [10] also reported the high dielectric permittivity of PVA/CdS nanocomposites. The dielectric permittivity of ZnS nanoparticles is high at

Dielectric Materials and Applications: ISyDMA'2016   Materials Research Forum LLC
Materials Research Proceedings **1** (2016) 1-4   doi: http://dx.doi.org/10.21741/2474-395X/1/1

lower frequencies [11-12]. This may be due to the contribution of the electronic, ionic, dipolar and space charge polarizations, which depend on the frequencies (Xue et al and Suresh et al). Space charge polarization is generally active at lower frequencies and indicates the purity and perfection of the nanoparticles [11-13].

In dielectric materials, dielectric losses (not shown in this paper) usually occur due to the absorption current. The orientation of the molecules along the direction of the applied electric field in polar dielectrics requires a part of the electric energy to overcome the forces of internal friction [13-14]. Another part of the electric energy is utilized for rotations of the dipolar molecules and other kinds of molecular transfer from one position to another, which also involve energy losses.

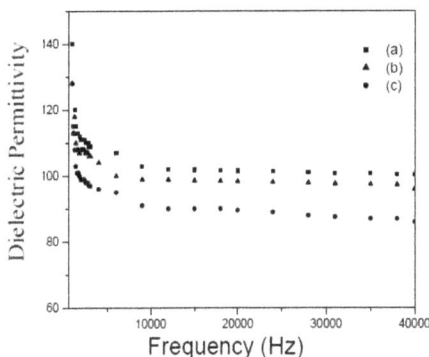

Figure 2   Variation of dielectric permittivity of nano-ZnS thin films with frequency for different deposition time (a) 10 min, (b) 20 min. and (c) 30 min.

Photoluminescence studies

Photoluminescence (PL) spectra, measured at room temperature (300 K) of the nanocrystalline ZnS thin film deposited on glass substrates are shown in Figure 3. All the plots contain two peaks centered at 315 nm and at 450 nm. The excitation wavelength was 260 nm. Appearance of the broad peak at 450 nm is comparable with the result obtained by Geng et al.[15] and Kovtyukhova et al [16]. It is evident from Figure 2 that the position of this peak is not shifted with the variation of deposition time i.e. with the variation of particle sizes. Hence the photoluminescence in this region is due to the presence of sulfur vacancies in the lattice, which is also previously reported by Sooklal et al.[17]] and Bol et al. [18]. The sulfur deficiency in the synthesized nanocrystalline ZnS films was also confirmed from EDX measurements of the samples. Also it was observed that with the increase of deposition time the intensity of the PL peak increased. This may be due to the fact that with the increase of deposition time sulfur deficiency increases which is supported by the EDX results. The broadness of the peak for our nanocrystalline ZnS can be explained [16] as follows: the photogenerated charge carriers trapped in shallow states are tunneled to each other to recombine. The emission from recombination in shallow traps appears at lower wavelength than deep traps. The broad emission band represents the superposition of wide distribution of traps distances (Spanhel et al.[19]). The other PL peak appeared at ~ 315 nm and from Figure 2 it is clear that the position of this peaks changes with the deposition time which indicate the peak position depends on the particle size. This peak is assigned due to the band-to-band transition of nanocrystalline ZnS. As the bandgap depends on the particle size, hence the position of this PL peak changes. The absorption edge of transmittance and reflectance spectra also appeared at around 315 nm which support the PL results.

Conclusions

Nanoparticles of ZnS have been successfully synthesized by rf-magnetron sputtering technique. The room temperature photoluminescence spectra of the films showed two peaks centered around 315 nm and 450 nm. EDX measurements confirmed the sulfur vacancy in the films. The dielectric property study showed high dielectric constant (85 to 100) at high frequency (>5 KHz).

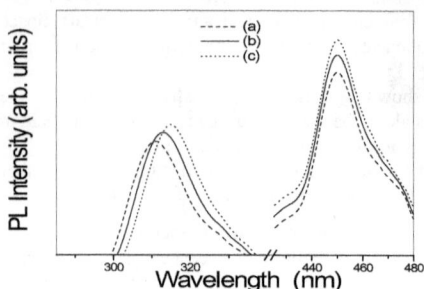

Figure 3 PL-spectra of nano-ZnS thin films deposited on glass substrates for different deposition time (a) 10 min., (b) 20 min. and (c) 30 min.

Acknowledgements: The author wishes to thank Jadavpur University and Professor K. K. Chattopadhyay of Jadavpur University for providing him the opportunity of performing this work in Thin Films and Nanoscience Laboratory and inspiring him for performing this work.

References

[1] H.J. Dai, E.W. Wong, Y.Z. Lu, S.S. Fan, C.M. Lieber, Nature (London) 375 (1995) 769. http://dx.doi.org/10.1038/375769a0

[2] P.D. Yang, C.M. Lieber, Science 273 (1996) 1836. http://dx.doi.org/10.1126/science.273.5283.1836

[3] M Bangal, S Ashtaputre, S Marathe, A Ethiraj, N Hebalkar, SW Gosavi, J Urban, SK Kulkarni, Semiconductor Nanoparticles. Hyperfine Interact. 160 (2005) 81-94. http://dx.doi.org/10.1007/s10751-005-9151-y

[4] CS Pathak, MK Mandal, V Agarwala. Mater. Sci.Semicond. Process. 16 (2013) 467-471. http://dx.doi.org/10.1016/j.mssp.2012.07.009

[5] Q.H. Wang, T.D. Corrigan, J.Y. Dai, R.P.H. Chang, A.R. Krauss, Appl. Phys. Lett. 70 (1997) 3308. http://dx.doi.org/10.1063/1.119146

[6] P.G. Collins, A. Zettl, Appl. Phys. Lett. 69 (1996) 1969. http://dx.doi.org/10.1063/1.117638

[7] R.G. Forbes, Solid State Electron. 45/6 (2001) 779. http://dx.doi.org/10.1016/S0038-1101(00)00208-2

[8] Landolt–Bronstein, Numerical Data and Functional Relationships in Science and Technology, vol. 22a, Springer Verlag, Berlin, 1987, p. 168.

[9] S. Bhattacharya, S.K Saha, D Chakravorty, Appl. Phys. Lett. 76 (26), (2000.) 3896.

[10] P. K. Ghosh, M K Mitra, K. K. Chattopadhyay, Nanotechnology, 16, (2005) 1-6. http://dx.doi.org/10.1088/0957-4484/16/1/022

[11] Xue D, Kitamura K. Solid State Commun.122 (2002) 537-541. http://dx.doi.org/10.1016/S0038-1098(02)00180-1

[12] Smyth CP Acta. Cryst. 9 (1956)838-839. http://dx.doi.org/10.1107/S0365110X56002382

[13] Sagadevan Suresh, International Journal of Physical Sciences, Vol. 8(21), pp. 1121-1127, 9 June, 2013.

[14] S. A. Mastia, A. K. Sharmab and P. N. Vasambekarc, Advances in Applied Science Research 4(4) (2013)335-339

[15] B. Y. Geng,L. D. Zhang, G. Z. Wang, T. Xie, Y. G. Zhang and G. W. Meng, Appl. Phys. Lett. 84 (2004) 2157. http://dx.doi.org/10.1063/1.1687985

[16] N. I. Kovtyukhova, E. V. Buzaneva, C. C. Waraksa and T. E. Mallouk, Materials Science and Engineering B 69-70, (2000) 411. http://dx.doi.org/10.1016/S0921-5107(99)00312-8

[17] K. Sooklal,B. S. Cullum, S. M. Angel and C. J. Murphy, J. Phys. Chem.100, (1996) 4551. http://dx.doi.org/10.1021/jp952377a

[18] A. A. Bol and A. Meijerink, Phys. Rev. B 58 (1998) 15997. http://dx.doi.org/10.1103/PhysRevB.58.R15997

[19] I. Spanhel and M. A. Anderson, J. Am. Chem. Soc. 113 (1991) 2826. http://dx.doi.org/10.1021/ja00008a004

Dielectric Materials and Applications: ISyDMA'2016   Materials Research Forum LLC
Materials Research Proceedings 1 (2016) 5-8   doi: http://dx.doi.org/10.21741/2474-395X/1/2

Studies of temperature-dependent behaviour and crystallization of $Ni_{36.3}Zr_{63.7}$ metallic glass determined by resistivity, thermopower and DSC

B. SMILI[*1,] M. MAYOUFI[1], I. KABAN[2], J G.GASSER[3], F.GASSER[3]

[1]Laboratory of Inorganic Materials Chemistry, Université Badji-Mokhtar Annaba, BP12, 23000 Annaba, Algeria.

[2]IIFW Dresden, Institute for Complex Materials, TU Dresden, Institute of Materials Science, P.O. Box 270116, 01171 Dresden, Germany

[3]Laboratoire de Chimie et Physique Approches Multiéchelles des Milieux Complexes LCP-A2MC Université de Lorraine, Institut de Chimie, Physique et Matériaux57078 cedex 3 Metz, France

*Corresponding author: E-mail: Smili.billel@yahoo.fr

Keywords: Metallic Glass, Electronic Transport Properties, Phase Transitions, Activation Energy

Abstract. In this paper, structural changes of amorphous $NiZr_2$ metallic glass will be characterized by thermal electrical resistivity, absolute thermoelectric power and by DSC measurements. A very good agreement between the phase transition temperatures determined using different techniques has been determined. In this context, the study of the amorphous $Ni_{36.3}Zr_{63.7}$ confirmed the potential of this means of investigation to study the kinetics, structural and thermal behavior of amorphous alloys. The crystallization kinetics of $Ni_{36.3}Zr_{63.7}$ metallic glass have been studied under non-isothermal and isothermal conditions using electrical resistivity measurement. The activation energies of crystallization E_x, for three measurements of "resistivity" as a function of temperature with different heating rate (0.5, 2.5 and 5) °C/min, is determined to be 334,2 kJ/mol and 344,6 kJ/mol using the Kissinger and Ozawa equations respectively. The Johnson-Mehl-Avrami equation has also been applied to the isothermal kinetics and the Avrami exponents are in the range of 2.97-3,33 with an average value of n=3,15 indicating the growth of small particles with an increasing nucleation rate. The activation energy calculated by the Arrhenius equation in the isothermal process (335°C/ 340°C/345°C) has been found to decrease with the transformed volume fraction between 10% and 90% of volume transformed, and the average value calculated to be E_x = 378,2 kJ/mol. Structural and morphology study after thermal treatment have been identified by X-ray diffraction (XRD) and scanning electron microscope (SEM).

Introduction

Metallic glasses (MG) are very important material since their discovery in 1960 by Klement and his team by rapid quenching technique of Au-Si binary alloy [1], an important research work has been developed on producing, structure, microstructure, mechanical, physical and thermal properties [2,3,4] and their applications in the industrial sector such the hydrogen storage and separation [5,6]. Amorphous membranes alloys become increasingly attractive due to their high thermal stability, excellent mechanical strength and soft magnetic properties [7,8] as compared to conventional crystalline alloys [9]. The separation of hydrogen through the amorphous membranes from contaminant gases such as carbon dioxide (CO_2) and methane (CH_4) becomes very important for the production of energy [10]. In recent years, amorphous alloy membranes Ni-Zr and Ni-Nb-Zr based alloys have been proposed to overcome the low cost of Pd-based alloys for hydrogen separation [11,12], and having excellent hydrogen permeation properties. Thermal and structural behavior studies of amorphous alloy membranes for the separation of hydrogen are focused on adapting and

Dielectric Materials and Applications: ISyDMA'2016   Materials Research Forum LLC
Materials Research Proceedings **1** (2016) 5-8   doi: http://dx.doi.org/10.21741/2474-395X/1/2

optimizing their functional properties for industrial applications. The operating temperature of amorphous alloy membrane is influenced by the glass transition temperature T_g and the crystallization temperature T_x, that must be studied to have clear and precise information on the behavior and the scope of these membranes, in order to prevent this crystallization phenomenon. The thermal behavior of MGs remained an interesting problem for basic and applied research. First, the research aims to answer fundamental questions about the nature of glass, glass transition, and devitrification. On the other hand, the understanding of the crystallization kinetics is very important with regard to the development and the thermal stability of the structure of MGs. The thermal stability of the amorphous state is a very important factor for using high temperature membranes. The resistivity and ATP (also known as or Seebeck coefficient) are very good and accurate to characterize the changes in the liquid and solid, especially the temperature dependent behavior and their kinetics of crystallization more accurate than the DSC measurement for the very large time-measuring with very small heating rate [13,14] In the present work, a ribbon of $Ni_{36.3}Zr_{63.7}$ amorphous alloy was prepared by the melt spinning technique were investigated in detail by electrical resistivity measurements, ATP, and differential scanning calorimetry under non-isothermal and isothermal conditions for the non-isothermal experiment. The activation energies were calculated by Kissinger [15] and Ozawa methods [16]. We applied Johnson-Mehl-Avrami (JMA) [17] model for the isothermal experiment to understand the mechanism of the crystallization process. The amorphous and crystalline structure in the $NiZr_2$ metallic glass is also confirmed by scanning electron microscopy (SEM) and by X-ray diffraction (XRD).

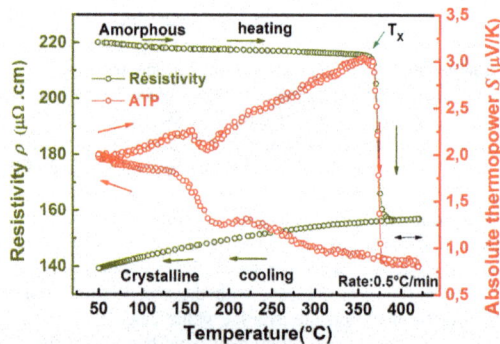

Fig. 1 Measurements of the electrical resistivity (left scale) and PTA   (right scale) of amorphous $Ni_{36.3}Zr_{63.7}$ metallic glass as function of temperature at the heating rate of 0.5 °C/min.

Results and discussion
The relaxation temperature modifies irreversibly the sample and shifts both the resistivity and the temperatures of the phase transitions. The activation energies of crystallization E_x , for resistivity as a function of temperature with different heating rate (0.5 , 2.5, and 5) °C/min , is determined to be 334,2 kJ/mol and 344,5 kJ/mol using the Kissinger and Ozawa equations respectively. The Johnson-Mehl-Avrami equation has also been applied to the isothermal kinetics and the Avrami exponents are in the range of 2.97-3,33 with an average value of n=3,15 indicating the crystallization process is mainly governed by three-dimensional growth. The activation energy calculated by the Arrhenius equation in the isothermal process has been found to decrease with the transformed volume fraction between 10% and 90% of volume transformed. Due to contributions of E_g with the progression of crystallization to a growth in three dimensions, the average value calculated to be E_x =

378,2 kJ/mol. The transformation from the amorphous state to the crystalline state has been reached completely during 90 min of treatment at 335 ° C, which correspond to the $NiZr_2$ phase tetragonal structure, (S.G I4 / mcm) with a small change in the lattice parameter, and a particle size of about 10 μm.

Conclusion

The high sensitivity of the electronic transport properties of the structural changes enabled us to find the phase transitions where the amorphous glass passes from the disordered state to a crystalline state, and study the kinetics of this passage in isothermal regime. The characterization by measuring resistivity and thermoelectric power is as an additional means of analysis equivalent to the DSC. In this context, the study of the amorphous $Ni_{36.3}Zr_{63.7}$ confirmed the potential of this means of investigation to study the kinetics and structural thermal behavior of Amorphous Alloy Membranes for hydrogen separation

References

[1] K. Klement, R.H. Willens, P. Duwez, Nature 187 (1960) 869–870. http://dx.doi.org/10.1038/187869b0

[2] M. Iqbal, W.S. Sun, H.F. Zhang, J.I. Akhter, Z.Q. Hu, Mater. Sci. Eng. A 447 (2007) 167. http://dx.doi.org/10.1016/j.msea.2006.10.039

[3] A. Inoue, A. Takeuchi, Mater. Sci. Eng. A 375–377 (2004) 16. http://dx.doi.org/10.1016/j.msea.2003.10.159

[4] M. Iqbal, Z.Q. Hu, H.F. Zhang, W.S. Sun, J. I Akhter, J. Non-Cryst.Solids 352 (2006) 3290. http://dx.doi.org/10.1016/j.jnoncrysol.2006.05.010

[5] G. Sandrock, J. Alloys Compd. 293e295 (1999) 877. http://dx.doi.org/10.1016/S0925-8388(99)00384-9

[6] V. Gorokhovsky, K. Coulter, T. Barton, S. Swapp, Thermal stability of Magnetron Sputter Deposited NiZr Alloys for Hydrogen Gas Separation, in: MS&T 2010: Proceedings from the Materials Science&Technology Conference,Houston, Texas, October 17e21, 2010, pp. 135e145. Inoue A, Koshiba H, Zhang T, Makino A. Wide supercooled liquid region and soft magnetic properties of Fe56Co7Ni7Zr0-10Nb (or Ta)0-10B20amorphous alloys. J. Appl.Phys. 1998;83(4):1967e74.

[7] Kimura H, Inoue A, Yamaura S-I, Sasamori K, Nishida M, Shinpo Y, et al. Thermal stability and mechanical properties of glassy and amorphous Ni-Nb-Zr alloys produced by rapid solidification. Mater. Trans., JIM 2003;44(6):1167-71.

[8] V. S. Vasantha, H.-S. Chin and E. Fleury, "Corrosion properties of Ni-Nb & Ni-Nb-M (M = Zr, Mo, Ta & Pd) metallic glasses in simulated PEMFC conditions" inJ Phys.: Con! Ser. 144 (2009) 012008. http://dx.doi.org/10.1088/1742-6596/144/1/012008

[9] Dolan MD, Dave NC, Ilyushechkin AY, Morpeth LD, McLennan KG. Composition and operation of hydrogen-selective amorphous alloy membranes. J. Membr. Sci. 2006; 285:30-55. http://dx.doi.org/10.1016/j.memsci.2006.09.014

[10] Hara, S. et aI, "An amorphous alloy membrane without noble metals for gaseous hydrogen separation" in J Membr. Sci (2000), vol. 164, pp.289-294. http://dx.doi.org/10.1016/S0376-7388(99)00192-1

[11] S.1.Yamaura et aI., "Hydrogen permeation and structural features of melt-spun Ni-Nb-Zr amorphous alloys" in Acta Materialia (2005), vol. 53, pp. 3703-3711. http://dx.doi.org/10.1016/j.actamat.2005.04.023

Dielectric Materials and Applications: ISyDMA'2016   Materials Research Forum LLC
Materials Research Proceedings 1 (2016) 5-8   doi: http://dx.doi.org/10.21741/2474-395X/1/2

[12] L. Abadlia, F. Gasser, K. Khalouk, M. Mayoufi, and J. G. Gasser, "New experimental methodology, setup and LabView program for accurate absolute thermoelectric power and electrical resistivity measurements between 25 and 1600 K: Application to pure copper, platinum, tungsten, and nickel at very high temperatures," Rev. Sci. Instrum., vol. 85, no. 9, 2014. http://dx.doi.org/10.1063/1.4896046

[13] O. Haruyama, N. Annoshita, H. Kimura, N. Nishiyama, A. Inoue, J. Non-Cryst, Solids 312–314 (2002) 552–556.

[14] H.E. Kissinger, Anal. Chem. 29 (1957) 1702–1706. http://dx.doi.org/10.1021/ac60131a045

[15] T. Ozawa, J. Therm. Anal. 2 (1970) 301–324. http://dx.doi.org/10.1007/BF01911411

[16] M. Avramin, Kinetics of phase change I: general theory, J. Chem. Phys. 7 (1939) 1103-1112. http://dx.doi.org/10.1063/1.1750380

Dielectric Materials and Applications: ISyDMA'2016
Materials Research Proceedings 1 (2016) 9-12

Materials Research Forum LLC
doi: http://dx.doi.org/10.21741/2474-395X/1/3

Study of physicochemical and dielectric proprieties of dicationic ionic liquids: the effect of the nature of the cation

Boumediene HADDAD*[1], El-habib BELARBI[2], Taqiyeddine MOUMENE[2],
Mustapha RAHMOUNI[2], Didier VILLEMIN[3]

[1]Department of Chemistry, Dr Moulay Tahar Univ Saida , Algeria.

[2]Synthesis and Catalysis Laboratory LSCT, Univ Tiaret, Algeria.

[3]LCMT, ENSICAEN, UMR 6507 CNRS, University of Caen, 6 bd MI Juin, 14050 Caen, France.

*Corresponding author. E-mail: haddadboumediene@yahoo.com

Keywords: Dicationic Ionic Liquids (DILs), Tetrafluoroborate, Dielectric Relaxation Spectroscopy (DRS), Conductivity

Abstract. In this paper a comparative study of three dicationic ionic liquids (DILs) is presented, these latter are characterized by sharing tetrafluoroborate [2BF$_4$$^-$] as anion and have three different structure piperidinium based cations [R$_1$(CH$_2$)$_n$PPI^{2+}]; namely: bis-methyl piperidinium propylidene [MPrPPI^{2+}], bis-methyl piperidinium butylidene [MBPPI^{2+}]and bis-ethyl piperidinium butylidene [EBPPI^{2+}], their structures were characterized by using ^{1}H, ^{13}C, ^{19}F-NMR and FT-IR techniques. In the frequency range [from 10^{-2} to 10^6 Hz] and over the temperature range from [212 to 248 K], real dielectric (ε'), imaginary dielectric (ε''), conductivity (σ) and complex electrical modulus (M*)-frequency (ω) relationships of these three ILs were investigated using dielectric spectroscopy and electric modulus formalism. In this context, the dielectric permittivity, conductivity and the activation energy values are observed to be increased respectively; on one hand, with the spacer length from [MPrPPI^{2+}] to [MBPPI^{2+}] and on the other hand, with the alkyl chain length attached to the cationic from [MBPPI^{2+}] to [EBPPI^{2+}] of the dicationic ionic liquids. Also, the results suggest that the effect of alkyl chain length is more important to the spacer length effect on the dielectric permittivity and conductivity proprieties. Finally, the activation energy is determined in order to understand the relaxation processes in our DILs.

Introduction

Nowadays, either in industry or in academic research field, the highlights are put on the new type of dicationic ionic liquids. Compared to the traditional monocationic ionic liquids, the dicationic DILs are new structures characterized by the presence of a doubly charged cation that is composed of two singly charged cations linked by an alkyl chain and paired with two singly charged anions [1-2]. This work is a follow up of our previous study [3-5], in which we focused on the understanding of chemical structure effect on the dielectric studies of three tetrafluoroborate dicationic ionic liquids (DILs) containing two central cationic unit , propylidene and butylidene capped by a basic functionality (piperidine), with methyl and ethyl as length of the side alkyl chain which share tetrafluoroborate [2BF$_4$$^-$] as anion. The structures of these compounds were identified by using ^{1}H, ^{13}C, ^{19}F, NMR and FT-IR spectroscopy.

The real permittivities, the modulus analysis and the conductivity have been studied using dielectric spectroscopy in the temperature range from 212 to 248 K and in the frequency range between 10^{-2} to 10^6 Hz. Moreover, the activation energy for the relaxation process is determined.

Experimental

2.1. Reagents and materials

The reagents used in this study are: 1,3-dibromopropane (98 wt.%), 1,4-dibromobutane (99.5 wt.%), N-methyl piperidine (95 wt.%), N-ethyl piperidine (97 wt.%), ammonium tetrafluoroborate (99.5

wt.%), diethyl ether and N,N-dimethylformamide. They were purchased from Fluka and used as received. Deionized H_2O was obtained by using a Millipore ion-exchange resin deionizer.

2.2.1. Synthesis and characterization by using NMR and infrared spectroscopy measurements:
All these details measurements have been reported in [4].
Figure 1 shows the chemical structure of the three DILs investigated:

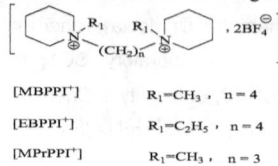

[MBPPI$^+$] $R_1=CH_3$, n = 4

[EBPPI$^+$] $R_1=C_2H_5$, n = 4

[MPrPPI$^+$] $R_1=CH_3$, n = 3

Figure 1. Chemical structure of the three DILs investigated.

A metathesis reaction of [MBPPI^{2+}][2Br$^-$] (2.07 g, 5 mmol) with ammonium tetrafluoroborate (1.04 g, 10 mmol) gives the desired compound using domestic microwave oven. The spectra details are given below:

[MPrPPI^{2+}][2BF$_4^-$]: ^{1}H-^{13}C NMR reported in [4]; IR ν(cm^{-1}) : νCH$_3$(C-H) : 2948, δ(C-H) : 1459,1427, ν(C-N) : 1285, ν(**B-F**) : 1051, ν(C-H) : 763; ^{19}F-RMN (D$_2$O) δ_F (ppm) : -150.73 (s, BF$_4^-$).

[MBPPI^{2+}][2BF$_4^-$]: ^{1}H-^{13}C NMR reported in [4]; IR ν(cm^{-1}): 2952, ν (C-H) : 1461, ν(C-N) : 1280, ν(**B-F**) : 1023, ν(C-H) : 764; RMN ^{19}F (DMSO) δ_F (ppm) : -148.25 (s, BF$_4^-$). δ

[EBPPI^{2+}][2BF$_4^-$]: ^{1}H-^{13}C NMR reported in [4]; IR ν(cm^{-1}) : νCH$_3$(C-H) : 2951, δ(C-H) : 1457, ν(C-N) : 1273, ν(**B-F**) : 1027, ν(C-H) : 763; ^{19}F-RMN (DMSO): -149.60 (s, BF$_4$).

2.3. Measurement of dielectric properties
These details have been described elsewhere [4].

The complex electric modulus can be represented as the reciprocal of the dielectric permittivity as given by the following equations:

$$M^*(\omega, T) = \frac{1}{\varepsilon^*} = j(\omega C_0)Z^* = M'(\omega, T) + j M''(\omega, T) \quad (1)$$

The real component M' and the imaginary component M'' were calculated from ε' and ε'' using the following expressions:

$$M'(\omega, T) = \frac{\varepsilon'}{(\varepsilon'^2 + \varepsilon''^2)} \quad (2)$$

$$M''(\omega, T) = \frac{\varepsilon''}{(\varepsilon'^2 + \varepsilon''^2)} \quad (3)$$

These complex functions are mutually related according to the equation
$1/\varepsilon^*(\omega) = j\omega C_0 Z^*(\omega) = M^*(\omega) = j\omega\varepsilon_0/\sigma^*(\omega).$ (4)

Results and discussion
3.1. Frequency dependent permittivity studies
In the order to understand the origin of dielectric losses and relaxation time, the study of the dielectric properties analysis was used.

Figure 2 (a and b) represent typical frequency dependences spectra of real $\varepsilon'(\omega)$ and imaginary part $\varepsilon''(\omega)$ of the dielectric permittivity measured for [MBPPI^{2+}][2BF$_4^-$] DILs at various temperatures. The real part of the dielectric function ε' at $f_c = (\omega_c/2\pi)$ turns from the high frequency limit to the static value ε_s [6].

We observed that ε' increases respectively with both; the spacer length from propylidene to butylidene in [MBPPI^{2+}] and [MPrPPI^{2+}] and the alkyl chain length from methyl to ethyl in [MBPPI^{2+}] and [EBPPI^{2+}]. In general, this cation sequence is [EBPPI^{2+}] > [MBPPI^{2+}] > [MPrPPI^{2+}]. For example, a close inspection of the variation of the real permittivity $\varepsilon'(\omega)$ for three DILs (fig 2 (a) at temperature T=212 k and at lower frequencies (f=0.1Hz) as can be seen, at ε' of [MPrPPI^{2+}] is (18.585), lower than (30.12) that of ε' of [MBPPI^{2+}], the latter is about 6.3 times lower than

Dielectric Materials and Applications: ISyDMA'2016
Materials Research Proceedings 1 (2016) 9-12

Materials Research Forum LLC
doi: http://dx.doi.org/10.21741/2474-395X/1/3

(202.87). of [EBPPI^{2+}]. These results suggest that the effect of alkyl chain length is more important than the spacer length effect on the increase of ε'.

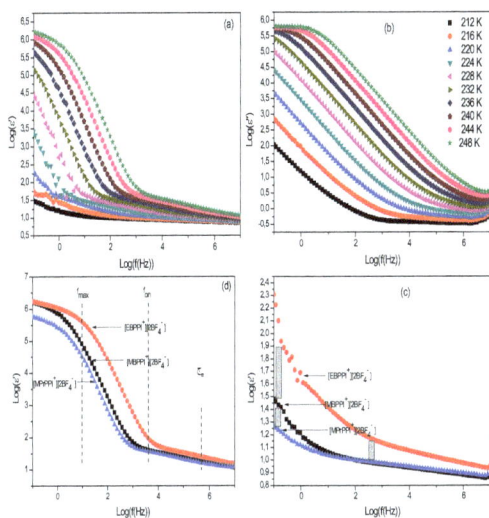

Figure 2. Frequency dependence of (a) real permittivity ε'(ω)and (b) dielectric loss ε'' (ω) at several temperatures of [MBPPI^{2+}][2BF$_4^-$]. Comparison of the frequency dependence of (c) real permittivity ε'(ω) at 246 K and (d) real permittivity ε'(ω) at 212 K of the three ILs.

3.2. Frequency dependent conductivity and Electric modulus studies

In order to understand and obtain information to their ion conductivity, the relaxation mechanism and study the space charge relaxation phenomena in our DILs, the real and imaginary conductivity (σ' and σ''), complex modulus formalism has been used vs. the frequency are given at various temperatures in Fig. 3. For each temperature, M' reaches a constant value at high frequencies. At low frequencies, M' approaches to zero, fact that confirm the presence of an appreciable electrode and/or ionic polarization in the studied temperature range.

For the data analysis, we observe that the conductivity increases as the temperature is elevated whatever applied frequency.

Moreover, when the frequency increases, M' increases too, wherein this latter attains its maximum value, M$_\infty$, at higher frequencies due to relaxation processes. It is also observed that the value of M$_\infty$ decreases with an increase in temperature. These results are similar with those reported by Saroj et al [7] about studying the behavior of PVA/ [EMIM][EtSO4] ionic liquids. They explained the increasing of the real and imaginary parts of modulus by the bulk effect, and indicated that the electrode polarization phenomenon makes small contribution.

Conclusion

In this work, three dicationic ionic liquids having tetrafluoroborate ([2BF$_4^-$]) anions and three different structures cations ([MPrPPI^{2+}], [MBPPI^{2+}], [EBPPI^{2+}]) were successfully synthesized.

The results of dielectric relaxation experiments on three dicationic ionic liquids, performed between 212 K and 248 K in the frequency range (10^{-2} to 10^6 Hz) are presented.

The influence of the effect of alkyl chain length compared with the spacer length effect on the conductivity and dielectric proprieties was investigated by dielectric measurements.

The increases of the spacer length from bis-methyl piperidinium propylidene [MPrPPI^{2+}] to bis-methyl piperidinium butylidene [MBPPI^{2+}] and in the alkyl chain length from this latter to bis-ethyl piperidinium butylidene [EBPPI^{2+}], leads to increase of the dielectric real permittivity and the conductivity. Especially, the values of the real dielectric permittivity and the conductivity of [EBPPI^{2+}] shows the best values compared with those of [MPrPPI^{2+}] and [MBPPI^{2+}] due to the important of the effect of alkyl chain length. Finally, the temperature dependence of the relaxation times was shown to be governed by the VFT equation in all three DILs and the activation energy

Dielectric Materials and Applications: ISyDMA'2016   Materials Research Forum LLC
Materials Research Proceedings 1 (2016) 9-12   doi: http://dx.doi.org/10.21741/2474-395X/1/3

values deduced from the VFT curves confirm our previous hypothesis about the link between the conductivity and the length of the spacer and the alkyl chains in the studied DILs.

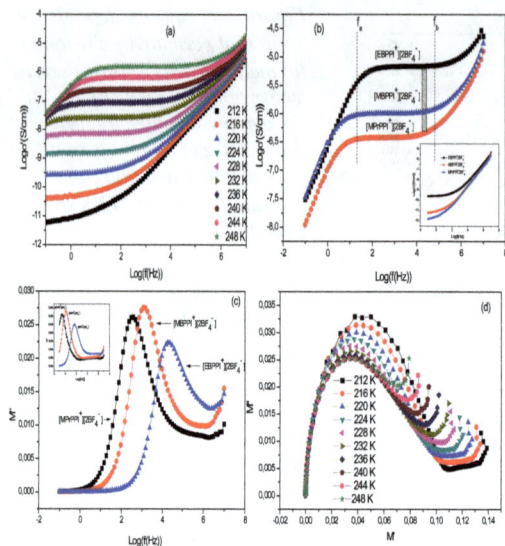

Figure 3. *Frequency dependence of the conductivity ((a) real conductivity σ'(ω) of [MBPPI^{2+}][2BF$_4^-$] at various temperatures. (b)Frequency dependence of the real conductivity σ'(ω) of three DILs at 246 K. (c) Frequency dependencies of the real part (M') of electrical modulus at T=248 K, (d) Complex modulus spectrum at different temperatures.*

Acknowlegment

The authors thank professor Kremer from Institute of Experimental Physics I, University of Leipzig for the dielectric measurements.

References

[1] Welton, T. (1999). Room-temperature ionic liquids. Solvents for synthesis and catalysis. *Chemical reviews, 99*(8), 2071-2084. http://dx.doi.org/10.1021/cr980032t

[2] Kim, J. Y., Kim, T. H., Kim, D. Y., Park, N. G., & Ahn, K. D. (2008). Novel thixotropic gel electrolytes based on dicationic bis-imidazolium salts for quasi-solid-state dye-sensitized solar cells. *Journal of Power sources, 175*(1), 692-697. http://dx.doi.org/10.1016/j.jpowsour.2007.08.085

[3] Haddad, B., Villemin, D., Belarbi, E. H., Bar, N., & Rahmouni, M. (2014). New dicationic piperidinium hexafluorophosphate ILs, synthesis, characterization and dielectric measurements. *Arabian Journal of Chemistry, 7*(5), 781-787. http://dx.doi.org/10.1016/j.arabjc.2011.01.002

[4] Haddad, B., Villemin, D., & Belarbi, E. H. (2012). Synthesis, Differential Scanning Calorimetry (DSC) and Dielectric Relaxation Spectroscopy (DRS) Studies of N-methyl-N-propylpiperidinium Bis (trifluoromethylsulfonyl) imide. *Environ. Sci, 3*, 312-319.

[5] Haddad, B., Moumene, T., Villemin, D., Lohier, j. f., & Belarbi, E. H.(2016). Bis-methyl imidazolium methylidene bis (trifluoromethanesulfonyl) imide, crystal structure, thermal and dielectric studies. *Bulletin of Materials Science*, 39, 797-801. http://dx.doi.org/10.1007/s12034-016-1193-z

[6] Sangoro, J. R., & Kremer, F. (2011). Charge transport and glassy dynamics in ionic liquids. *Accounts of chemical research, 45*(4), 525-532. http://dx.doi.org/10.1021/ar2001809

[7] Saroj, A. L., & Singh, R. K. (2012). Thermal, dielectric and conductivity studies on PVA/ionic liquid [EMIM][EtSO 4] based polymer electrolytes. Journal of Physics and Chemistry of Solids, 73(2), 162-168. http://dx.doi.org/10.1016/j.jpcs.2011.11.012

Dielectric Materials and Applications: ISyDMA'2016   Materials Research Forum LLC
Materials Research Proceedings 1 (2016) 13-16   doi: http://dx.doi.org/10.21741/2474-395X/1/4

Complex impedance study of carbon nanotubes/polyester polymer composites

Z. SAMIR*[1], Y. EL MERABET[1], M. P. F.GRAÇA[2], S. SORETO TEIXEIRA[2],
M. E. ACHOUR[1], L. C. COSTA[2]

[1]LASTID Laboratory, Department of Physics, Faculty of Sciences, University Ibn Tofail, BP 133, 14000 Kenitra, Morocco

[2]I3N and Physics Department,University of Aveiro, 3810-193 Aveiro, Portugal

*Corresponding author e-mail:zineebsamir@gmail.com

Keywords: Carbon Nanotubes, Electrical Conductivity, Cole-Cole

Abstract. The dielectric relaxation characteristics of different concentrations of carbon nanotubes loaded in polyester polymer matrix has been studied as a function of frequency over a wide range (100 Hz–10^6Hz). The effect of filler loading and frequency dependence on the real and imaginary parts of the impedance was explained by the relaxation process. Spectral curve of impedance was fitted using the Cole-Cole dielectric relaxation function. The calculated relaxation parameters of this model and the distribution function of relaxation times are calculated and presented.

Introduction

With great technologic and scientific progress in the twentieth century, the need for materials with high-properties has become prominent. During the last decade, carbon nanotube (CNT) has attracted interest of researchers due to the fact that it is a promising nanosized filler which enhances the thermal, mechanical and electrical properties of different materials [1-5]. Discovered by Iijima [1-5], CNT enable the dissipation of electrostatic charges for antistatic applications [6] at a very low filler concentration. Indeed, the insulator–conductor transition, also called percolation threshold [7], can be reached with this type of particles in a semi-crystalline polymer matrix with less than 1% in weight [8]. This percentage depends mainly on the polymer matrix, particle dispersion state and particle apparent aspect ratio [9].

In this paper, we report the dielectric relaxation using the impedance spectroscopy technique for different concentration of CNT loaded in polyester polymer matrix. The experimental data have been analyzed using the complex impedance Cole-Cole formalism [10].

Experimental

Multiwalled CNT was purchased from Cheap-Tubes, USA, presenting average diameter about 50 nm, length in the range from 10 to 20 μm and purity higher than 95 wt%. The matrix used in this study was unsaturated polyester resin154TB, containing 31 wt% of styrene monomer, requiring 30 min for gelation at room temperature (298 K), and was obtained from Cray Valley/Total, USA. The multiwalled carbon nanotubes/polyester composite (CNT/PS) has been prepared by mixing in beaker, with 5.87 g of liquid polyester resin, and 0.2% weight of cobalt octanone as a reaction activator. Multiwalled CNT was introduced in different fractions, before adding 1% of hardener to make each mixture cohesive. Each sample of the CNT/PS was stirred at room temperature and then placed into a mold for 5 minutes to obtain a gel. After a few hours we have unmolded the samples, however it took several weeks for the complete polymerization. Several samples, with nominal CNT volume concentrations in the range of 0.0% up to 3.0%, were prepared.

Impedance spectroscopy measurements were carried out in the frequency range from 100 Hz to 10^6Hz, in a helium atmosphere, using an Agilent 4294A precision impedance analyzer, in the Cp-Rp configuration, at room temperature.

Dielectric Materials and Applications: ISyDMA'2016   Materials Research Forum LLC
Materials Research Proceedings 1 (2016) 13-16   doi: http://dx.doi.org/10.21741/2474-395X/1/4

The samples were placed between two parallel plate electrodes and the impedance $Z^*(\omega) = Z'(\omega) - iZ''(\omega)$ was measured.

Results and discussion

Fig 1 shows the variation of the imaginary part (Z'') of the impedance as a function of frequency for the seven samples (0.0, 0.2 and 0.5 % below the percolation threshold x_c, and 0.8, 1.0, 2.0 and 3.0 above x_c). Based on our previous work [11], it was found that the percolation threshold is about 0.6 %. From this figure, two important phenomena have been detected. First, for the concentration 0.0 and 0.2 % there is an absence of any relaxation peak in the measured frequency range. The second, (Z'') values reach a maximum peak (Z''_{max}) and then decreases with increasing frequency indicating the presence of dipolar relaxation in the polymer nanocomposite [12-15].

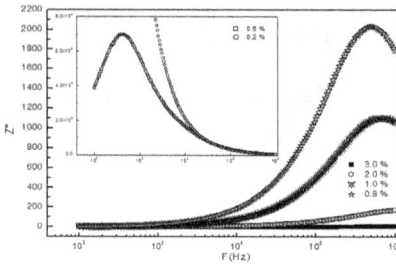

Fig. 1. Imaginary part of the complex impedance, for different CNT concentrations.

In order to study the behaviour of the relaxation frequency, we used the Cole-Cole representation, as depicted in Fig. 2, for x=0.8 %. The observed arcs are not centered on the Z'-axis, suggesting that a single relaxation time, as described by the Debye equation, cannot account for this electrical relaxation. We have analyzed the data with the Cole–Cole function [10]:

$$Z^*(\omega) = Z_\infty + \frac{Z_S - Z_\infty}{1 + (i\omega\tau)^{1-\alpha}} \tag{1}$$

Z_∞ is the high frequency resistance, Z_S is the low frequency resistance, τ the relaxation time and α a parameter between 0 and 1 that reflects the dipole interaction. When α is close to 0, the material is more homogeneous. An angle of depression can be defined as [16].

$$\phi = \alpha\pi/2 \tag{2}$$

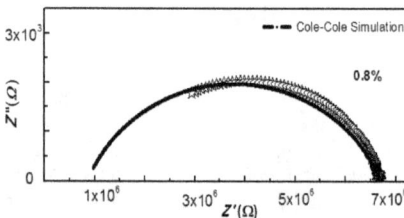

Fig. 2 Nyquist plot for (x=0.8%) concentration of CNT.

In Fig 3, we present the calculated relaxation parameters, $\Delta z = z_S - z_\infty$, ϕ and the relaxation frequency F_{max}, as a function of CNT concentration. It can also be seen from the figure, that the strength of resistance, Δz, decreases with the concentration x, indicating the increase in the electrical conductivity. At the same time, the relaxation frequency increases. The calculated values of ϕ are between 0.40 for a concentration of 0.5% and 0.75 for the concentration 3.0%, responsible for the depressed semicircle in impedance plane, which indicates an increase in heterogeneity with the concentration of the conducting particles[18]. The values of the relaxation frequencies obtained by simulation for all concentrations above x_c are between 5×10^5 and 10^6 Hz .

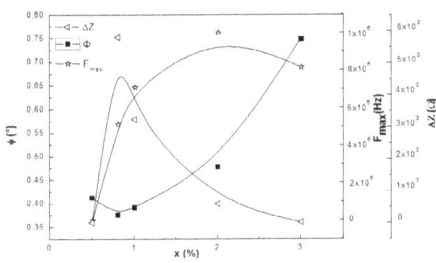

Fig. 3. Dielectric relaxation parameters obtained by simulation of the Cole-Cole equation of different CNT concentrations.

The Departure from the ideal Debye relaxation observed in the relaxation process can be construed as due to a distribution of relaxation times. Considering τ the relaxation time, we can write:

$$Z^*(\omega) = Z_\infty + (Z_S - Z_\infty) \int_0^\infty \frac{G(\tau)}{1+i\omega\tau} d\tau \tag{3}$$

where $G(\tau)$ is the distribution function of time constants, and $G(\tau)d(\tau)$ is the probability of finding a Debye element in the differential time element. The distribution must be normalizable:

$$\int_0^\infty G(\tau)d\tau = 1 \tag{4}$$

For the Cole–Cole model, the derived distribution function is [15]:

$$G(\tau) = \frac{1}{2\pi} \frac{\sin(\alpha\pi)}{\cosh[(1-\alpha)\log(\frac{\tau}{\tau_{cc}})]-\cos(\alpha\pi)} \tag{5}$$

We present in Fig 4 the calculated distribution functions for three concentrations of the conducting filler. It is more suitable to use the variable $\log(\tau / \tau_{cc})$ in the abscissa axis, where this model presents a symmetry with respect to the central frequency or relaxation time. With the increasing concentration of nanotube carbon particles, the system becomes more heterogeneous, with a progressive departure from the Debye relaxation process.

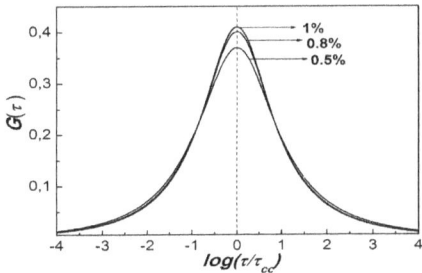

Fig. 4. Relaxation times distribution functions for different concentration of CNT.

Conclusion

The impedance spectroscopy technique was used to interpret the relaxation processes for different concentrations of CNT loaded in polyester polymer matrix, in the frequency range from (100 Hz–10^6 Hz). The results were analyzed using the Cole-Cole model. The calculated relaxation parameters and the distribution function of relaxation times show that when the filler concentration increases, the system becomes more heterogeneous, with a progressive departure from the Debye relaxation process.

Acknowledgment

The authors acknowledge the support from National Center for Scientific and Technical Research (CNRST) and FCT-CNRST bilateral cooperation and FEDER by funds through the COMPETE 2020

Programme and National Funds through FCT - Portuguese Foundation for Science and Technology under the project.

References

[1] K. Zhang, Davis M, Qiu JJ, Hope-Weeks L, Wang SR. Nanotechnology 2012:23.

[2] GT . Pham, YB. Park, SR . Wang, ZY. Liang, B. Wang, C. Zhang, et al. Nanotechnology2008:19.

[3] K. Zhang, SR. Wang. Carbon 2014;69:46. http://dx.doi.org/10.1016/j.carbon.2013.11.055

[4] K. Zhang, Y. Zhang, SR. Wang. Carbon 2013;65:105. http://dx.doi.org/10.1016/j.carbon.2013.08.005

[5] Li, Wang SR, D. Hui, JJ. Qiu. Compos Part B Engineering 2015;71:40.S. Iijima, Nature 354 (1991) 56.

[6] J.-M. Thomassin, I. Huynen, R. Jerome, C. Detrembleur, Polymer 51 (2010) 115. http://dx.doi.org/10.1016/j.polymer.2009.11.012

[7] G. Stauffer, Introduction to the Percolation Theory, Taylor and Francis, London, 1985. http://dx.doi.org/10.4324/9780203211595

[8] D. Carponcin, E. Dantras, G. Aridon, F. Levallois, L. Cadiergues, C. Lacabanne, Compos. Sci. Technol. 72 (2012) 515. http://dx.doi.org/10.1016/j.compscitech.2011.12.012

[9] Balberg, C.H. Anderson, S. Alexander, N. Wagner, Phys. Rev. B 30 (1984) 3933. http://dx.doi.org/10.1103/PhysRevB.30.3933

[10] K.S. Cole, R.H. Cole, J. Chem. Phys. 9 (1941) 341. http://dx.doi.org/10.1063/1.1750906

[11] H.H.S. Javadi, F. Zuo, K.R. Cromack, M. Angelopoulos, A.G. Mac, J. Epstein Synth. 279 Met. 29 (1989) 409–416. 280. http://dx.doi.org/10.1016/0379-6779(89)90326-3

[12] F. Zuo,M. Angelopolous, A.G.Mac, J. Epstein Phys. Rev. B 39 (1989) 3570–3578. 281. http://dx.doi.org/10.1103/PhysRevB.39.3570

[13] A.N. Papathanassiou, J. Grammatikakis, S. Sakkopoulos, E. Vitoratos, E. Dalas, J. 282 Phys. Chem. Solids 63 (2002) 1771–1778. 283. http://dx.doi.org/10.1016/S0022-3697(01)00264-5

[14] S. Banerjee, A. Kumar, J. Non-Cryst. Solids 358 (2012) 2990–2998. http://dx.doi.org/10.1016/j.jnoncrysol.2012.07.033

[15] A.K. Jonscher, Dielectric Relaxation in Solids, Chelsea Dielectric Press, London, 1983. p. 98.

[16] L.C. Costa, M.E. Achour, M.P.F. Grac¸a, M. El Hasnaoui, A. Outzourhit, and A. Oueriagli, Non-Cryst. Solids., 356, 270 (2010). http://dx.doi.org/10.1016/j.jnoncrysol.2009.11.008

Dielectric Materials and Applications: ISyDMA'2016   Materials Research Forum LLC
Materials Research Proceedings 1 (2016) 17-20   doi: http://dx.doi.org/10.21741/2474-395X/1/5

Numerical simulation of the microwave complex permittivity of carbon black particles filled polymer composites

Yassine NIOUA[*1], Salahddine EL BOUAZZAOUI[1], Mohammed Essaid ACHOUR[1], Fouad LAHJOMRI[2]

[1]LASTID Laboratory, Physics department, Faculty of Science Ibn Tofail University, B.P:133, 14000 Kenitra, Morocco

[2]ENSA, Abdelmalek Essaadi University, LabTIC, Tanger, Morocco

*Corresponding author: E-mail: yassinenioua200@gmail.com

Keywords -Dielectric properties, Complex permittivity, Mixing laws, Composites

Abstract. This article reports on a study of the dielectric properties of carbon black particles in a polymer matrix at microwave frequencies (0.5- 9.5 GHz). The experimental data was compared to the calculated values obtained by McLachlan and Lichteneker and Rother's Laws. These both laws did not fit accurately the experimental data for the high concentrations of carbon particles. To overcome this difficulty, these models with adjustable parameters were used with good agreement between the experimental results and calculated values. Nevertheless, the adjustable parameter is not universal and it depends on the frequency.

Introduction

Modeling the dielectric response of conductor-insulator composites materials is one of challenge facing research working in the area of the heterogeneous materials. For answer this problem many theoretical model have been proposed for predicting the complex permittivity [1]. In our previous work we show that effective medium theories correctly account for the experimental results at low conducting particle concentrations [2]. However at concentrations higher than a few percent, these laws fail to interpret experimental results and all tentative results must take into account parameters such as the particle size, their distribution, and the existence of agglomerates. In this study, we present the results of an experimental study on the dielectric behavior of the carbon black (CB)/epoxy composites materials at the microwave frequencies (0.5–9.5 GHz.). The variations of the complex permittivity with the conducting particles concentration is analyzed by the prediction of the general effective medium (GEM) equation proposed by McLachlan [3] which takes into account the morphology parameters such as a critical volume fraction and exponents α and β of the equation (1), and the Lichteneker and Rother's equation. We show that the proposed models give good results and prove a better fits with the experimental data.

Mixture Laws

Characterization of a non-homogeneous dielectric medium by the dielectric functions is not so obvious. In this direction many theoretical approaches derived from the mixing laws and effective medium theories have been developed and proposed to predict the complex permittivity $\varepsilon^* = \varepsilon'- i\varepsilon''$ like the Maxwell-Garnett [4] and Bruggeman [5] equations. Thus these approaches are strictly valid just for small volume fraction, but have a drawback when the volume fraction of spheres increases [6]. For this paper and in order to minimize the discrepancies observed between experimental data and the calculated values for our sets of sample we consider two models used in the literature for predict the complex permittivity. The first model given by McLachlan, and nouns as general effective medium (GEM) [3] that depends on α and β for characterizing the effective complex permittivity ε^*_{eff} .

Dielectric Materials and Applications: ISyDMA'2016   Materials Research Forum LLC
Materials Research Proceedings 1 (2016) 17-20   doi: http://dx.doi.org/10.21741/2474-395X/1/5

$$(1-\phi).\frac{\varepsilon_m^{1/\alpha}-\varepsilon_{eff}^{1/\alpha}}{\varepsilon_m^{1/\alpha}-A\varepsilon_{eff}^{1/\alpha}}+\phi.\frac{\varepsilon_c^{1/\beta}-\varepsilon_{eff}^{1/\beta}}{\varepsilon_c^{1/\beta}-A\varepsilon_{eff}^{1/\beta}}=0 \qquad (1)$$

The notation used in this equation for two phases of composite where ϕ is the volume fraction of the inclusion , ε_c the complex permittivity for the filler phase and ε_m the complex permittivity for the matrix. The exponent α describe the divergence behavior of dc conductivity when $\phi \leq \phi_c$ [6], the exponent β characterizes the ac and dc conductivity for $\phi \geq \phi_c$ and $A=(1-\phi_c)/\phi_c$. The second model is the lichteneker and Rother's law [7], given by the following equation.

$$\varepsilon_{eff}^k = \phi.(\varepsilon_c^*)^k + (1-\phi).(\varepsilon_m^*)^k \qquad (2)$$

k is a constant within the range [-1,1] , and describes a specific topology of a composite materials. For k= ± 1, one recovers the well known Wiener's bounds which represent the harmonic and arithmetic means of the intrinsic permittivities. If k is set to 1/3, one recovers an equation developed by Looyenga, independently [6].

Experemental
The conductor-insulator composites that have been examined in this study consisted of conducting particles (produced by CABOT Co.) embedded in an insulating epoxy resin matrix DGEBF: Diglycidylic Ether of Bisphenol F (from CIBA GEIGY Co), whose DC conductivity is $1.40.10^{-14}$ ($\Omega.m$)$^{-1}$ and density is 1.19 (g.cm^{-3}). The series of samples: Monarch 700/Epoxy resin polymer composites were prepared by mixing the desired volume concentration ϕ of carbon powder with the fluid resin. This series have a static percolation threshold $\phi_c = 8$ %. The scattering parameters of the sample corresponding to reflection (S11) and transmission (S21) of a transverse electromagnetic (TEM) wave were measured by the coaxial method using a Hewlett-Packard HP8753 A network analyzer in the frequency range of 0.5–8.0 GHz.The real ε' and imaginary ε'' parts of the complex permittivity, $\varepsilon = \varepsilon'-i\varepsilon''$, were determined from the relative complex scattering. In the other side the measurements of the complex permittivity at 9.5 GHz were Characterizing by the cavity resonator. The experimental measurement procedure was described earlier in our previous work [2].

Results and Discussion
For predict correctly the complex permittivity we consider the McLachlan equation reported in equation (1), with the complex permittivity of the resin epoxy is $\varepsilon_m^* = \varepsilon_m' - j\varepsilon_m''$ has a value for each frequency and the complex permittivity for the carbon Black (CB) was estimated by Drude model $\varepsilon_c^* = 0 - j\sigma_c / \varepsilon_0.\omega$ [8][9] in which σ_c is the electrical conductivity $\sigma_c = 1500(\Omega.m)^{-1}$. Our method is to apply the TEPPE at a given frequency. In the above prescription, the only remaining free parameters are α and β that give the best fit. This problem was solved by using the nonlinear Levenberg - Marquardt iterative algorithm with the Newton– Gauss type of a constrained least-square minimization. Figure 1 report the experimental and calculated values using the McLachlan model for different frequencies. We show that the McLachlan model gives a good agreement with experimental data.We notice that α and β depends strongly with the frequency and the best fits can be obtained for $3.2 \leq \alpha \leq 5.1$ and $4.3 \leq \beta \leq 5.24$. These values are very different to those proposed by McLachlan s ≈ 0.8 and t ≈ 2 which make the description of the dielectric properties using the McLachlan equation is inappropriate.

Dielectric Materials and Applications: ISyDMA'2016   Materials Research Forum LLC
Materials Research Proceedings 1 (2016) 17-20   doi: http://dx.doi.org/10.21741/2474-395X/1/5

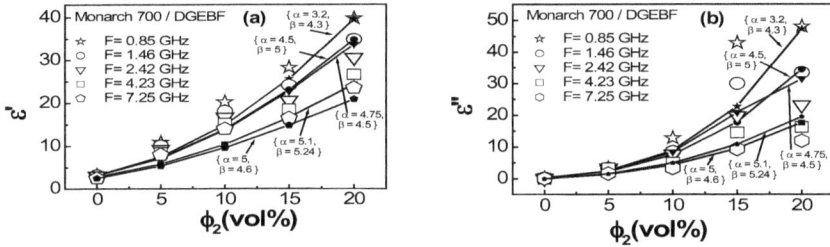

Fig 1: Real part ε' (a) and the imaginary part ε'' (b) of the effective complex permittivity as a function of CB volume fraction lines are the calculated values using the McLachlan equation and symbols are the experimental data for different frequencies respectively.

Lichteneker and Rother's model given in equation 2 is also used to calculate the complex permittivity of the composites materials with different volume fraction of CB for five microwave frequencies. The exponent k is used as an adjustable parameter. In figures 2 and 3 we present the values of ε' and ε'' at the different frequencies, solid lines reports the calculated value using the proposed model. We show that the Lichteneker and Rother's model gives a good agreement with the experimental data and the best fits can be obtained for $k = 0.15$ concerning the frequency range 0.5 – 8 GHz and $k = 0.02$ for 9.5 GHz. These values show that k depends on the frequency and that is a contradiction with the literature because the k value describes a specific morphology [7] and for that the Lichteneker and Rother's model fail to explain the experimental data. Overall these models derived from mixing laws and effective medium theories fail to interpret experimental data because many parameters such as particle size, their distribution, and also the interaction between the components of composites system [10] [11] that are not taken into consideration. Thus it is important to revisit this issue, to minimize this discrepancy observed between the experimental data and theoretical results by proposed models which provides physical insight to the complex permittivity of composite systems.

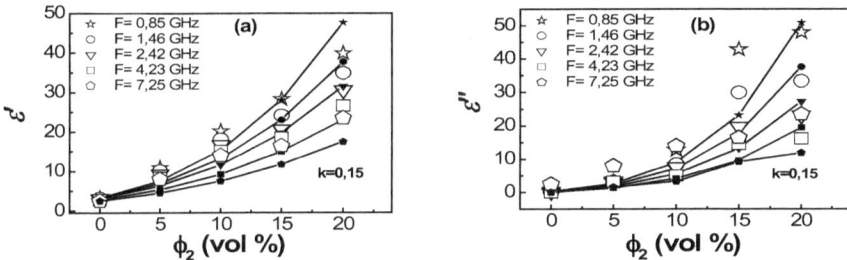

Fig 2: Real part ε' (a) and the imaginary part ε'' (b) of the effective complex permittivity as a function of CB volume fraction, line are the calculated values using the Lichteneker and Rother' equation and symbols are the experimental data for the range of frequencies

Dielectric Materials and Applications: ISyDMA'2016
Materials Research Proceedings 1 (2016) 17-20

Materials Research Forum LLC
doi: http://dx.doi.org/10.21741/2474-395X/1/5

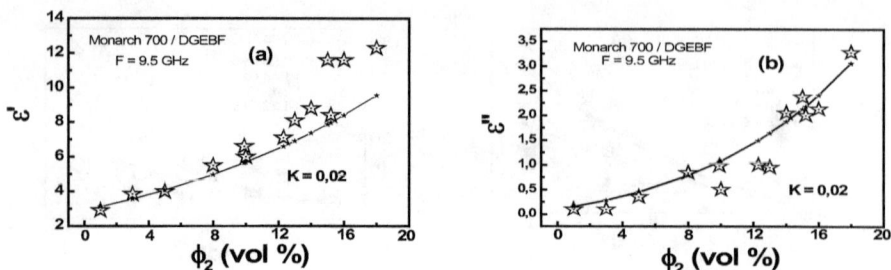

Fig 3: Real part ε' (a) and the imaginary part ε'' (b) of the effective complex permittivity as a function of CB volume fraction, line are the calculated values using the Lichteneker and Rother' equation and symbols are the experimental data at 9.5 GHz.

Conclusion

This work reports on a study of the dielectric properties of carbon black particles in a polymer matrix at microwave frequencies using the McLachlan and Lichteneker and Rother's models. These both models given by the Equations (1) and (2), with adjustable parameters, were used, with good agreement between the experimental results and calculated values. Nevertheless, the exponent parameters are not universal and depending on the frequency. For that the proposed laws are inappropriate and fail to interpret experimental results and all tentative results must take into account parameters such as the particle size, their distribution, and the existence of agglomerate.

References

[1] M. Sahimi, Heterogeneous Materials I: Linear Transport and Optical Properties (Springer, New York, 2003).

[2] M.E.Achour " Electromagnetic properties of carbon black filled epoxy polymer composites", in/Prospects in filled polymers engineering: mesostructure, elasticity network, and macroscopic properties Brosseau, C. ed. Transworld Research Network. (2008)

[3] D.S. McLachlan, "The complex permittivity of emulsions: an effective media-percolation equation", Solide State Comunications, Vol 72 ,p. 831 (1989) .

[4] M.Garnett, J.C.Philos, 'Colours inMetal Glasses and in Metallic Films'Trans.R.Soc.London., Vol. 203, pp. 385-420 . (1904). http://dx.doi.org/10.1098/rsta.1904.0024

[5] D.A.G.Bruggeman, Ann phys, Vol. 24, P. 636. (1935).

[6] S. El Bouazzaoui, M.E. Achour,C. Brosseau, 'Microwave effective permittivity of carbon black filled polymers: Comparison of mixing law and effective medium equation predictions'. J.Appl. Phys., Vol 110, 074105, (2011). http://dx.doi.org/10.1063/1.3644947

[7] A.V. Goncharenko, V.Z. Lozovski, E.F. Venger "Lichtenecker's equation: applicability and limitations," Optics communications, Elsevier, Vol. 174 , pp. 19-32, 15 junuary 2000.

[8] M. Essone Mezeme,S. El Bouazzaoui,M. E. Achour,2 C. Brosseau, JOURNAL OF APPLIED PHYSICS 109, 074107 (2011). http://dx.doi.org/10.1063/1.3556431

[9] Achour, M.E., El Malhi,M., Miane, J.L., Carmona, F., Lahjomri, F. (1999). *J. Appl. polym. Sci.*, Vol .73, pp. 969-973. http://dx.doi.org/10.1002/(SICI)1097-4628(19990808)73:6<969::AID-APP14>3.0.CO;2-1

[10] H.T.Vo, F.G.Shi, Microelectronic journal. 33, 409-415 (2002).

[11] M.G.Todd, F.G.Shi, J.Appl. Phys. 94, 7 (2003). http://dx.doi.org/10.1063/1.1604961

Dielectric Materials and Applications: ISyDMA'2016
Materials Research Proceedings 1 (2016) 21-24

Materials Research Forum LLC
doi: http://dx.doi.org/10.21741/2474-395X/1/6

Dielectric properties of high density polyethylene-nano-composites

Ossama GOUDA[1], Ahmed S. HAIBA[2]

[1]Professor of High Voltage, Cairo University, Faculty of Engineering, Cairo, Egypt

[2]Lecturer Assistant, National Institute of Standards (NIS), Cairo, Egypt

Corresponding author: E-mail: Prof_ossama11@yahoo.com

Keywords: Polymer Nano-Composites, High Density Polyethylene, Permittivity, Tan Delta

Abstract. Recently, polymer Nano-composites have been used for various important industrial applications. This paper studies the dielectric properties of high density polyethylene HDPE composed with different concentrations of Clay-Nano-Filler over a frequency range of 200 Hz-2 MHz. The experimental results show that 6% filler concentration is the optimum clay content for HDPE/Clay Nano-Filler material.

Introduction

It is concluded by several researches that Polymer Nano-composites have better dielectric and electrical insulation properties [1-3]. These composites consist of more than one material and at least one of them must be a Nano-filler having Nano-size dimensions (1-100) nm which completely dispersed into the industrial polymers. A small amount of Nano-materials increases the surface area of the material and creates a large interfacial interaction between them and the neat polymer [1]. Consequently, Nano-fillers are able to change the properties of the dielectrics to be improved and enhanced such as increasing thermal stability, decreasing the permittivity and decreasing the conductivity, in addition for improving the mechanical properties at low Nano-filler concentrations (1-10%) .

X- ray diffraction analysis (XRD) for prepared samples

The XRD analysis is divided into two categories. One of them shows the modification process by comparison between inorganic and modified clay as shown in Fig. 1. The other category shows the dispersion of modified clay-nanoparticles into the polymer matrix by comparison between modified clay and HDPE/Clay composites as shown in Fig. 2. XRD patterns in Fig. 1 and Fig. 2 show that all samples have a diffraction peak after $°2$ theta = 20. In Fig. 1, montmorillonite clay (MMT) has a characteristic peak at $°2$ theta = 26.57 with d-spacing $2.573°A$ while modified clay has a characteristic peak at $°2$ theta = 26.7045 with d-spacing 3.355 $°A$. From the above results, it is concluded that clay is successfully modified.

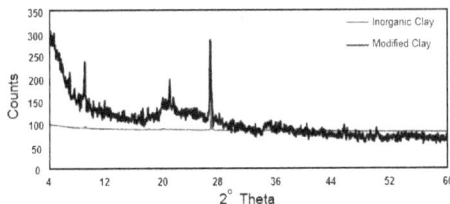

Fig. 1. XRD pattern of clay and modified clay

Dielectric Materials and Applications: ISyDMA'2016   Materials Research Forum LLC
Materials Research Proceedings 1 (2016) 21-24   doi: http://dx.doi.org/10.21741/2474-395X/1/6

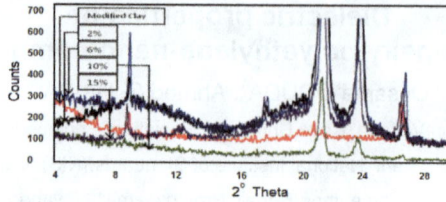

Fig. 2. XRD patterns of modified clay and other Nano-composites

Scanning electron microscopy (sem)

The thermal stability of the prepared samples was measured using thermo-gravimetric analyzer (TGA) as given in Fig. 3. The morphology of the SEM images for HDPE with 2% clay, 6% clay, 10% clay, and 15 % clay composites is shown in Fig.4. As it is noticed SEM images for samples revealed that, clay was dispersed in polymer matrix very well and there was not any accumulation of clay Nano-filler in it.

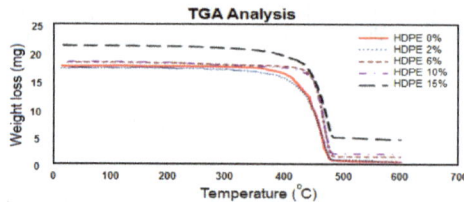

Fig. 3 *TGA curves for HDPE and HDPE/clay composites*

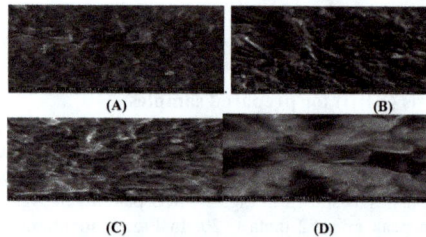

Fig. 4 SEM images for HDPE at (200x & 40000x) magnifications for (A) 2% clay, (B) 6% clay, (C) 10% clay and (D) 15% clay

High density polyethylene-nano-composites performance

Dielectric Breakdown Strength

From the tests carried out on samples of HDPE at 25 °C room temperature and 50 % relative humidity it is noticed that the optimum value of breakdown voltage (36.1 kV) is recorded at clay-Nano-filler concentration of 6% then, the breakdown voltage decreases at 10%, and 15% clay-Nano-fillers. The results are given in Fig.5

Dielectric Materials and Applications: ISyDMA'2016   Materials Research Forum LLC
Materials Research Proceedings **1** (2016) 21-24   doi: http://dx.doi.org/10.21741/2474-395X/1/6

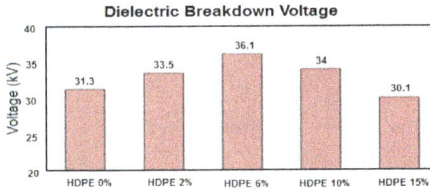

Fig. 5 Dielectric breakdown strength measurements for HDPE/clay composites

HDPE Relative Permittivity (εr)

Testing of 100 samples of HDPE, shows that εr decreases considerably with the addition of clay-Nano filler up to 6% filler concentration, and then it increases at 10% and 15% filler concentrations. The value of εr at HDPE 10% is still lower than that of pure HDPE and its value at HDPE 15%, is higher than the pure material. The results are given in Fig. 6

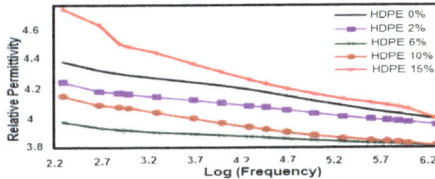

*Fig. 6 Relative permittivity variation with frequency at 23.7 °C room temperature and 50 %
relative humidity*

HDPE Dissipation Factor (Tan δ)

100 samples of HDPE are tested to study the effect of Clay-Nano-Filler concentrations on Tan δ at different frequencies at room temperature 23.7 °C and 50 % relative humidity. As given in Fig. 7. Tan δ decreases with increasing the concentrations of clay Nano-filler incorporated in polymeric material up to 6% filler concentration. On the other hand, the values of Tan δ for 15% filler concentration are higher than that of pure material. This means that clay Nano-filler improves the dissipation factor for HDPE polymeric material.

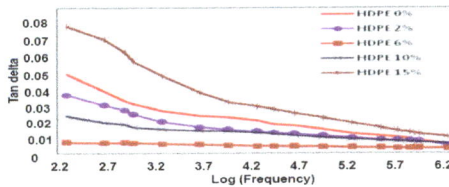

Fig. 7 Frequency dependence of tan delta at room temperature

HDPE Volume Impedance

100 HDPE samples are tested for measuring the insulation impedance with different frequencies at 23.7 °C room temperature and 50 % relative humidity. An important observation is that the insulation impedance increases at the same frequency with the addition of clay-Nano filler up to 6% filler concentration, and then it decreases at 10% and 15% filler concentrations.

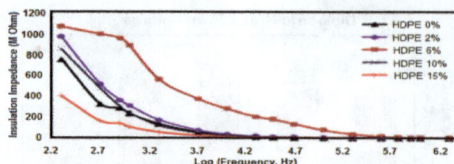

Fig. 8 The insulation impedance in ohms variation with variable frequencies at 23.7 °C room temperature and 50 % relative humidity

Conclusions

HDPE/Clay composite is prepared by melting flow rate method. Its dielectric properties over a frequency range of 200 Hz to 2 MHz have been investigated. From the investigation of obtained results, it is concluded that 6% filler concentration is the optimum clay content for HDPE/clay system.

References

[1] F. Carmona, "Conducting Filled Polymers", Physica A, Vol. 157, pp. 461-469, 1989. http://dx.doi.org/10.1016/0378-4371(89)90344-0

[2] Y. Bai, Z. -Y. Cheng, V. Bharti, H. S. Xu and Q. M. Zhang, "High dielectric-constant ceramic-powder polymer composites", Appl. Phys. Letters, Vol. 76, No. 25, pp. 3804-3806, 2000. http://dx.doi.org/10.1063/1.126787

[3] S. S. Ray and M. Okamoto, "Polymer/layered silicate nanocomposites: a review from preparation to processing", Prog. Polym. Sci., Vol. 28, pp. 1539-1641, 2003. http://dx.doi.org/10.1016/j.progpolymsci.2003.08.002

Dielectric Materials and Applications: ISyDMA'2016
Materials Research Proceedings 1 (2016) 25-30

Materials Research Forum LLC
doi: http://dx.doi.org/10.21741/2474-395X/1/7

Vegetable oil electrical properties analysis for power insulation use

O. IDIR*[1], A. REFFAS[1], S. AMEUR[1], F. CHETIBI[1], H. MOULAI[1], I. KHELFANE[2],
M. OUAGUENI[2], A. BEZAZ[2]

[1]USTHB University of Sciences and Technology Houari Boumediene; LSEI/FEI, BP 32, Bab Ezzouar, Algiers, Algeria

[2]Centre of Research and Development of Electricity and Gas (CREDEG)-SONELGAZ, 36 Route de Ouled Fayet, Algiers, Algeria

* Corresponding author: E-mail :omaranubis@hotmail.com

Keywords: Olive Oil, Transformers, Dielectric Liquids, Thermal Aging

Abstract. Transformer industry called upon insulating oils, among them mineral oils used since more than one century, and PCBs which are prohibited because of their high toxicity and non-biodegradability and it could be source of contamination of soils and waterways, in the other hand petroleum based mineral oils are going to run out according to many studies. For these reason many studies undertaken by researchers and industrial groups allowed improvements of the performances of mineral oils, and the design of new liquids of substitution to replace the PCBs such synthetic esters and silicones. Although they present good properties, they remain nonrenewable sources and poorly biodegradable substances, hence the necessity to seek other alternative oils that resolve both problems of biodegradability and durability, criteria that vegetable oils correspond perfectly. In this paper, we report results of the behavior of a vegetable oil (olive oil) subjected to a thermal stress of 110°C in an oven for 716 hours duration by checking the evolution of the electric and physicochemical properties in regular intervals of time. The results so obtained show a decrease in flashpoint and water content inducing an increase of breakdown voltage. However, a trend of increase in total acidity and viscosity are observed.

Introduction

Power transformers are key elements in the transmission and distribution of electrical energy, and thermal and electrical stresses which they are subjected to cause generally faults which will have a direct impact on the economy, in addition to those on the environment due to insulating oils of petroleum origin whose composition is toxic and non-biodegradable.

With this in mind, many studies have been carried out in view of a progressive or total replacement of mineral oils [1-4] currently in service with more environmentally friendly oils.

Vegetable oils benefiting from a high moisture saturation point, high flash point and virtually no toxicity are potential candidates to spend mineral oils [5].

The purpose of this work is to study the physical and chemical properties of a non-inhibited vegetable oil (olive oil) subject to thermal accelerated aging at 110 ° C for 716 hours.

Experimental setup

The vegetable oil used in this work is a virgin quality olive oil, produced and commercialized by the food group IFRI. All samples have the same initial properties, and then they were heated at 110°C using an oven. At regular intervals of time, a sample is taken for analysis of physicochemical properties according to international standards, They are carried out according to IEC 156 as regards the measurement of dielectric strength, the IEC 247 regarding the loss factor Tanδ, the relative permittivity ε_r and the resistivity ρ, the IEC 814 as regards the water content, ASTM D1500 regarding color index, kinematic viscosity is measured according to ASTM D445, flash point according to ASTM D93, and acidity according to NF ISO 6618.

Dielectric Materials and Applications: ISyDMA'2016   Materials Research Forum LLC
Materials Research Proceedings **1** (2016) 25-30   doi: http://dx.doi.org/10.21741/2474-395X/1/7

Results

Color Index

The color index provides information on changing the visual appearance of the oil due to the presence of degradation by-products. Figure 1 shows a slight decrease from 1.5 to 1.2 of the color index. This change can be in part due to the elimination of absorbing substances of light spectrum including chlorophyll and carotene [5, 6].

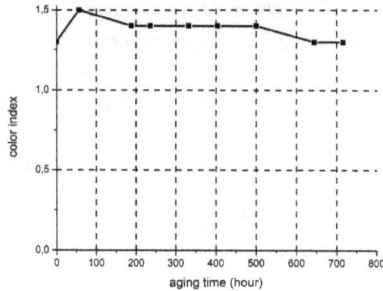

Fig. 1. Variation of color index as a function   of aging time

Breakdown voltage

At the beginning of the aging time, the breakdown voltage increases slightly to reach a value of 38.9 kV and decreases until 22.5 kV, then it increases again to reach a value of 79,9kV. After 716 hours, the breakdown voltage has increased from 27 kV to more than 79kV as shown in Figure 2. This can be the consequence of contaminating agents removal, in particular water which is a very important factor in determining the breakdown voltage [8, 9].

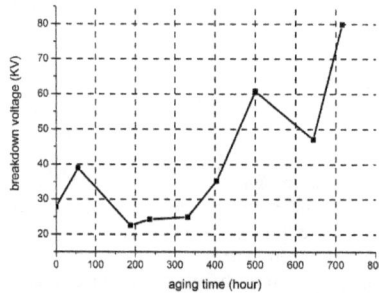

Fig. 2. Variation of breakdown voltage as a function of aging time

Water content

There has been a decrease in water content from 752 ppm to 105 ppm after 716 hours of aging as shown in Figure 3. The origin of this decrease in water may be due to the heat at which the oil was subjected during the tests, and the hydrolysis reaction (saponification) of triglyceride esters under the catalytic action of acids and temperature [10].

Dielectric Materials and Applications: ISyDMA'2016   Materials Research Forum LLC
Materials Research Proceedings 1 (2016) 25-30   doi: http://dx.doi.org/10.21741/2474-395X/1/7

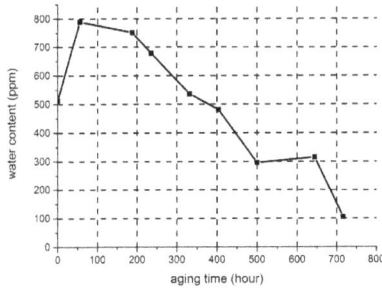

Fig. 3. Variation of water content as function of time aging

Permittivity
The permittivity is a fundamental property of dielectric materials; it mainly depends on the structure and chemical composition of the considered dielectric. Figure 4 shows the variations of the permittivity as a function of the aging time. This property, reflecting the polar aspect of materials, shows a decrease from 3.28 to 2.8. It provides useful information on the contamination of polar molecules. It can be correlated with the decrease in water content previously found as water is a highly polar liquid. Its elimination can induce a decrease in permittivity.

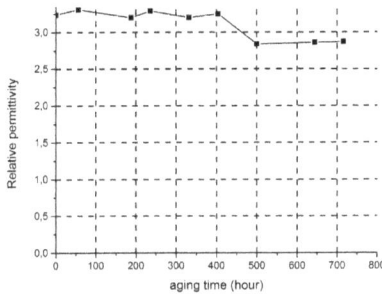

Fig. 4. Variation of permittivity as function of time aging at 50 Hz

Dissipation factor
The dissipation factor (tan δ) is expressed by a dimensionless number. It depends mainly on the conductivity of the oil. For the investigated samples as shown in figure 5, the measured loss factors are relatively high compared to the recommended ones for mineral oils. This is mainly due to the presence of contaminating agents resulting from different mechanisms of degradation of the vegetable oil and the presence of high amounts of water which initiates conduction processes [11].

Conductivity/Resistivity
Insulations are supposed to not conduct electrical currents, but by applying low voltages, current appears in phase with the applied voltage due to the movement of free charges. This characteristic is the conductivity and inversely the resistivity. The latter varies between 3 and 6.5 GΩm versus time aging without clear trend. If we consider the conductivity (the inverse of resistivity) variations depicted in figure 6, the tendency is perfectly the same than the factor dissipation one. This clearly indicates that these losses are due essentially to conduction.

Dielectric Materials and Applications: ISyDMA'2016   Materials Research Forum LLC
Materials Research Proceedings **1** (2016) 25-30   doi: http://dx.doi.org/10.21741/2474-395X/1/7

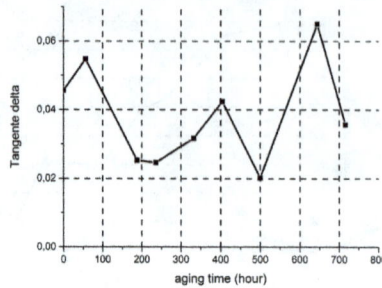

Fig. 5. Variation of loss factor (tangent delta) as function of time aging at 50 Hz

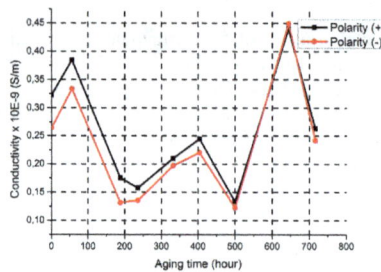

Fig. 6. Variation of conductivity as a function of time aging.

Acidity number

An increase of the acidity number according to the aging time is observed from 0.4 to 1.9 at the end of the tests as shown in the figure 7. This is associated to the hydrolysis of triglyceride esters [12]. A sharper increase between 200 to 400 hours is recorded, which is partly due to the consumption of natural antioxidant and also to the hydrolysis mechanism catalyzed by the presence of metals traces in the chlorophyll [7] and acidity of the medium (autocatalysis) [13].

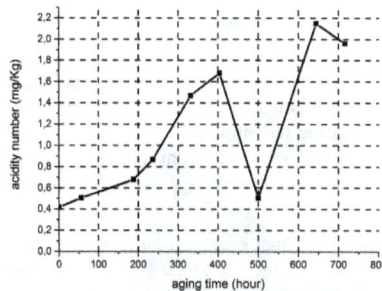

Fig. 7. Variation of acidity number as a function of time aging

Materials Research Forum LLC
doi: http://dx.doi.org/10.21741/2474-395X/1/7

Viscosity

Figure 8 shows the variations of viscosity as a function of the aging time. The curve shows an increase from 36 to 41 cSt. It is mainly due to the crosslinking phenomenon of fatty acids catalyzed by the temperature and the acidity of the medium [14].

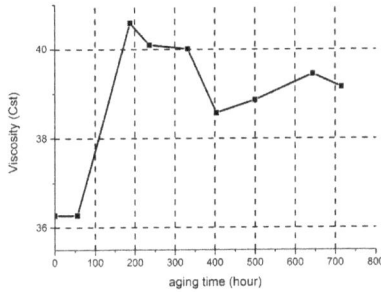

Fig. 8. Variation of viscosity as a function of time aging

Flash point

The flash point is a property that must be taken into account, and supervising its evolution helps to protect against fire disaster of the high voltage equipments including transformers. The vegetable oil flash point is higher than that of mineral oils, but according to the results reported in figure 9, it has decreased from 228°C to 130°C after aging time. This decrease can be due to the presence of volatile particles, low molecular weight and easily flammable as methane CH_4, ethane C_2H_6 and the acetylene C_2H_2 [15, 16]. In the case of vegetable oils, the presence of aldehydes as oxidation products of unsaturated fatty acids in high quantities can also affect the flash point [17].

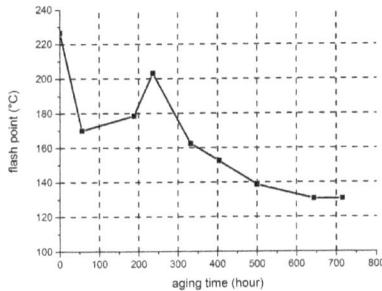

Fig. 9. Variation of flash point as a function of aging time.

Conclusion

The global result of this work is that the submission of the oil to temperature 110°C for a period of 716 hours affects significantly all the physical and chemical properties, which reduces the water content in the oil and increases the voltage breakdown. But in opposition, the temperature promotes chemical reactions that can affect heavily the other physicochemical properties (acidity, viscosity, dissipation factor and flash point).

References

[1] U. U. Abdullahi, R. Yunus, and A. Nordin, "Use Of Natural Vegetable Oils As Alternative" *The Institution of Engineers Journal*, vol. 67, no. 2, pp. 4–9, 2006.

[2] I. L. Hosier, A. Guushaa, A. S. Vaughan, and S. G. Swingler, "Selection Of A Suitable Vegetable Oil For High Voltage Insulation Applications" *Journal of Physics*. 2009. http://dx.doi.org/10.1088/1742-6596/183/1/012014

[3] C. TRAN DUY, "Propriétés Diélectriques De Liquides Isolants D'origine Végétale Pour Applications En Haute Tension,"Thèse de doctorat, Université JOSEPH Fourier-Grenoble 1, 2009.

[4] C. P. Mcshane, J. Corkran, K. Rapp, J. Luksich "Natural Ester Dielectric Fluid Development Update"in Power and Energy Society General Meeting, pp. 1–5, 2009.

[5] IEEE Std C57.147, IEEE Guide for Acceptance and Maintenance of Natural Ester Fluids in Transformers,July2008.

[6] I. L. Hosier, A. Guushaa, E. W. Westenbrink, C. Rogers, A. S. Vaughan, and S. G. Swingler, "Aging of Biodegradable Oils and Assessment of their Suitability for High Voltage Applications," *IEEE Transactions on Dielectrics and Electrical Insulation*, vol. 18, no. 3, pp. 728–738, 2011. http://dx.doi.org/10.1109/TDEI.2011.5931059

[7] H. Benabid, "Caracterisation De L'huile D'olive Algerienne, Apports Des Méthodes Chimiométriques" These de Doctorat, Universite Mentouri De Constantine, 2009.

[8] T. Suzuki, R. Oba, A. Kanetani, T. Kano, and T. Tamura, "Consideration on the Relationship between Dielectric Breakdown Voltage and Water Content in Fatty Acid Esters," in *IEEE conference on dielectric liquids*, pp. 10–13, 2011. http://dx.doi.org/10.1109/icdl.2011.6015424

[9] M. Amanullah, S. M. Islam, C. Sammer, and I. Gary, "Evaluation of Several Techniques and Additives to De-moisturise Vegetable Oils and Bench Mark the Moisture Content Level of Vegetable Oil-based Dielectric Fluids." In international conference of dielectric liquids, 2008.

[10]J. T. Mullen and N. Carolina, "Biodegradable electrical insulation fluids," in *Electrical Insulation Conference*, pp. 465–468, 1997

[11]T. Iwasawa, K. Ohota, H. Mitsui, and M. Sone, "Effect of molecular hindrance on dc conductivity in dielectric liquids including water," in *Conference on Electrical Insulation and Dielectric Phenomena*, pp. 709–712, 1996. http://dx.doi.org/10.1109/CEIDP.1996.564570

[12]M. Monfreda, L. Gobbi, and A. Grippa, "Blends Of Olive Oil And Seeds Oils : Characterisation And Olive Oil Quantification Using Fatty Acids Composition And Chemometric Tools . Part II," *Food Chemistry*, vol. 145, pp. 584–592, 2014. http://dx.doi.org/10.1016/j.foodchem.2013.07.141

[13]A. Bendin, L. Cerretani, M. Salvador, G. Fregapane, and G. Lercker, "Stability Of The Sensory Quality Of Virgin Olive Oil During Storage," *Italian Food and Beverage Technology*, pp. 5–18, 2010.

[14]Y. Xu , S. Qian, "Oxidation Stability Assessment of a Vegetable Transformer Oil under Thermal Aging," *IEEE Transactions on Dielectrics and Electrical Insulation*, vol. 21, no. 2, pp. 683–692, 2014. http://dx.doi.org/10.1109/TDEI.2013.004073

[15]Y. Liu, J. Li, and Z. Zhang, "Gases Dissolved In Natural Ester Fluids Under Thermal Faults In Transformers," in *2012 IEEE International Symposium on Electrical Insulation*, 2012, pp. 223–226. http://dx.doi.org/10.1109/ELINSL.2012.6251462

Dielectric Materials and Applications: ISyDMA'2016
Materials Research Proceedings 1 (2016) 31-34

Materials Research Forum LLC
doi: http://dx.doi.org/10.21741/2474-395X/1/8

Interphase effect on the complex permittivity for composite materials at low frequencies

Salahddine EL BOUAZZAOUI[*1], Yassine NIOUA[1],
Mohammed Essaid ACHOUR[1a], Fouad LAHJOMRI[2]

[1]LASTID Laboratory, Physics department, Faculty of Science Ibn Tofail University, B.P:133, 14000 Kenitra, Morocco

[2]ENSA, Abdelmalek Essaadi University, LabTIC, Tanger, Morocco

*Correpondng Author: Email: s_elbouazzaoui@yahoo.fr, [a]sachoum@yahoo.fr

Keywords: Dielectric Properties Complex Permittivity, Mixing Laws, Interphase Region

Abstract. This article aims to present a study of the dielectric properties of the epoxy polymer /carbon black nanocomposites, with different concentrations of the dispersed conducting particles in frequencies between 180 Hz and 15 MHz. Measurements showed that the complex permittivity of the composites depends strongly on the volume concentration of the conducting medium. The dielectric behavior of these nanocomposites is compared with that obtained by the calculation of the effective permittivity based on theoretical model taking into account of the effect of the interphase region on the dielectric property of filler/polymer composites. We also attempted to model the dielectric behavior of these environments from a model that takes into account the interphase region. The first results show that the calculated values of the complex permittivity of the latter model, taking some coefficients as adjustable parameters, are in good agreement with experimental values.

Introduction

The complex permittivity of composite materials composed from filler embedded in a matrix is defined as $\varepsilon^* = \varepsilon' - i\varepsilon''$, and the prediction of the complex permittivity is one of challenge facing the research. In this direction there are many models that have been derived and proposed to predict the effective permittivity of composite materials, such as the Maxwell-Garnett [1] and the Bruggeman [2] laws. However, these models have a drawback when filler concentrations are higher than few percent these laws generally disagree with the experiments [3]. The discrepancy observed between the experimental data and calculated values can be attributed to the simplicity of the mixing laws that do not take into consideration many parameters like the morphology parameters such as particle size, their distribution. Indeed recent studies [4] showed the effect of the interphase region between the filler and the matrix on the complex permittivity of these materials. A model derived from the quasi-crystalline approximation with coherent potential model (QCA-CP) [5-6] including the interphase between filler and matrix is established for predict the complex permittivity. In our previous work [7] we found that the interphase between the matrix and the filler may place a significant effect on the effective permittivity for composite materials at microwave frequencies. In this paper as we report a study of the effect of the interphase region on the effective complex permittivity at low frequencies (180 Hz – 15 MHz) to understand the interphase effect on transport mechanism in composites materials.

Models for Effective Dielectric Constant

There have been several outstanding reviews of mixing laws and effective media theories for the prediction of the effective complex permittivity ε^* of two-component composite materials formed by conducting particles in insulating matrix. The first mixture law used in this study is the Maxwell Garnett equation (MG) which is expressed by the following equation [1]:

$$\frac{\varepsilon^* - \varepsilon^*_m}{\varepsilon^* + 2.\varepsilon^*_m} = \phi.\frac{\varepsilon^*_c - \varepsilon^*_m}{\varepsilon^*_c + 2.\varepsilon^*_m}$$

(1)

Where ε^*_c $(\varepsilon^*_c = \varepsilon'_c - j\varepsilon''_c)$ and ε^*_m $(\varepsilon^*_m = \varepsilon'_m - j\varepsilon''_m)$ are the complex permittivities of the conducting and insulating phases and ϕ is the volume fraction of the inclusion. A classic example of effective medium equations is the symmetric Bruggeman equation (BG) [2] that can be written, for spherical particles, in the form :

$$\phi.\frac{\varepsilon^* - \varepsilon^*_m}{\varepsilon^* + 2\varepsilon^*_m} + (1-\phi).\frac{\varepsilon^*_c - \varepsilon^*_m}{\varepsilon^*_c + 2\varepsilon^*_m} = 0$$

(2)

We are interested by the interphase effect on the complex permittivity of composite materials using another model that is a simple extension of the quasi-crystalline approximation with coherent potential (QCA-CP) model [5-6]. In the model, expressed by (3) the interphase region is described as a material having dielectric properties different from that of the polymer and filler phases:

$$\varepsilon^* = \varepsilon^*_m + 2\phi'.\frac{\varepsilon^*(\gamma\varepsilon^*_c - \varepsilon^*_m)}{3\varepsilon^* + (1-\phi')(\gamma\varepsilon^*_c - \varepsilon^*_m)}$$

(3)

with

$$\gamma = \frac{\beta(1+2\beta) + 2\alpha\beta(1-\beta)}{(1+2\beta) - \alpha(1-\beta)}$$

(4)

$$\alpha = \frac{a^3}{b^3} \ (5), \quad \beta = \frac{\varepsilon^*_{int}}{\varepsilon^*_c}$$

(6)

and

$$\phi' = \frac{\phi + k.\phi}{1 + k.\phi}$$

(7)

Where the permittivity of the interphase region is given by ε^*_{int} and the radii of the core particle and the interfacial shell are a and b thus the thickness of the interfacial layer is $t = b - a$ [4]. ϕ' is the volume fraction of the equivalent solid particle which contains the core particle and the shell, its strongly depend on the volume fraction of the inclusion ϕ. The value of k reflects the inclusion and matrix interaction strength, $k = 0$ indicates that the interaction between the filler and the matrix is negligible and a large positive value of k indicates a strong filler matrix interaction [5-6].

Experemental
The conductor-insulator composites that have been examined in this study consisted of conducting particles (produced by CABOT Co.) embedded in an insulating epoxy resin matrix DGEBF: Diglycidylic Ether of Bisphenol F (from CIBA GEIGY Co). The average size of the primary Raven 2000 particles is about 9.6 nm [8]. The epoxy has a DC conductivity of $1.40.10^{-14}$ $(\Omega.m)^{-1}$ and a density of 1.19 (g.cm^{-3}). The series of samples: Raven 2000/Epoxy resin polymer composites were prepared by mixing the desired volume concentration ϕ of carbon powder with the fluid resin. This series have a static percolation threshold $\phi c = 3.6$ vol%. The experimental setup used to measure the effective complex permittivity, $\varepsilon = \varepsilon' - i\varepsilon''$ of the composite samples relative to the vacuum permittivity $8.85*10^{-12}$ C^2 J^{-1}m^{-1} has been described elsewhere [8]. The permittivity was measured in

the frequency range from 180 to $1.5*10^7$ Hz using an impedance analyzer (HP 4194 A). The real ε' and imaginary ε'' parts were deduced from the measurement of the capacitance and resistance of the sample, respectively. All the measurements were performed at room temperature.

Results and Descussion
Firstly we used the Maxwell-Garnett equation which are proposed in equation 1 to predict both the real and imaginary parts of the complex permittivity for the composite materials Raven 2000 / DGEBF, in the range of frequencies 83 KHz to 15 MHz. The respective permittivities for components used in the calculation are, the complex permittivity of the matrix defined as $\varepsilon_m^* = \varepsilon_m' - j.\varepsilon_m''$ and is given for each frequency. Concerning the carbon black the complex permittivity was estimated by the Drude Model $\varepsilon_c^* = 0 - j\sigma_c / \varepsilon_0.\omega$ in which σ_c represent the electrical conductivity [9]. Figure 1 report the experimental data for the Raven 2000/epoxy as a function to the carbon black volume fraction and the calculated values using the Maxwell-Garnett model. It's clear that's the model gives a good agreement just for low concentration but when the volume fraction of the filler increases the difference between experimental data and calculated values become large and the model underestimate the experimental data. These results reflect the simplicity of the mixing laws that do not note take into consideration many important parameters like the morphology, the distribution of the particle in the matrix and mostly the interphase region between the matrix and the filler.

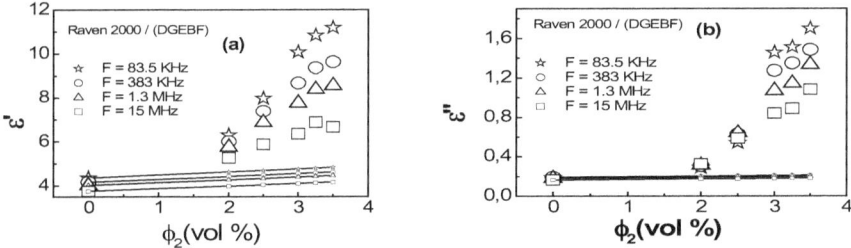

Fig 1. Real and imaginary parts of the effective complex permittivity as a function of CB volume fraction, line are the calculated values using the Maxwell Garnett equation and symbols are the experimental data for different frequencies respectively.

As a further step and in order to minimize the discrepancies observed between the theoretical and experimental data for our series of samples, we take the properties of the interphase region like the complex permittivity and the volume fraction of the interphase region into consideration and we attempt to fit the dielectric behavior using the proposed model derived from the QCA-CP model includes the interphase surrounding the filler which is supposed in equation (3). Figure 2 illustrates the calculation values, using the interphase approach with k and the interphase permittivity an as adjustable parameters. It's clear that the model gives a good agreement with the experimental data. Table 1 report the values of k and the interphase permittivities which gives the best fits for the different frequencies.

Table 1: k and the interphase permittivities for the different frequencies.

frequency	K	ε_{int}'	ε_{int}''
83 KHZ	7.9	47	18
383 KHZ	6.5	44	22
1.3 MHZ	5.9	39	19
15 MHZ	5.7	28	13

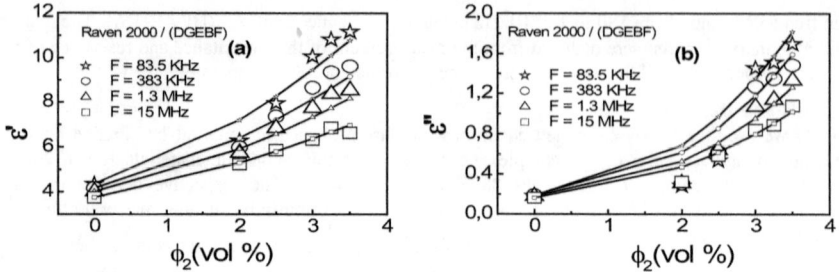

Fig 2. The real and imaginary parts of the effective complex permittivity as a function of CB volume fraction, line are the calculated values using the QCA-CP) model equation and symbols are the experimental data for different frequencies respectively.

Conclusion

In this work we studied the interphase effect on the complex permittivity predictions for composite materials made up of carbon black inclusion in a polymer of the epoxy resin type. The measurements conduced at a range of frequencies between 83 KHz and 15 MHz show that the QCA-CP with interphase model gives a good results and prove a better fits with the measurement. It indicate that the interphase between the matrix and the filler may place a significant effect on the effective permittivity for heterogeneous materials.

References

[1] M.Garnett, J.C.Philos, 'Colours inMetal Glasses and in Metallic Films'Trans.R.Soc.London., Vol. 203, pp. 385-420 . (1904). http://dx.doi.org/10.1098/rsta.1904.0024

[2] D.A.G.Bruggeman, Ann phys, Vol. 24, P. 636. (1935). http://dx.doi.org/10.1002/andp.19354160705

[3] N. Aribo, S. Elbouazzaoui, M. E. Achour, C. Brosseau," Investigating the Dielectric Properties of Carbon Black-Epoxy Composites", Spectroscopy Letters 47 (2014) 336. http://dx.doi.org/10.1080/00387010.2013.829103

[4] H. Vo and F. G. Shi, J. Microelectronics, 33, 5, (2012) 409

[5] Liu, X., Youpeng, W., Chun, W, Zeyang , Z. (2010) 'Study on permittivity of composites with core-shell particle' Physica B., Vol. 405, pp. 2014-2017. http://dx.doi.org/10.1016/j.physb.2010.01.093

[6] Xiangxuan Liu, YoupengWu, XuanjunWang, Rong Li and Zeyang Zhang,"Effect of interphase on effective permittivity of composites", J. Phys. D: Appl. Phys. 44 (2011) 115402. http://dx.doi.org/10.1088/0022-3727/44/11/115402

[7] Y. Nioua, S. El Bouazzaoui , M. E. Achour , F. Lahjomri, Microwave dielectric properties modeling of filled polymer composites using the interphase approach; Polymer Composites (Submitted).

[8] M.E Achour, "Electromagnetic properties of carbon black filled epoxy polymer composites", in Prospects in Filled Polymers Engineering: Mesostructure, Elasticity Network, and Macroscopic Properties, Brosseau C. Transworld Research Network, p. 129, (2008)

[9] M. Essone Mezeme, S. Elbouazzaoui, M. E. Achour, C. Brosseau, "Uncovering the intrinsic permittivity of the carbonaceous phase in carbon black filled polymers from broadband dielectric relaxation", Journal of Applied Physics. 109, 074107 (2011). http://dx.doi.org/10.1063/1.3556431

Dielectric Materials and Applications: ISyDMA'2016
Materials Research Proceedings 1 (2016) 35-37

Materials Research Forum LLC
doi: http://dx.doi.org/10.21741/2474-395X/1/9

The study of gate length AlGaAs/InGaAs/GaAs pseudomorphic high electron mobility transistor

Fatima Zohra BECHLAGHEM, Ahlam GUEN BOUAZZA, Benyounes BOUAZZA, Badia BOUCHACHIA

Research Unit of Materials and Renewable Energies, Faculty of Science and Engineering, University, Abou-Bakr Belkaid, Tlemcen, Algeria

* Corresponding author: E-mail: bechlaghem_f@yahoo.fr

Keywords: Component, Pseudomorphic HEMT, Silvaco, Gate Length Scaling

Abstract. In this paper a detailed simulation study is presented, the device characteristics of AlGaAs/InGaAs/GaAs pseudomorphic high electron mobility transistors (PHEMTs) with different gate lengths sheets are comprehensively and systematically investigated. Based on a two-dimensional simulator of Atlas, the detailed calculations and studies including, carrier distributions, and DC and microwave performances are reported . Due to the use of InGaAs DC structure, good pinch-off and saturation characteristics, higher current drivability, larger and linear transconductance correlation has been analyzed for prediction of the impact on device performances.

Introduction

Field-effect transistors (FETs) performing with excellent linear transfer characteristic and low distortion have been applied for amplifiers and microwave circuits [1–3]. Though high electron mobility transistors (HEMTs) have well performed, these devices show a sharp peak in transconductance as a function of gate voltage due to the onset of parallel conductance in doping high-band gap materials. Researchers worldwide have attached considerable importance to the pseudomorphic high electron mobility transistor (pHEMT) which utilizes the high carrier mobility property of a twodimensional electron gas (2DEG). The channel modulation efficiency of heterojunction devices are highly dependent on the quality of Schottky contact, which implies a strong impact on many critical device parameters, such as gate leakage current, transconductance and threshold voltage, etc.[4] We present a detailed simulation study of the expected improvements in the performance of PHEMT devices scaled to de nano dimensions. The simulated PHEMTs are scaled down in proportion to gate lengths of 300 and 50 nm. Important to this paper is that the devices are shrunk not only in the lateral but also in the vertical dimensions. The real device geometry and the corresponding parasitics are also properly included in the simulations.

Devise modeling

An accurate knowledge of the electrical properties of the semiconductor material as well as the nature of the physical contact to the material allows a reliable determination of the electrical characteristics of GaAs devices [5]. The physical modelling software used in this work is ATLAS, The commercially produced pseudomorphic high electron mobility transistors (**P**HEMT) is shown in Fig. 1 which is the two-dimensional physical device simulator by SILVACO. Physical modelling uses physical parameters describing the device geometry, material and performance characteristics.

Dielectric Materials and Applications: ISyDMA'2016   Materials Research Forum LLC
Materials Research Proceedings **1** (2016) 35-37   doi: http://dx.doi.org/10.21741/2474-395X/1/9

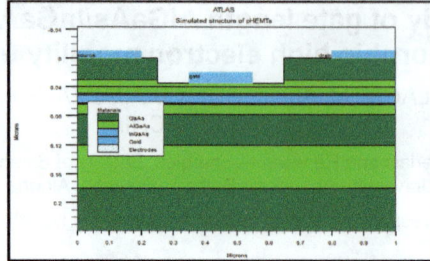

Figure 1. Simulated structure of PHEMTs

Resultats and discussions

Thow different gate length (300, 50 nm) PHEMTs have been simulated and their logic characteristics have been determined. The drain characteristics of the three devices at VGS=0 V is shown in figures(2),and (3) .It was observed that the maximum drain current obtained at the bias of Vd=2 V and Vg=0V for PHEMT   (589.94 mA/mm) for Lg=50nm and.(431.87 mA/mm) Lg=300nm respectively.

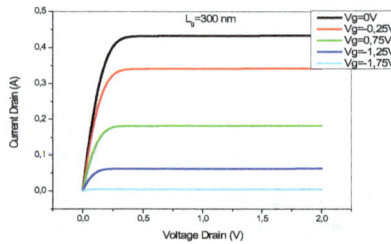

Figure 2. Drain characteristics Of AlGaAs/InGaAs PHEMT for Lg=300 nm.

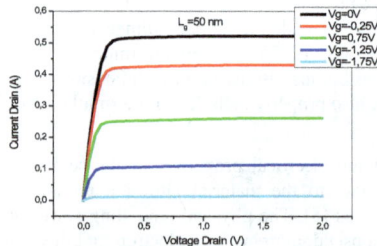

Figure 3. Drain characteristics Of AlGaAs/InGaAs PHEMT for Lg=50 nm

The transfer characteristics at Vd=2V for PHEMT differentgate lengths of AlGaAs/InGaAs PHEMT (Fig.4, Fig.5). The linearity and maxima of transconductance characteristics increase as the basic PHEMT device is modified the various gate length. The maximum transconductance (gm,max) values are 442.92,   and 372.47 , mS/mm for device 300 nm,and 50 nm, respectively. It was observed that the threshold voltage (Vth), which is defined as the gate-bias intercept by linearly extrapolating drain current curve from the peak gm position in transfer characteristics, is nearly -1.8V, and -2.1V for AlGaAs/InGaAs PHEMT for Lg=300 nm and Lg=50nm respectively.

Dielectric Materials and Applications: ISyDMA'2016   Materials Research Forum LLC
Materials Research Proceedings 1 (2016) 35-37   doi: http://dx.doi.org/10.21741/2474-395X/1/9

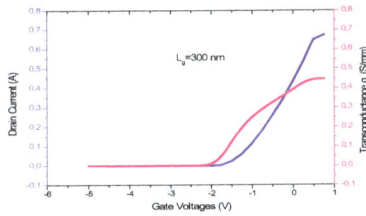

Figure 4. Drain characteristics Of AlGaAs/InGaAs PHEMT for Lg=300 nm.

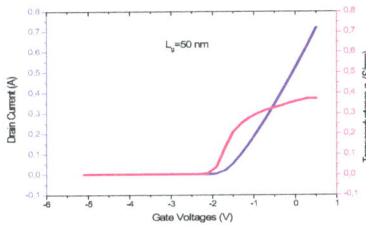

Figure 5. Drain characteristics Of AlGaAs/InGaAs PHEMT for Lg=50 nm

Conclusion

In summary, device   for various gate length is simulated and analyzed. The two dimensional simulator, obtains optimal designs of pHEMTs. Various gate length AlGaAs / InGaAs pHEMTs has decreased maximum drain current (Idmax) and and higher transconductance, the augmentation of drain current, the maximum drain-source current (Ids) was 589.94 mA/mm, for Lg=50 nm and and transconductance (Gm) of the pHEMT is 442.92 mS/mm for Lg=300nm. This highlights the significance of maintaining an optimized value of aspect ratio as gate length is reduced for digital applications.

References

[1] Jung-lui Tsai King-Poul Zhu "Eletrical proprieties of single deltat-doped InGaP/InGaAs/GaAs pseudomorphic HEMT wit canal-like gate structure" Materials Chemistry and Physics 82 (2003) 501–504. http://dx.doi.org/10.1016/S0254-0584(03)00327-4

[2] W.C. Liu, W.C. Hsu, L.W. Laih, J.-H. Tsai, W.S. Lour, Appl. Phys. Lett. 66 (1995) 1524. http://dx.doi.org/10.1063/1.113634

[3] M.J. Kao, H.M. Shieh, W.C. Hsu, T.Y. Lin, Y.H. Wu, R.T. Hsu, IEEE Trans. Electron Dev. 43 (1996) 1181. http://dx.doi.org/10.1109/16.506766

[4] Muniba Ahmad Reg. No. 2010-NUST-MS-EE(S)-48 student of SEECS have submitted the final draft copy of his/her thesis titled: "physical device modelling and simulations of phemts.

[5] Ka Wa Ian,Mohamad Adzhar Md Zawawiand and Mohamed Misous "Termally stable In0.7Ga0.3As/In0.52/Al0.48As phemt using thermally evaporated palladium gate metallization",Semicond.Sci.Technol.29(2014)035009(spp). http://dx.doi.org/10.1088/0268-1242/29/3/035009

Dielectric Materials and Applications: ISyDMA'2016   Materials Research Forum LLC
Materials Research Proceedings 1 (2016) 38-41   doi: http://dx.doi.org/10.21741/2474-395X/1/10

Dielectric investigation of polytetrafluoroethylene manufactured by a newly spark plasma sintering (SPS) technique

H. LAHLALI*[1], A. MDARHRI[1], I. EL ABOUDI[1], M. EL AZHARI[1], M. ZAGHRIOUI[2],
C. HONSTETTRE[3]

[1]Condensed Matter and Nanostructures Laboratory, Faculty of Sciences and Techniques. Cadi Ayyad University, BP 549, 40 000 Marrakesh, Morocco

[2]Laboratoire GREMAN CNRS-UMR 7347Blois, Université François Rabelais, CS 2903, 41029 Blois Cedex, France

[3]Laboratoire GREMAN CNRS-UMR 7347 Blois, Université François Rabelais, Parc de Grandmont, 37200 Tours, France

*Corresponding author: E-mail: hind.lahlali@edu.uca.ma

Keywords: PTFE, SPS Process, Dielectric Properties, Extrusion, Molding, Temperature Effect, Microstructure

Abstract. A new processing method named Spark Plasma Sintering (SPS) is particularly adapted to process a dense polymer polytetrafluoroethylene (PTFE) which is difficult to sinter by conventional techniques. It consists in applying simultaneous electric current and pressure directly on the micrometric particles powder. The aim of the present work is to highlight the dielectric properties response of PTFE synthesized by the SPS technique in a low frequency domain typically between 100Hz and 10 MHz and over a temperature range from 313 to 553 K. In a first step, the dielectric constant measured at room temperature indicates a slight decrease when the frequency increases reflecting the general trend for this kind of materials. By growing the temperature up to 553 K, the dielectric constant spectra show a notable increase and the obtained values can reach four times more than of those measured at room temperature. This unexpected behavior is then compared to the dielectric spectra response highlighted by commercially PTFE samples manufactured by using two conventional methods i.e. extrusion and molding. It is found that the processing parameters affect greatly the dielectric properties measured either at room temperature or in 313-553 Ktemperature range. Thermogravimetry Analysis (TGA) is used to probe the thermal stability of different PTFE samples used in this study. From this experimental investigation, it is concluded that the SPS technique can be seen as an efficient and rapid route to manufacturing polymer materials with tailored dielectric properties for many useful applications.

Introduction

Polytetrafluoroethylene (PTFE) is semi-crystalline in nature with complex structures even at ambient conditions. Due to the C-F bonds, it exhibits special properties surpassing those of most polymers such as very high melting temperature (600 K) when compared to other crystalline polymers, good chemical resistance, a low friction coefficient and good dielectric properties [1]. In particular, its dielectric properties have always attracted much interest for electronic applications such as a high voltage equipment, insulator cablesand printed circuit boards to cite a few[2]. Nevertheless, the high melting temperature and the very high melt viscosity (10^{11}P.s at 653 K), as well as its insolubility rule out conventional processing routes[3]. In this regard, Spark Plasma Sintering (SPS) has been shown to be an effective unconventional sintering method for obtaining fully dense materials at lower sintering temperatures and shorter holding times (few minutes). It consists in applying simultaneous electric current and pressure directly on the micrometric particles powder. Our central

Dielectric Materials and Applications: ISyDMA'2016   Materials Research Forum LLC
Materials Research Proceedings 1 (2016) 38-41   doi: http://dx.doi.org/10.21741/2474-395X/1/10

goalis to study the dielectric properties response of PTFE synthesized by the SPS technique in a low frequency domain typically between 100Hz and 10 MHz and over a temperature range from 313 to 553 K. Then, the results were compared to the dielectric spectra response highlighted by commercially PTFE samples manufactured by using two conventional methods i.e. extrusion and molding.

Materials and techniques

For comparison with PTFE fabricated by a new non-melting process namely Spark Plasma Sintering (SPS) technique, two series of commercial samples made by Extrusion and molding melting methods have been used. The extruded PTFE samples (PTFE-Ext) and the molded PTFE samples (PTFE-Mol) were obtained from PETRO-MAT company. According to their different fabrication processes,the initial form of thethree samples under study is not the same. Indeed, thePTFE-Ext sampleswere produced with the shape of a plate with thickness of about 2mm, while the second commercial samples;PTFE-Mol,have a cylindrical shape with a diameter of 16 mm. The sintered PTFE (PTFE-SPS) samples have a disk shape with a diameter and thickness of 40 and 4 mm respectively. The sintering of the PTFE micrometric powder was carried out in a vacuum using a SPS system (Model SPS-FCT HP D25/1). To sinter a circular disk with the above dimensions a constant mass (15g) of powder is used. The processing conditions that enable us to fabricate our PTFE-SPS samples from the PTFE powder are detailed in a previous study [4-6].

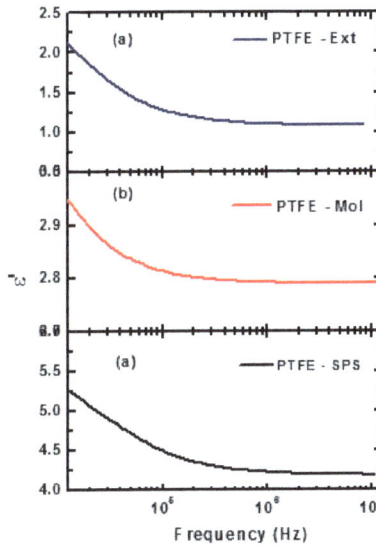

Fig. 1. Dielectric constant spectra measured at room temperature of PTFE samples made by using three processes: **(a)**: Extrusion, **(b)** Molding, and **(c)** a non-conventional SPS technique.

To conduct the dielectric measurements, standard dimensions of about $8.5 * 8.5 * 0.6 \ mm^3$were used for all samples carefully cradled with smooth and parallel surfaces from their initial shapes. Moreover, Au electrodes with thickness of about $100nm$were sputtered onto the faces of thespecimens prior to electrical characterization. The dielectric constant was measured at room temperature in a frequency range from 100 Hz to 10 MHz using an Agilent 4294A impedance analyzer. The thermal stability of PTFE was investigated using a thermogravimetric analyzer

Dielectric Materials and Applications: ISyDMA'2016   Materials Research Forum LLC
Materials Research Proceedings 1 (2016) 38-41   doi: http://dx.doi.org/10.21741/2474-395X/1/10

(Perkin-Elmer TGA-7) thermograms were obtained in the temperature range 323-1073K. The heating runs were made at a constant heating rate of 293K/min under nitrogen atmosphere.

Resultsand discussion

The dielectric constant spectra for PTFE samples according to the three technological processes are depicted in Fig. 1. At room temperature, all samples show a typical behavior characterized by a decrease of dielectric constant ε' with increasing the frequency and followed by approximately constant value in the frequency range 0.1 MHz up to 10 MHz.

As can be seen from Fig.1 high values of ε' are measured in PTFE-SPS samples. This behavior indicates clearly the effect of the technological process that affects significantly the microstructure of resulting samples. The unusual high values of the dielectric constant can be due to the polar impurities as is suggested by Eby and Sinnott [7]. This trend is more pronounced in the cases of sintered PTFE samples for which the graphic papers are placed between the die and the powder to guarantee electrical contact. At high electrical current (2kA), a possible migration of slight traces graphic papers may be involved. In addition, the high values of both density and crystallinity of PTFE-SPS compared to the commercial PTFE may partially explain the difference in dielectric responses for these three samples as reported in an early work [5]. For instance the crystallinity, deduced from the density measurements, of sintered PTFE is found to be highest than in the extruded and molded PTFE samples with respective values of 82, 66, and 61%. In that report, other techniques, i.e. m-IR and Raman spectroscopies, confirm this trend. These results are consistent with the mechanical behaviors of those samples demonstrating the superiority of SPS samples.

By growing the temperature up to 553K, the dielectric constant spectra show a notable increase and the obtained values can reach four times more than of those measured at room temperature (not shown here). Other investigations are needed to best understand the temperature effect. By using thermogravimetric analysis (TGA) the weight change of the PTFE samples as function of temperature is probed. Fig. 2 shows that the TGA curves of the PTFE-SPS samples exhibit a high stability compared to samples elaborated by the conventional methods. For SPS samples, the weight loss firstly became apparent about 773Kwith less than 1% and reaches 80% when the temperature exceeds 873K.

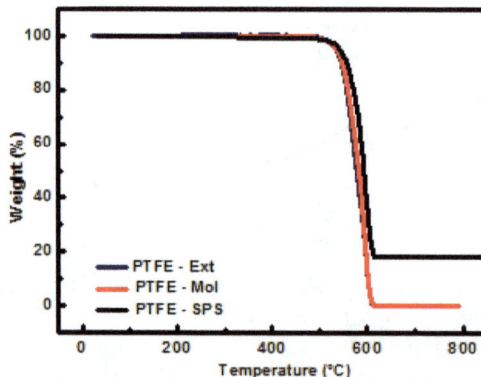

Fig. 2. Thermogravimetric curves of the PTFE samples synthesized by three technological processes.

Conclusion

This experimental investigation deals with the dielectric properties of PTFE synthesized by three technological processes. The dependence of the typeof process on dielectric constant ε' is clearly highlighted indicating that the microstructure of different samples is strongly affected. Compared to the conventional techniques, the SPS technique permits to sinter fully dense PTFE with relatively high dielectric constant associated with improved stability. These results are encouraging when taking in account the advantages of the SPS technique. More investigations are needed to understand the apparent dielectric response and its relation to the SPS's parameters, and the possible growth of PTFE particles size under this fast and efficiency process.

Acknowledgements

The authors and especially H. Lahlali thank CNRST-Morocco for the Excellence Scholarships.

References

[1] S. Ebnesajjad and P. R. Khaladkar " Fluoropolymers applications in chemical processing industries : the definitive user's guide and databook", edited by PDL Flurocarbon Series/ William Andrew Publising, Norwich, NY (2005).

[2] J. G. Drobny, Technology of Fluoropolymers Secon Edition, vol. 71, no. 1–2. 2010.

[3] C. A. Sperati and H. W. Starkweather, "Fluorine-containing polymers. II. Polytetrafluoroethylene," in Fortschritte Der Hochpolymeren-Forschung, Berlin/Heidelberg: Springer-Verlag, 1961, pp. 465–495. http://dx.doi.org/10.1007/BFb0050504

[4] H. Lahlali, A. Mdarhri,F. El Haouzi, M. El Azhari, I. El Aboudi O. Lame D. Fabregue, and G. Bonnfont.Verres, Céramiques & Composites, Vol. 4, N°2 (2015), 5-8

[5] H. Lahlali, A. Mdarhri, M. Zaghrioui, F. El Haouzi, I. El Aboudi, and M. El Azhari.Journal of Advances in Materials and Processing Technologies (2016). Accepted.

[6] A. Mdarhri, H. Lahlali, O. Lame, M. El Azhari, I. El Aboudi and D. Fabregue (Umpublished).

[7] R. K. Elby and K. M. Sinnott. Journal of Applied Physics. Vol 32, 1765-1771 (1961). http://dx.doi.org/10.1063/1.1728433

Dielectric Materials and Applications: ISyDMA'2016   Materials Research Forum LLC
Materials Research Proceedings 1 (2016) 42-44   doi: http://dx.doi.org/10.21741/2474-395X/1/11

Influence of DC voltage on the dielectric properties of nematics

N. ÉBER[1*], B. FEKETE[1], P. SALAMON[1], Á. BUKA[1], A. KREKHOV[2]

[1]Institute for Solid State Physics and Optics, Wigner Research Centre for Physics, Budapest, Hungary

[2]Laboratory for Fluid Dynamics, Pattern Formation and Biocomplexity, Max Planck Institute for Dynamics and Self-Organization, Göttingen, Germany

*Corresponding author: E-mail: eber.nandor@wigner.mta.hu

Keywords: Liquid Crystals, Conductivity, Anisotropy

Abstract. We report on precise impedance measurements, with the aim of exploring the influence of a dc bias voltage on the dielectric permittivity and electrical conductivity of liquid crystals and on their anisotropies. We prove that the dielectric permittivity is not affected by a dc bias; however, the electrical conductivity suffers a substantial reduction upon increasing the superposed dc voltage. Moreover, we show that the relative conductivity anisotropy also diminishes at increasing dc bias.

Introduction

Nematic liquid crystals are anisotropic dielectric materials with a low electrical conductivity, which originates from a small concentration of ionic contaminants. Experimental studies of the electro-optical properties of nematics, as well as most display applications of them utilize ac driving voltage; investigations under dc voltage are rather scarce. Recently, electroconvection and flexoelectric patterns occurring under superposed action of ac and dc voltages have been studied and, unexpectedly, an inhibition of the pattern formation at high voltages (an increase of the thresholds) was found [1-2]. The fact that conductivity is an important factor determining the threshold voltage, motivated the present testing of the effect of dc bias on the electrical conductivity.

Experimental Method

Measurements were performed on the compound 4-n-octyl-oxyphenyl 4-n-methyloxybenzoate (1OO8), having a nematic phase in the temperature range between 53 °C and 77 °C and a chemical structure shown in Fig. 1. It is a material used for studying pattern formation in electric field [2-3].

Precise impedance measurements were carried out with the dielectric analyzer Novocontrol Alpha equipped with a ZG4 test interface. The liquid crystal was filled into a plane condenser (glass plates with transparent electrodes of $A = 1$ cm^2 size, separated with a gap of $d = 1$ mm). The electrodes were coated with polyimide layer, in order to mimic the electrical boundary conditions of the thinner commercial cells employed in pattern formation studies. The cell was placed in a shielded chamber, with its temperature kept at $T = 58 \pm 0.05$ °C.

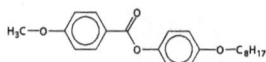

Fig. 1. The chemical structure of the nematic 4-n-octyloxyphenyl 4-n-methyloxybenzoate (1OO8).

As liquid crystals are anisotropic materials, the dielectric permittivity (electrical conductivity) can be characterized by the quantities $\varepsilon_\parallel$ ($\sigma_\parallel$) and $\varepsilon_\perp$ ($\sigma_\perp$), measured with an electric field **E** parallel with and perpendicular to the director **n**, respectively. These two orthogonal experimental geometries can be easily realized, if the chamber is put into a magnetic field **H**. For the large cell thickness used, a magnetic induction value of $B = 1$ T is large enough to align the director parallel to **H**. Thus applying

Dielectric Materials and Applications: ISyDMA'2016   Materials Research Forum LLC
Materials Research Proceedings 1 (2016) 42-44   doi: http://dx.doi.org/10.21741/2474-395X/1/11

a magnetic field parallel to the electrode surface ($\mathbf{H}$, $\mathbf{n} \perp \mathbf{E}$), the perpendicular component is obtainable; when the chamber is rotated so that the magnetic field becomes normal to the electrodes ($\mathbf{H}$, $\mathbf{n} \parallel \mathbf{E}$), the parallel component can be measured.

The complex impedances of the empty and the liquid- crystal-containing cell were measured at the frequency of 1 kHz, with a probing ac rms voltage of $U_{ac} = 0.2$ V. The liquid crystal sample was interpreted as a parallel RC circuit, with

$$R_i = \left(\sigma_i \frac{A}{d} \right)^{-1} ; C_i = \varepsilon_0 \varepsilon_i \frac{A}{d}, \text{ where } i = \parallel, \perp . \qquad (1)$$

The dc bias voltage was incremented from $U_{dc} = 0$ up to $U_{dc} = 40$ V in 1 V steps, in each 10 minutes. At a fixed bias voltage, measurements were performed repeatedly several times, alternately in the parallel and perpendicular geometries. Permittivity and conductivity values were evaluated using (1).

Results

Measurements proved that the dielectric permittivity values do not depend on the dc bias voltage: we obtained $\varepsilon_\parallel = 4.96 \pm 0.01$ and $\varepsilon_\perp = 5.45 \pm 0.01$; i.e., the dielectric anisotropy, $\varepsilon_a = \varepsilon_\parallel - \varepsilon_\perp$, is negative, as expected from the molecular structure in Fig. 1. The small deviations of about 0.2% are attributed to temperature fluctuations. The constancy of $\varepsilon_\parallel$ and $\varepsilon_\perp$ indicate that the dc bias did not alter the director alignment created by the magnetic field; i.e., no pattern formation occurred at the applied voltages.

In contrast to the permittivity, the electrical conductivity substantially depends on the bias voltage. Figure 2(a) exhibits the dc bias dependence of $\sigma_\parallel$ and $\sigma_\perp$, as well as that of the conductivity anisotropy $\sigma_a = \sigma_\parallel - \sigma_\perp$; the symbols show the values averaged over a 10 minutes interval. It is clearly seen that the bias dependence is nonlinear. For $U_{dc} < 4$ V the change is small, there is even a slight increase of σ; however, for $U_{dc} > 4$ V, increasing the dc bias reduces strongly the conductivities. The maximal reduction is by about a factor of 6. The conductivity anisotropy also diminishes, though remains positive in the whole tested voltage range. It is important to note that the relative conductivity anisotropy, defined as $\delta\sigma = \sigma_a/\sigma_\perp$, is decreasing with the dc bias voltage too, as depicted in Fig. 2(b); at the highest bias it is only about 60% of its initial value at $U_{dc} = 0$.

The temporal dynamics of how the conductivity responds to voltage variations was also tested. Unsurprisingly, σ does not change instantaneously after a switching on/off the dc bias voltage; the voltage jump initiates a relaxation process instead. Measurements proved that this relaxation is rather long; moreover, it can be characterized by several relaxation times, from seconds up to several hours, indicating the presence of various ionic processes (e.g., migration, diffusion, association, recombination, and adsorption).

Discussion

By its original concept, a constant electrical conductivity describes the linear, voltage independent current response of the system to the electric excitation; i.e., it implicitly assumes ohmic conductivity. Though in liquid crystals conductivity is mostly due to ions, under ac voltage the assumption of ohmic conductivity is usually an acceptable approximation. This fails, however, in the presence of dc voltage. A dc voltage induces migration of ions leading to a partially ion-depleted region in the bulk, the formation of a charged double layer with high electric field gradients near the electrodes, and adsorption of ions at the substrates. It is not surprising that these phenomena result in a reduction of the number of active charge carriers and thus in dc voltage dependent, diminishing electrical conductivity. On the other hand, the relative conductivity anisotropy is assumed to depend rather on the structure of the nematic matrix, which is not expected to change with the number of charge carriers. Thus the dc- voltage-induced reduction of the relative conductivity anisotropy is an unexpected founding; it may perhaps be related with a change in the type of charge carriers.

Electrical conductivity, as well as its (relative) anisotropy is a key, though hardly controllable material parameter for governing the formation of electroconvection patterns. Usual theoretical modelling of electroconvection phenomena assumes ohmic conductivity with constant $\sigma_\parallel$ and $\sigma_\perp$. The results shown above clearly demonstrate that this assumption fails and should be given up if the applied voltage has a dc component. Therefore, if it is not done so, a mismatch between theoretical predictions and experimental observations about the onset characteristics (threshold voltage, critical wave number) of electroconvection patterns is practically unavoidable and actually has been reported for patterns induced by superposed ac and dc voltages [1-2]. A precise description of all ionic processes mentioned above and thus the understanding of the dc-voltage-induced conductivity changes would be a huge challenge for the future development of the theory of electric-field-induced pattern formation.

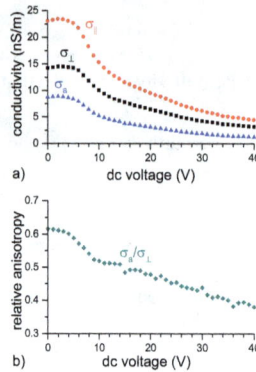

Fig. 2. Dc voltage dependence of a) the electrical conductivity parallel with ($\sigma_\parallel$) and perpendicular to ($\sigma_\perp$) the director, and the anisotropy $\sigma_a = \sigma_\parallel - \sigma_\perp$; b) the relative anisotropy $\sigma_a/\sigma_\perp$.

Some qualitative features of the pattern formation at superposed ac and dc voltages can, however, be estimated by testing the effect of varying $\sigma_\perp$ and $\delta\sigma$ on the pattern onset voltages. Recent simulations have shown that reduction of $\delta\sigma$, that occurs upon increasing the dc bias voltage, results in the increase of the threshold voltage of electroconvection, independently of the frequency of the ac voltage component [2]. This offers an explanation for the experimental observations and thus resolves the mismatch with theory.

Acknowledgment

Financial support by the Hungarian Scientific Research Fund (OTKA) grant NN110672 and the Moroccan (CNRST) – Hungarian (NKFIH) bilateral project TÉT_12_MA-1-2013-0010 is gratefully acknowledged.

References

[1] P. Salamon, N. Éber, B. Fekete, and Á. Buka, "Inhibited pattern formation by asymmetrical high-voltage excitation in nematic fluids,". Phys. Rev. E, vol. 90, 022505/1-5, August 2014. http://dx.doi.org/10.1103/PhysRevE.90.022505

[2] N. Éber, P. Salamon, B. A. Fekete, R. Karapinar, A. Krekhov, and Á. Buka, „Suppression of spatially periodic patterns by dc voltage," Phys. Rev. E, in press. http://dx.doi.org/10.1103/physreve.93.042701

[3] P. Salamon, N. Éber, A. Krekhov, and Á. Buka, "Flashing flexodomains and electroconvection rolls in a nematic liquid crystal," Phys. Rev. E, vol. 87, 032505/1- (2013).

Dielectric Materials and Applications: ISyDMA'2016
Materials Research Proceedings 1 (2016) 45-48

Materials Research Forum LLC
doi: http://dx.doi.org/10.21741/2474-395X/1/12

Effect of temperature and filler concentration on the electrical parameters of a dispersion of carbon nanotubes in an epoxy matrix

S.BOUKHEIR[1,2,*], M.E. ACHOUR[2], OUERIAGLI[1], A. OUTZOURHIT[1], N.ÉBER[3], L.C. COSTA[4]

[1]LN2E Laboratory , Faculty of Sciences, Sémlalia, Cadi Ayyad University, Marrakech, Morocco

[2]LASTID Laboratory , Faculty of Sciences, Ibn Tofail University, Kenitra, Morocco

[3]Institute for Solid State Physics and Optics, Wigner Research Centre for Physics, Hungarian Academy of Sciences Budapest, Hungary

[4]I3N and Physics Department, University of Aveiro, Aveiro, Portugal

Corresponding author : E-mail: boukheirsofia@gmail.com

Keywords: Electrical Properties, Relaxation, Impedance Spectroscopy

Abstract. We have investigated the electrical properties of carbon-nanotubes-loaded DGEBA polymer composites in the frequency range between 1Hz and 10 MHz and temperature range between 25°C and 105°C. The frequency dependence of electrical data have been analyzed in two frameworks: the electrical modulus formalism with the Kohlrausch-Williams-Watts stretched exponential function (KWW) and the electrical conductivity by using the Jonscher's power law. The stretching exponent β_{KWW} and the Jonscher exponent n are found to be temperature dependent for all carbon nanotubes concentrations and show a very slight variation with increasing the amount of filler percentage at room temperature.

Introduction

Multiwall carbon nanotubes (MWCNTs) have several characteristics (such as flexibility, low mass density, and large aspect ratio) that make them excellent for improving the electrical properties of polymers [1]. Due to their reduced size and dimensionality, carbon nanotubes form complex networks of aggregates within the composite materials [2]. Samples containing different volume concentrations (0.2% to 5%) of MWCNT were studied by impedance spectroscopy. Complex impedance spectroscopy is a nondestructive method that allows us to study the electrical properties of multiwall-carbon-nanotubes-based polymer composite samples. This technique can correlate the structural and electrical characteristics of composites in a wide range of frequencies and temperatures, as we already reported in previous works [3]. It is by now well demonstrated that carbon nanotubes and carbon-nanotube-based polymer composites have the potential to be applied as effective sensing devices in various applications [4]. In the present paper we report the electrical conductivity and modulus properties of epoxy polymer composites loaded with multiwall carbon nanotubes, investigated in the frequency range of 1 Hz to 10 MHz and temperature range between 25°C and 105°C. The obtained spectra were analyzed by appropriate theoretical models, as discussed below.

Methodology

The composite samples investigated were MWCNTs (Cheap-Tubes, USA Laboratories) with the diameter of the primary CNT about 50 nm, the length in the range of 10 – 20 µm and the purity higher than 95wt%, dispersed in an insulating epoxy matrix DGEBA (diglycidylic ether of bisphenol A) with a density of 1.19 (g/cm^3), a conductivity of the order of $\sigma_{DC} = 1.4*10^{-14}$ (Ωm)$^{-1}$ and a glass transition temperature about T_g = 60 °C. We added the MWCNTs to the epoxy resin in different concentrations, before adding 1% of hardener to make each mixture cohesive. The mixture was

Dielectric Materials and Applications: ISyDMA'2016   Materials Research Forum LLC
Materials Research Proceedings **1** (2016) 45-48   doi: http://dx.doi.org/10.21741/2474-395X/1/12

stirred at room temperature. Each sample of the MWCNTs /DGEBA took 5 min to gelate after pouring it into the mold. After a few hours we unmolded our samples, which took several weeks for reaching a complete polymerization. The percolation threshold for this series of samples is about ϕ_C =2.7%. The temperature dependent AC impedance spectra were measured with a Novocontrol Alpha-A Analyzer combined with the impedance interface ZG4 in a 4 wire arrangement, in the frequency range of 1 Hz to 10 MHz. Measurements and data recording were performed with the WinDeta software [5].

The bulk AC conductivity was determined by nonlinear mean-square-deviation curve fitting of the impedance spectrum using the WinFit program provided by Novocontrol, Hundsagen, Germany [5].

Theoretical models
The electrical modulus, M*, is defined as:
$$M^*(\omega) = M'(\omega) + iM''(\omega) \tag{1}$$

where M'(ω) and M''(ω) are the real and imaginary parts of the electrical modulus. The non-exponential conductivity relaxation could be described by using the Kohlrausch-Williams-Watts (KWW) function [6], which represents the distribution of the relaxation times in charges conducting materials [7]. The frequency dependent complex modulus can be given as:

$$M^*(\omega) = M_\infty \left[1 - \int_0^\infty \exp(i\omega t)\,(-d\emptyset/dt)dt \right] \tag{2}$$

$$with \quad \emptyset(t) = \emptyset_0 \exp\left(-\frac{t}{\tau_\sigma}\right)^{\beta_{KWW}} \tag{3}$$

where M_∞ represents the asymptotic value of $M'(\omega)$ when $\omega \to \infty$, τ_σ is the conductivity relaxation time and β_{KWW} is the Kohlrausch exponent; its value is located in the range $0 < \beta_{KWW} \le 1$. Furthermore, the total conductivity at a given temperature over a wide range of frequencies can be written as [8]:
$$\sigma_{tot}(\omega, T) = \sigma_{DC}(T) + \sigma_{AC}(\omega, T) \tag{4}$$

Here σ_{DC} is the DC conductivity and
$$\sigma_{AC}(\omega, T) \propto \omega^{n(T)} \tag{5}$$

is the AC electrical conductivity following Jonscher's power law [9]; $n(T)$ is the power exponent depending on the temperature, which fulfils $0 \le n(T) \le 1$.

Results
Figure 1 shows the variation of the conductivity as a function of the frequency at room temperature, for three different volume concentrations of MWCNT. At low frequencies, for the concentrations above ϕ_C, the AC conductivity is almost independent of frequency, approaching the DC conductivity in the plateau region, while at high frequency the conductivity has a dispersion that shifts to higher frequencies with increasing MWCNT. Several studies show that the dielectric response of the composites conductor/insulator obeys the Jonscher power laws [9]:
$$\sigma_{AC}(\omega) = A\,\omega^{n(T)}.$$

The relaxation behavior is analyzed using the complex electric modulus, which reflects only the dynamic properties of the sample without the polarization effects at the interface, with a Kohlrausch-Williams-Watts (KWW) distribution of relaxation times. The variation of imaginary M'' (ω) part of the electrical modulus as a function of frequency at room temperature is shown in Figure 2. the asymmetric M''(ω) is immediately suggestive of stretched exponential relaxation behaviour.

Dielectric Materials and Applications: ISyDMA'2016   Materials Research Forum LLC
Materials Research Proceedings 1 (2016) 45-48   doi: http://dx.doi.org/10.21741/2474-395X/1/12

The parameters n and β_{KWW}, which were obtained from the analysis by fitting to the Jonscher's power law and the Kohlrausch-Williams-Watts (KWW) function, respectively, are depicted in Figure 3 as the function of the concentration of the carbon nanotubes at room temperature. The parameter β_{KWW} is calculated by using the relation $\beta_{KWW} = 1.14/FWHM$, where $FWHM$ is the full width at half height of the frequency dependent modulus spectrum. The exponent n was measured as the slope of the frequency dependent AC conductivity spectrum in the high frequency region. We found n values ranging between 0.84 and 0.94, and β_{KWW} values ranging between 0.40 and 0.50. Below a concentration of 2.0% of MWCNT, the two parameters do not show any significant change with increasing the concentration, but we can observe that they slightly decrease above this MWCNT concentration of 2.0%.

The temperature dependence of the stretched exponent β_{KWW} and the Jonscher exponent n is depicted in figure 4; $n(T)$ decreases slightly with the temperature increasing from 25°C to 105°C, whereas β_{KWW} increases significantly with temperature. The variation of n with temperature can be related to the existence of a distribution of the relaxation parameters [10].

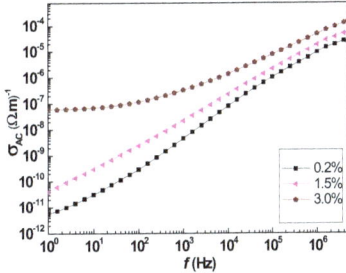

Fig 1. σ_{AC} versus frequency at room temperature for three selected carbon nanotube concentrations.

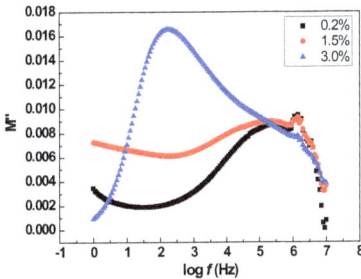

Fig 2. Imaginary part of the complex modulus at room temperature for three selected carbon nanotube concentrations.

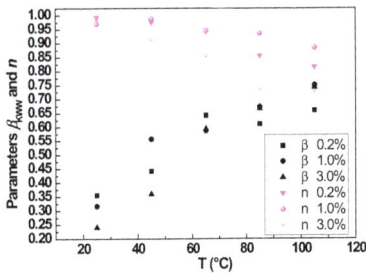

Fig. 3. Electrical parameters, β_{KWW} (obtained from fitting by KWW function) and n (from the conductivity power law), as a function of volume concentrations of carbon nanotubes, at room temperature.

Dielectric Materials and Applications: ISyDMA'2016   Materials Research Forum LLC
Materials Research Proceedings 1 (2016) 45-48   doi: http://dx.doi.org/10.21741/2474-395X/1/12

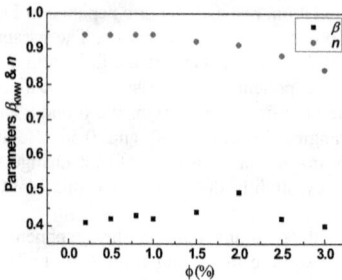

Fig. 4. The temperature dependence of the stretched exponent β_{KWW} and the Jonscher exponent n.

Conclusion

The complex electric modulus model and the Jonscher's power law have been used to investigate the electrical properties of carbon-nanotubes-loaded polymer composites. The stretching exponent β_{KWW} representing the degree of interaction and the exponent n obtained by Jonscher's power law are found to be temperature-dependent. This result confirms the fact that the dielectric response show significant change with temperature.

Acknowledgment

Financial support by the Moroccan (CNRST) - Hungarian (NKFIH) bilateral project TÉT_12_MA-1-2013-0010 and the Budapest Neutron Centre (www.bnc.hu) are gratefully acknowledged. We also thank FEDER funds through the COMPETE 2020 Program and National Funds through FCT - Portuguese Foundation for Science and Technology under the project UID/CTM/50025/2013.

References

[1] C. McClory, S. J. Chin, and T. McNally, Australian Journal of Chemistry, vol. 62, no. 8. pp. 762–785, 2009. http://dx.doi.org/10.1071/CH09131

[2] Z. Spitalsky, D. Tasis, K. Papagelis, and C. Galiotis, Prog. Polym. Sci., vol. 35, no. 3, pp. 357–401, 2010. http://dx.doi.org/10.1016/j.progpolymsci.2009.09.003

[3] S. Boukheir, A. Len, J. Füzi, V. Kenderesi, M.E. Achour, N. Éber, L. C. Costa, and A. Outzourhit., J. Appl. Polym. Sci. 133, 44514, 2016.

[4] K. S. N. A.K. Geim, Nat. Mater., vol. 6, pp. 183–191, 2007. http://dx.doi.org/10.1038/nmat1849

[5] "WinData and Winfit are software trademarks of Novocontrol." http://www.novocontrol.de/html/winfit.htm , (accessed June 2016).

[6] J. Trzmiel, K. Weron, J. Janczura, and E. Placzek-Popko, J. Phys. Condens. Matter, vol. 21, no. 34, p. 345801, 2009. http://dx.doi.org/10.1088/0953-8984/21/34/345801

[7] Z. Zallen, The Physics of Amorphous Solids. New York: Wiley, 1985.

[8] A. K. Jonscher, J. Phys. D. Appl. Phys., vol. 32, no. 14, pp. R57–R70, 1999. http://dx.doi.org/10.1088/0022-3727/32/14/201

[9] A. K. Jonscher, IEEE Trans. Electr. Insul., vol. 27, no. 3, pp. 407–423, 1992. http://dx.doi.org/10.1109/14.142701

[10] J. N. Jain, H.; Mundy, Solid State Ionics, vol. 91, pp. 3–15, 1987.

Dielectric Materials and Applications: ISyDMA'2016
Materials Research Proceedings 1 (2016) 49-52

Materials Research Forum LLC
doi: http://dx.doi.org/10.21741/2474-395X/1/13

Mineralogical, chemical and textural study of roman mortars for their restoration case of the historic site of Volubilis (Morocco)

Abdelmalek AMMARI[1*], Abdelilah DEKAYIR[2], M.A.DOS-BENNANI[3]

[1]LMM, Dept. of physics, Mohammed V University, Rabat, Morocco

[2]Equipe Géotech, Dept. of geologie, BP 11201Zitoune, Meknes, Morocco

[3]LPEE, Bassatine, Meknes, Morocco

Corresponding author: E-mail: Abdelmalek09@gmail.com

Keywords: Mortar Mosaic, Characterization, Catering, Mechanical Testing, Volubilis

Abstract. The objective of this study is to conduct a mineralogical and chemical characterization of the mortars Roman archaeological site of Volubilis to rebuild spare mortars for restoration. Samples of mortar, broken tile palate garden, and pavement mosaic Falavius Germanus. The analysis by X-ray diffraction reveals that the coarse mortar Flavius Germanus is made of quartz and calcite, with feldspar and probably, mica and dolomite in small amounts. The binder end is formed calcite and quartz. However, the broken tile mortar is formed by coarse particles, clay base mixed with a binder phase dominated by calcite. These results allowed us to reformulate spare mortars for the restoration of damaged Roman mosaics, the mortars are made up by 63.6% of lime and 36.4% of sand(with (4.19% of large grain, 71, 04% of coarse sand, 24.22%, of fine sand and 0.55% fines parts). The performance of these mortars was tested by mechanical testing.

Introduction

Roman mosaics [1-2] Volubilis are precious works of leaders. Since its ranking on the list of UNESCO World Heritage in 1987, this city continues to receive special attention. Roman artists have exploited the geological formations around the city for the construction and decoration of mosaics, through different materials. Building mortars, including those of the mosaics and the baths built pavements show the very advanced states of deterioration related to the phenomena of weathering and soil instability of the city. The purpose of this research is first to proceed to a mineralogical and chemical characterization of a mean particle size of mortar samples from the Roman mosaic (Flavius Germanus) and another from the Baths of Galliens. Then there is the alternative of recreation mortars with similar characteristics to the former moat.

Materials and methods

In the mosaic of Flavius Germanus (FG), mortar samples were collected in the nucleus layer (FGMG) and end binder (FGLF), while the mortar broken tile comes from basins Gallien baths. The latter was used by the Romans in coating walls thermal baths.

The mineralogy of the mortars is analyzed by X-ray powder (XRD) and the Fourier transformed infrared spectroscopy (FTIR) [2].

A. Mineralogy and chemical composition of Roman mortars

A.1 Mineralogy and chemistry of the nucleus and binding end mineralogy

The XRD spectra of the nucleus FGMG show in (Fig.1a). It is consisted mainly of quartz and calcite, with traces of feldspar and probably mica and dolomite. FTIR spectra of this mortar is presented in (Fig. 1 :b). The bands located at (1797, 1427, 874, 712 cm-1) are related to the calcite. The bands situated at (1034, 778, 796 cm-1) are assigned to the vibrations bands of quartz. The bands located at (3695, 3620, 1034, 796, 767 cm-1) are related to the kaolinite. The last bands located at (2982, 2924,

Dielectric Materials and Applications: ISyDMA'2016   Materials Research Forum LLC
Materials Research Proceedings 1 (2016) 49-52   doi: http://dx.doi.org/10.21741/2474-395X/1/13

2913, 2874 cm-1) are due to the presence of OH groups probably related to humic and fulvic acids and caused by root activity [2, 3].

While the analysis of the end binder (FGLF) shows spectra superimposed with those of coarse mortar (Fig. FGLF 1.a and 1.b FGLF) with very small quartz grains cemented by calcite. Unlike the coarse mortar, this binder is richer in quartz, calcite that, as evidenced by the chemical analysis.

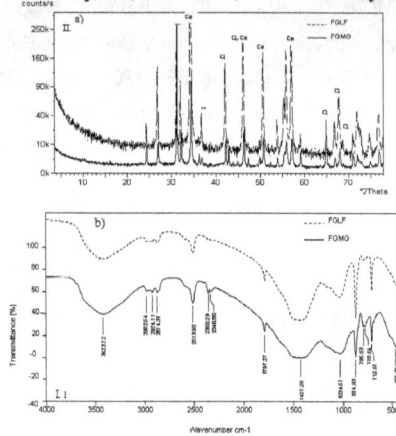

Fig.1: a) XRD spectra of the coarse mortar (FGMG) and binder end (FGLF). (Q: quartz; Ca: calcite; D: dolomite; F: feldspar M: Mica). b) FTIR spectra of: Coarse mortar (FGMG) and the end binder (FGLF) [2]

A.2 Chemical analysis and mortar outlet
The chemical analyzes of the coarse mortar (FGMG) show a SiO2 content (35 .63%) high compared with that of the end binder (FGLF) due to the quartz concentration. Conversely, the binder end shows an enrichment in CaO linked to the abundance of carbonate matrix. The heat losses are to be bonded with the presence of water.

The matrix serves as a binder for both the coarse and fine mortar binder, is essentially represented by calcite. It is the product of the carbonation of the lime used in the preparation of the mortar.

Lime wasted with water and mixed with sand has the property of setting rapidly agglomerating inert particles, such as quartz [2].

B.   Broken tile mortar Analysis
XRD shows that the binder of broken tile mortar is composed mainly of calcite (Fig. 2). The following bands of vibrations, indexed (1972.8217; 2098.1721; 2231.2363; 2281.3765; 2408.6553, 2665.1414, 3802.9368, 3926.3586, 838.8831; 1770.3326 cm-1) are typical   of calcite (Fig. 3).

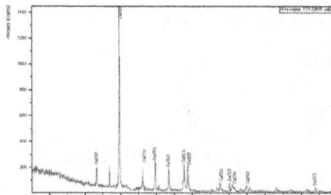

Fig.2: Diffractogram binder, mortar broken tile

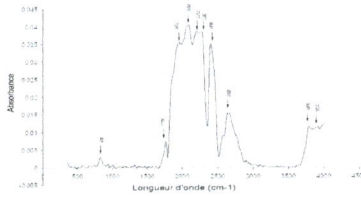

Fig.3: FTIR Analysis of the binder mortar broken tile

Preparation of spare mortar prisms for restoration

The preparation of the mortar through a mixture of lime, water, sand, with very precise proportions. Indeed, the lime mixed with water is mixed and left for several hours. Once the mixture is ready, a well-defined amount of sand is added. Prisms of 4 cm x 4 cm x 16 cm were made with different percentages of sand and lime (Table.1), the prisms are placed in the oven for 72h at a temperature of 50 ° C for drying.

Table.1: Percentages of the used sand and lime mortar in preparation

Mortars	Percentage of sand (%)	lime (g)	sand (g)	water (ml)
M1	20	400	100	800
M2	30	350	150	700
M3	40	300	200	600
M4	50	250	250	500
M5	66	170	330	300

A.1 Mortar of Replacement

Table.2, below, gives the percentages of lime and sand used in the mortar spare and the drying period in the oven. We use this mortar the same refusal percentages of sand that are already found (table.2)

Table.2: percentages of sand and lime mortar of replacement

Mortar of replacement	Lime (g)	Sand (g)	Lime (%)	Sand (%)	Drying time (months)
FGR-1	636	364	63 .6	36.4	2
FGR-2	636	364	63 .6	36.4	3

A.2 The Mechanical tests on mortar spare FG-R

The mechanical tests [5, 6] aim to characterize the laws of behavior [4] materials (continuum mechanics). This law establishes the relationship between stress (pressure = force / area) and deformations (unit lengthening without dimensions). Table.3 summarizes the results of different mechanical tests (bending Fig.4 and Compressive Fig.5) conducted on the different parts of the mortar prisms.

Fig.4: bending test   *Fig.5: Compressive strength*

Table.3: values of flexural and compression of the Replacement mortar MR

Prism	Drying time	Mass of the prism (g)	Flexion (Ff) in KN	Flex ural MPa	Compre ssive (Fc) in KN	Compre ssive n MPa
MR-1-1	2 month	279.3	0.15	0.35	1.45 1.63	0.91 1.02
MR-1-2	2 month	278.7	0.16	0.37	1.22 1.73	0.76 1.08
MR-1-3	2 month	259.8	0.12	0.28	1.49 1.59	0.93 0.99
MR2-1	3 month	310.4	0.32	0.75	4.68 3.27	2.92 2.04

Conclusion

This study allowed us to characterize the mineralogy and chemistry of Roman mortar (the nucleus of the mosaic of broken tile and mortar Roman baths).

According to chemical analyzers, we found that the coarse and fine mortar matrix binder are represented mainly by calcite. The mineralogical analysis by X-ray diffraction of the coarse mortar and binder FG end, shows that they consist mainly of quartz and calcite with different proportions (with traces of feldspar and probably mica and dolomite).

The replacement mortar (FG) is comprised of 63.6% lime and 36.4% sand (4.19% of the coarse component, 71.04% coarse sand, fine sand 24.22%, and 0 55% fine fraction).

Mechanical tests exerted on the mortar spare FG show that the prism (MR2) exhibits good Compressive compared to other prisms because of a very slow setting.

Spare mortars developed in this study should be tested in order to ensure their performance.

References

[1] F. Davidovits, Les Mortiers de pouzzolanes artificielles chez Vitruve évolution et historique architecturale, Mémoire de D.E.A, Université Paris X-Nanterre, 1993.

[2] Dekayir, M.Amouric, J.Olives, C.Parron, A. Nadiri, A. Chergui, M. A. El Hajraoui, Structure and characterisation of the materials used in the building of Roman mosaics in Volubilis City (Morocco), C. R. Géoscience, vol.336, 2004, pp. 1061–1070.

[3] K.Bouassria, A.Ammari, A.Tayyibi, H.Bouabid, J.Zerouaoui,M. Cherraj, S. CharifD'ouazzane, The effect of lime on alumino-silicate and cement on the behavior of compressed earth blocks, J. Mater. Environ. Sci. 6 (12) (2015) 3430-3435.

[4] H. Bouabid, K. .Zine-dine, M. El Kortbi, S. Charif d'Ouazzane, O. Fassi-Fehri, Comportement mécanique non linéaire du mortier de terre stabilisée, Rev. Mar. Gén. Civ., n° 81, 1999, pp : 16-19.

[5] Hakimi, N. Yamani, H. Ouissi, Résultats d'essais de résistance mécanique sur échantillon de terre comprimée, Matériaux et Construction, Vol.29, 1996, pp. 600-608.

[6] H.B. Nagaraj, M.V. Sravan, T.G. Arun, K.S. Jagadish, Role of lime with cement in long-term strength of Compressed Stabilized Earth Blocks, International Journal of Sustainable Built Environment (2014) 3, 54–61. http://dx.doi.org/10.1016/j.ijsbe.2014.03.001

Dielectric Materials and Applications: ISyDMA'2016   Materials Research Forum LLC
Materials Research Proceedings 1 (2016) 53-59   doi: http://dx.doi.org/10.21741/2474-395X/1/14

Analytical estimation of flashover voltage based on simplified formulation of the resistance of the pollution layer

S.A. BESSEDIK*[1], H. HADI[2], R. DJEKIDEL[1]

[1]Laboratoire d'Analyse et de Commande des Systèmes d'Energie et Réseaux Electriques (LACoSERE), Université Ammar Telidji Laghouat, PO BOX 37G, Laghouat (03000), Algeria

[2] Laboratoire de Génie Electrique d'Oran (LGEO), Université des Sciences et de la Technologie Mohamed Boudiaf d'Oran (U.S.T.O), BP1505 El Mnouar Oran, Algeria

Corresponding author: E-mail: s.bessedik@mail.lagh-univ.dz

Keywords: High Voltage Insulator, Flashover Voltage, Resistance of the Pollution Payer, Open Model

Abstract. High voltage insulators are supposed to provide a continuous and safety insulation in energy transmission systems. However, pollution layer that occurs on the surface of insulator decreases the performance of insulators and causes surface flashovers and therefore short circuit failures. Numerous works have been developed to calculate the flashover characteristics of polluted insulators, these works were carried out on models with simple geometry. These models are advantageous for both experimental measurements and theoretical modelling. However, these models negligent the effects of the complex insulator geometry. In this paper, an analytical method based on the open model knows as AR model is proposed to calculate the flashover voltage of polluted insulator. The proposed method takes into account the new simplified formulation of the resistance of the pollution layer. The validity of the method was verified by comparing the computed results with the experimental results of previous researchers and good correlation has been shown.

Introduction

High voltage insulators form an essential part of high voltage electric power transmission systems. Any failure in the satisfactory performance of high voltage insulators will result in considerable loss of capital, as there are numerous industries that depend upon the availability of an uninterrupted power supply. Outdoor insulators are being subjected to various operating conditions and environments. Contamination on the surface of the insulators enhances the chances of flashover. Under dry conditions the contaminated surfaces do not conduct, and thus contamination is of little importance in dry periods. In cases when there is light rain, fog or dew, the contamination on the surface dissolves. This promotes a conducting layer on the surface of the insulator and the line voltage initiates the leakage current. High current density near the electrodes results in the heating and drying of the pollution layer. An arc is initiated if the voltage stress across the dry band exceeds the withstand capability. The extension of the arc across the insulator ultimately results in flashover. It has often been suggested that the discharge elongation is caused by drying of the pollution layer around the discharge root and the extension of the dry zone across which the discharge burns [1-3].

Several methods have been proposed to calculate the flashover of polluted insulators [4-9].But some difficulties still exist. Usually, real insulators with complex shapes need to be converted to equivalent rectangles [10-13]. The rectangular model due to its uncomplicated geometry is helpful for experimental measurements. However, this model cannot show the effects of the complex shape of insulator. Consequently, using the open model, which displays the open shape of the insulator surface and corresponds completely to the actual insulator geometry, is a more suitable approach [4, 14-17]. In addition, equivalent electrical circuit based on the flat plate model and circular strips model have been developed [6, 18-20].

The resistance of the pollution layer plays a significant role in calculation of flashover voltage of polluted insulators. Therefore, a considerable number of approaches have been proposed in literature

Dielectric Materials and Applications: ISyDMA'2016   Materials Research Forum LLC
Materials Research Proceedings 1 (2016) 53-59   doi: http://dx.doi.org/10.21741/2474-395X/1/14

to simplify their expression [12, 14-15 21-23]. For this purpose, an analytical method based on new formula simplifying the resistance of the pollution layer is proposed to estimate flashover voltage of polluted insulators.

Analytical modelling

In this study the cap-and-pin insulator and its open model are used (see figures 1 and 2 for the geometry and open model of the U40 insulator). An open model of an insulator can be obtained from the following equation [4]:

$$y(x) = \theta r(x) \tag{1}$$

where x is the distance from the cap to the point which is on the leakage distance, $y(x)$ is the vertical distance on the open model, $r(x)$ is the radius which has been determined from the distance of a node to the insulator axis, and θ an angle which varies from 0 to 2π.

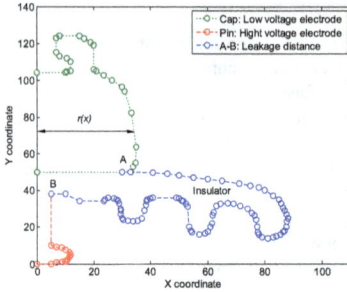

Figure 1. Half of U40 cap-and-pin insulator

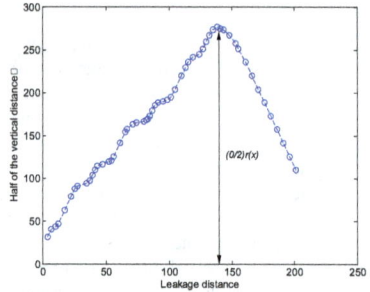

Figure 2. Half of the open model of U40 insulator

Resistance of the pollution layer

From the profile of the insulator given in figure 1, we obtain the coordinates (x_i, y_i) of the maximum number of the points on the surface of the insulator to create the table of the geometrical data of the insulator. Knowing coordinates x_i, y_i, we can calculate the elementary leakage distance between two successive rings i and $i+1$ on the surface of the insulator to find the leakage distance of the insulator as follows:

$$l_i = \sqrt{(x_{i+1} - x_i)^2 + (y_{i+1} - y_i)^2} \tag{2}$$

$$L = \sum_{i=1}^{m} l_i \tag{3}$$

where L is the leakage distance of the insulator, m is the number of points on the leakage distance and l_i is a length of the leakage distance.

We can easily calculate the elementary resistances R_i and the total resistance of the pollution layer R_p by the following relations:

$$R_i = \frac{1}{\sigma_s} \int_{l_i}^{l_{i+1}} \frac{dl}{\theta r(x)} \tag{4}$$

Dielectric Materials and Applications: ISyDMA'2016   Materials Research Forum LLC
Materials Research Proceedings **1** (2016) 53-59   doi: http://dx.doi.org/10.21741/2474-395X/1/14

$$R_p = \frac{1}{\sigma_s} \sum_{i=1}^{m} \int_{l_i}^{l_{i+1}} \frac{dl}{\theta r(x)} \tag{5}$$

where $r(l)$ is the radius to the point of curvilinear abscissa l, σ_s the surface conductivity and dl is the incremental leakage length,

The resistance of the pollution layer of equation (5) (for $\theta=2\pi$) used by most researchers is not correct in the presence of a partial arc established at the surface of the polluted insulator. However, it can only be validate in the case when the density current distribution is uniform on the entire pollution layer which is not the reel case for a real insulator shape. Therefore, an improved formulation of the resistance of the pollution layer based on the corrective angle $\theta<2\pi$ is adopted in this paper [21].

Flashover criteria

The simplest model that has been developed by Obenhaus consists of a partial arc bridging the dry zone and the resistance of the polluted wet zone in series. Therefore, the voltage across the insulator will be:

$$U = AXI^{-n} + r_y(L-X) \tag{6}$$

Where U is the voltage applied to the insulator, I the current passing through the surface of the insulator, r_y is the resistance per unit leakage length , X is the arc length considered to burn from one electrode and to elongate towards the other, and A, n the constants of the arc characteristics.

To determine conditions in which the discharge will extinguish, consider first the voltage required to maintain conduction for any given value of X. This voltage depends on the current and the V/I curve has only one turning-point, which occurs when $dU/dI = 0$ leading to a current and corresponding voltage given as [14]:

$$I_m = \left(\frac{nAX}{r_y(L-X)} \right)^{\frac{1}{n+1}} \tag{7}$$

$$U_m = (1+n)(AX)^{\frac{1}{n+1}} \left(\frac{r_y(L-X)}{n} \right)^{\frac{n1}{n+1}} \tag{8}$$

The critical voltage, current and critical length are obtained when $dU_m/dX = 0$, leading to a critical length of $X_c=L/(n+1)$ Substituting in Eqns. (7) and (8) for critical current and voltage given as [14]:

$$I_c = \left(\frac{A}{r_y} \right)^{\frac{1}{n+1}} \tag{9}$$

$$U_C = L r_y^{\frac{n}{n+1}} A^{\frac{1}{n+1}} \tag{10}$$

The resistance per unit leakage length r_y (Ω/cm) is given by following formula:

$$r_y = \frac{1}{\sigma_s(L-X_c)} \int_0^l \frac{dl}{\theta r(l)} \tag{11}$$

Dielectric Materials and Applications: ISyDMA'2016   Materials Research Forum LLC
Materials Research Proceedings 1 (2016) 53-59   doi: http://dx.doi.org/10.21741/2474-395X/1/14

Experimental setup

The experimental open model of the U40 insulator was constructed according to the calculated dimensions of the model presented on figure 2. This experimental model presents a length of 190 mm and a variable width defined using Eq. (1). The experimental setup is shown in figure 3.

Positive high voltage (U) is applied on the copper needle electrode. The gap distance h between the tip electrode and the solution surface is set up to 2 mm and kept constant during the experiments. The aqueous solution of (H2O+NaCl) was used to simulate the surface of wet polluted insulator.

Experiments of the study of the discharge propagation direction over electrolytic surface are carried out in air, at atmospheric pressure.

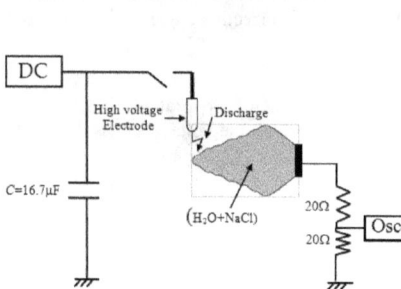

Figure 3. Experimental setup for DC systems

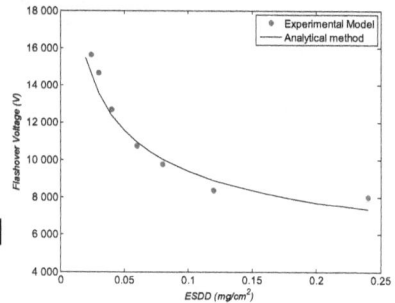

Figure 4. Comparison of critical flashover voltages obtained from experimental tests and from the analytical method

Application and validation

The proposed method is applied to estimate flashover voltage of polluted insulator, as it has been presented in previous section. The characteristic of the investigated insulators and the simulation parameters used in this study are summarized in table 1.

Table 1 Simulation parameters and Characteristic values of the insulators

	Characteristic of the investigated insulators		Simulation parameters
	U40 insulator	7k3 insulator	
D	175	288	$A = 63, \quad n = 0.76$
L	185	305	$\theta < 2\pi$
H	110	185	$m = 58$

The experimental critical flashover voltages obtained for different electrolyte solutions with the open model was compared with the results provided by the proposed method.

The results obtained for different ESDD presented in figure 4 shows that the analytical method presents a very good accuracy compared to experimental results. This results permit to conclude that the analytical method describes well the experimental device.

The proposed method is validated by comparing the results with the theoretical and experimental data of other researchers for the 7k3 cap-and-pin insulator (figure 5). Using the present analytical method, flashover voltage of this insulator has been computed and compared.

Dielectric Materials and Applications: ISyDMA'2016   Materials Research Forum LLC
Materials Research Proceedings 1 (2016) 53-59   doi: http://dx.doi.org/10.21741/2474-395X/1/14

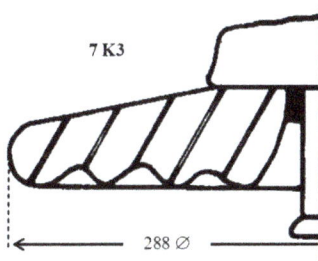

Figure 5. *7k3 insulator geometries used for the analytical method validation*

Figure 6 shows a comparison between the experimental results of Von Cron [3], theoretical results of Wilkins [12], Dynamic Model [21] and Sundarajan [24] with the present method for seven 7k3 insulator units.

The results demonstrated a good agreement between the proposed analytical method and the experimental results of Von Cron and theoretical results of Sundarajan.

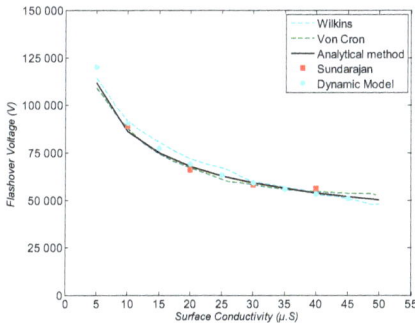

Figure 6. *Comparison of flashover voltages for 7K3 insulator (For a seven-unit)*

Conclusion

The basic difficulties in calculation of flashover voltage of polluted insulators come from the complex shape of the insulators and the complicated nature of the flashover phenomenon. However, with increasing knowledge on the mechanism of polluted flashover and computational facilities in recent years, encouraging progress has been achieved in the theoretical finding of flashover performances of polluted insulators. It is hoped that analytical techniques may gain importance as supplement to experimental methods in the evaluation of pollution flashover performances.

In this study, an analytical method based on simplified formulation of the resistance of the pollution layer has been successfully applied for the estimation of the flashover voltage on polluted insulators. The obtained results show a very good agreement compared with the experimental and theoretical results obtained by different authors. The proposed analytical method has fewer parameters (A, n and θ) to be adjusted and has been demonstrated to be competitive than other methods concerning pollution flashover performances.

References

[1] Gillem, G. H, "Report on the activities of study comitee Zo.5 Insulators, CIGRE", Report 234, 1960.

Dielectric Materials and Applications: ISyDMA'2016
Materials Research Proceedings 1 (2016) 53-59

Materials Research Forum LLC
doi: http://dx.doi.org/10.21741/2474-395X/1/14

[2] Korbut, N. V and Kerkhelev, S.F, "Investigation of the flashover characteristics of dirtly insulators", Elektrichestvo, 3, 76-81, 1962.

[3] Von Cron, H. and Estorff, W, "The HV insulator as a problem of extraneous films", E.T.Z, 1260-1266, 1952.

[4] Rumeli, M. Hızal and Y. Demir, "Analytical Estimation of Flashover Performances of Polluted Insulators", MADRAS, Vol.1.02,pp.01_06, 1981.

[5] R. Sundararajan, and R. S. Gorur, "Dynamic Arc Modeling of Pollution Flashover of Insulators under dc Voltage", IEEE Trans. Electr. Insul. Vol. 28, pp. 209-218, 1993. http://dx.doi.org/10.1109/14.212246

[6] N. Dhahbi-Megriche, A. Beroual and L. Krahenruhl, "A New Proposal Model for Flashover of Polluted Insulators", J. Phys.D: Appl. Phys., Vol. 30, pp. 889-894, 1997.

[7] N. Dhahbi-Megriche and A. B´eroual, "Flashover Dynamic Model of Polluted Insulators under ac Voltage", IEEE Trans.Dielectr. Electr. Insul., Vol.7, pp. 283-289, 2000. http://dx.doi.org/10.1109/94.841822

[8] F. V. Topalis, I. F. Gonos, I. A. Stathopoulos, Dielectric behaviour of polluted porcelain insulators, IEE Proceedings Generation Transmission and Distribution 148 (July (4)) (2001) 269–274. http://dx.doi.org/10.1049/ip-gtd:20010258

[9] Z. Aydogmus, M. Cebeci, "A New Flashover Dynamic Model of Polluted HV Insulators". IEEE Transactions on Dielectrics and Electrical Insulation Vol. 11, No. 4, pp. 577-584, August 2004. http://dx.doi.org/10.1109/TDEI.2004.1324347

[10] L. L. Alston, S. Zoledziowski, "Growth of Discharges on Polluted Insula-tion", PIEE, Vol. 110, pp. 1260-1266, 1963.

[11] B. F. Hampton, "Flashover Mechanism of Polluted Insulators", PIEE, Vol. 111, pp. 985-990, 1964.

[12] R. Wilkins, "Flashover Voltage of High Voltage Insulators with Uniform Surface-Pollution Films", Proc. IEE, Vol. 116, pp.457-465, 1969. http://dx.doi.org/10.1049/piee.1969.0093

[13] S. Flazi, "Etude du contournement électrique des isolateurs haute tension pollués, Critères d'élongation de la décharge et dynamique du phénomène"; thèse de doctorat, Université Paul Sabatier de Toulouse 1987.

[14] G. Zhicheng, and Z. Renyu, "Calculation of dc and ac Flashover Voltage of Polluted Insulators", IEEE.Trans.Electr.Insul.,Vol.25, pp.723-729, 1990. http://dx.doi.org/10.1109/14.57096

[15] Derkaoui, "Etude de la validité d'un nouveau modèle expérimental du contournement des isolateurs haute tension pollués"; memoire de Magistère, Université USTO, Algérie 2004.

[16] M. Marich, H.Hadi and R.Amiri. "New approach for the modeling of the polluted insulators". 2006 Annual Report Conference on Electrical Insulation and Dielectric Phenomena.2006. http://dx.doi.org/10.1109/CEIDP.2006.311964

[17] S. A .Bessedik, H. Hadi, M. Marich, A. Bouyekni, "Modélisation Dynamique du Phénomène de Contournement des Isolateurs Pollues".8ème Conférence Nationale sur la Haute Tension, CNHT'2011,Univ. Ibn Khaldoun de Tiaret, Mai 2011.

[18] N. Dhahbi.Megriche and A. Beroual, "Model for calculation of flashover characteristics on polluted insulating surfaces under DC stress", 1998 Annual Report Conference on Electrical Insulation and Dielectric Phenomena CEIDP, 1998. http://dx.doi.org/10.1109/CEIDP.1998.733855

[19] B. S. Ram, "Flashover Voltage of Contaminated Insulators", IEEE Int. Symp. On Electrical Insulation, pp. 39-43, 1988. http://dx.doi.org/10.1109/elinsl.1988.13862

[20] S. Flazi, N. Boukhennoufa and A.Ouis, "Critical Condition of DC Flashover on a Circular Sector ModelIEEE Transactions on Dielectrics and Electrical Insulation Vol. 13, No. 6; December 2006.

[21] S. A. Bessedik, H. Hadi, C. Volat and M. Jabbari, "Refinement of Residual Resistance Calculation Dedicated to Polluted Insulator Flashover Models", IEEE Trans. on Diel. and Elect. Insul., Vol. 21 pp .1207-1215, June 2014.

[22] S. Flazi, A. Ouis and N. Boukhennoufa, "Resistance of pollution in equivalent electric circuit of flashover", IET Gener. Transm. Distrib., Vol. 1, No. 1, January 2007.

[23] S. A. Bessedik, H. Hadi, "Etude expérimentale du modèle ouvert d'un isolateur réel haute tension", 4th International Conference on Electrotechnics, ICEL'2009, Univ.USTO Oran, Novembre 2009.

[24] G. G. Karady, F. Amarh, R. Sundararajan, "Dynamic modeling of AC insulator flashover characteristics". High Voltage Engineering Symposium, Conference Publication No. 467, 1999. http://dx.doi.org/10.1049/cp:19990804

Dielectric Materials and Applications: ISyDMA'2016   Materials Research Forum LLC
Materials Research Proceedings 1 (2016) 60-62   doi: http://dx.doi.org/10.21741/2474-395X/1/15

Advanced nanosized spinels for
energy conversion application

T. PETKOVA*[1], D. NICHEVA[1], B. ABRASHEV[2], K. PETROV[2], P. PETKOV[3]

[1]Solid state electrolytes department, Institute of electrochemistry and energy systems, Acad. Evgeni Budevsky, Bulgarian Academy of Sciences, 1113 Sofia, Bulgaria

[2]Department of electrocatalysts, Institute of electrochemistry and energy systems, Acad. Evgeni Budevsky, Bulgarian Academy of Sciences, 1113 Sofia, Bulgaria

[3]Departments of Physics, TFTLab, University of Chemical Technology and Metallurgy, 1756 Sofia, Bulgaria

*Corresponding author: E-mail: tpetkova@iees.bas.bg

Keywords: Nanomaterials, Properties, Energy, Catalysts

Abstract. Samples from $NixCo_{3-x}O_4$ system, with different dopant content, x, have been prepared by means of Pechini method. The materials are characterized by physical chemical methods, i.e. X-ray diffraction and infrared spectroscopy. The conductivity has been determined by impedance spectroscopy analysis. Electrochemical tests have been carried out to define the catalytic activity.

Introduction

During the last few years, synthesis of nano structured oxide materials had attracted considerable attention [1]. Oxides constitute a wide class of materials with good electrocatalytic activity [2].

Nickel cobaltite, $NiCo_2O_4$, is a mixed-metal oxide spinel that has shown exceptional ability to serve as an electrode for oxygen evolution reaction (OER) and has been studied quite extensively by electrochemical methods for the purpose of the Zn-air batteries. However, the oxygen reduction reaction (ORR) on cathode is sluggish which limits the rate performance of the battery. Developing of highly active catalysts for ORR is a key to improve the performance of zinc-air battery.

The aim of the present work is study of oxide materials from $NixCo_{3-x}O_4$ system, where x = 0; 0.7; 1.4; 2.1 as bifunctional catalysts for oxygen evolution and reduction electrodes. The nano-sized catalyst materials will be used to improve the efficiencies of carbon free bi-functional air electrodes, BAE-s.

Experimental

Nickel nitrate hexahydrate, cobalt nitrate hexahydrate, and sodium hydroxide were used as starting materials. Citric acid was used as stabilizer. The initial chemicals obtained from Alfa aesar were used for preparation of the nano particles of nickel oxide, cobalt oxide and nickel-cobalt oxide. The oxides were prepared by means of Pechini method. The powder obtained was sintered at 400 and 500°C for 6 hours in a furnace to obtain the respective oxide structure.

XRD study was carried out using Philips model powder diffractometer employing Cu- Kα radiation (λ=1.54060 Ao) operating at 40 kV, 30 mA. The XRD patterns were obtained in the range 2Theta = 20-80. The surface morphology of the powdered sample was studied by Scanning Electron Microscope (SEM) [JEOL/EO JSM-6390]. FTIR spectra were recorded using Brucker FTIR Spectro Photo Meter in the range from 400 cm^{-1} to 4000 cm^{-1}. AC conductivity studies were done using Gain Phase Impedance Analyzer, MODEL HP, 4294 A, 40 Hz-110 MHz.

Results and disscusion

The XRD patterns of Co_3O_4 and $NiCo_2O_4$ prove the crystal spinel structure of the prepared samples. The relative crystalline sizes determined from the XRD spectra using Debay-Scherrer equation demonstrate values between 50 and 100 nm. It has been found that the higher the sintering temperature is the bigger crystallites size is. The broadening of the XRD peaks is due to the

Dielectric Materials and Applications: ISyDMA'2016   Materials Research Forum LLC
Materials Research Proceedings 1 (2016) 60-62   doi: http://dx.doi.org/10.21741/2474-395X/1/15

nanocrystalline nature of the particles. This broadening could also appear from defects like dislocations and twinning grown unexpectedly during the chemical reaction. Most probably the reason is the insufficient energy of the atom to move and to locate at a proper site of the crystalline lattice.

The SEM pictures reveal the powder as black homogenous in color with particles mostly with spherical shape.

The IR spectra clearly confirm the presence of spinel structure by the modes detected which determine metal – oxygen bonds. The bond observed at 657 cm^{-1} is associated with Co-O bonds while that at 560 cm^{-1} is due to vibrations of Ni-O bonds. The IR spectra are typical for cubic spinels with space group Fd3m with most intensive bands in the range 550-660 cm^{-1}.

The AC conductivity measurements in the temperature range 25 – 250 °C show higher values in the sample with smaller amount of nickel (fig.1)l. Dopping cobalt oxide with nickel produces significant increase in the electrical conductivity while maintaining the spinel structure. Nickel cations are found most probably to reside in octahedral sites with a valence 2$^+$ and 3$^+$ represented by equation below:

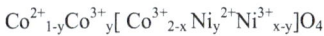

$$Co^{2+}_{1-y}Co^{3+}_{y}[\ Co^{3+}_{2-x}Ni_{y}^{2+}Ni^{3+}_{x-y}]O_4$$

The conductivity increases as Ni content decreases by 2 orders of magnitude however it decreases when the nickel proportion becomes higher than 1, probably because of appearance of NiO. The NiO phase, which is less conductive than the spinel phase "dilutes" the spinel composition and the measured conductivity progressively decreases [3]. Oxides prepared by sol-gel methods as a rule present high conductivity values. The suggested reason is the increase of the charge carriers concentration and growth in their mobility due to decrease in Co-Co distance which may be inferred to the evolution of the cell parameter [4].

The electrodes produced from Co$_3$O$_4$ and silver powder tested electrochemically in alkaline solution show good catalytic activity for OER (fig.2) and ORR (fig.3.).

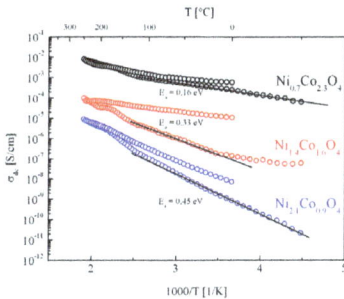

Fig. 1. Temperature dependence of conductivity for Ni$_x$Co$_{3-x}$O$_4$

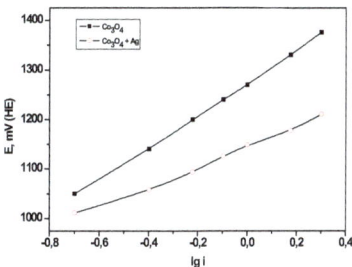

Fig. 2. Tafel plot for OER of pure Co$_3$O$_4$ and Co$_3$O$_4$ with Ag electrodes

Dielectric Materials and Applications: ISyDMA'2016   Materials Research Forum LLC
Materials Research Proceedings **1** (2016) 60-62   doi: http://dx.doi.org/10.21741/2474-395X/1/15

The activity is better in the catalyst consisting of oxide and silver powder. We suggest the phenomena is due to to a synergistic catalytic effect in both OER and ORR: Co_3O_4 cocatalyses oxygen reduction at Ag and Ag cocatalyses solution oxidation at Co_3O_4 [5].

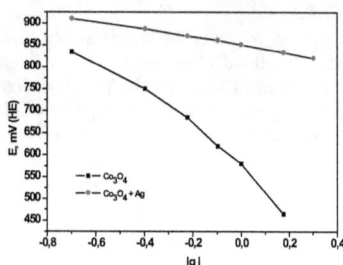

Fig. 3. Tafel plot for ORR of pure Co_3O_4 and Co_3O_4 with Ag electrodes

Conclusion

Cobalt oxide and cobalt nickel having the spinel crystal structure and the compositions Co_3O_4 and $Ni_xCo_{3-x}O_4$ were successfully prepared via sol-gel method. The study shows that mixed oxides prepared by sol gel route, Pechini method, possess good catalityc properties. The addition of silver improve the catalytic activity probably due to synergistic catalytic effect in both OER and ORR

Acknowledgment

The research leading to these results has received funding from EC, H2020, Zinc-Air Secondary batteries based on innovative nanotechnology for efficient energy storage.

References

[1] J.-S. Lee, S. Tai Kim, R. Cao, N.-S. Choi, M. Liu, K.T. Lee, J. Cho, Metal–Air Batteries, with High Energy Density: Li–Air versus Zn–Air, Advanced Energy Materials 1, 34 2011. http://dx.doi.org/10.1002/aenm.201000010

[2] Hamdani R.N. Singh, P. Chartier, Co3O4 and Co- Based Spinel Oxides Bifunctional Oxygen Electrodes, Int. J. Electrochem. Sci., 5 556 – 577, 2010

[3] MC.F Jr Windish,.; K.F Ferris,.; G.J Exarhos.,.; S.K Sharma,. Conducting spinel oxide films with infrared transparencyThin Solid Films, 420/421,89, 2002

[4] Kahoul, A.; Hammouche, A.; Nâamoune, F.; Chartier, P.; Poillerat, G.; Koenig, J.F. Mat.Res. Bull., 35, 1955, 2000

[5] F.H.B. Lima, J.F.R. de Castro, E.A. Ticianelli, Silver-cobalt bimetallic particles for oxygen reduction in alkaline media, J. Power Sources 161, 806. 2006. http://dx.doi.org/10.1016/j.jpowsour.2006.06.029

[6] H M.A. Amin, Ht Baltruschat , D. Wittmaier, K.A Friedrich, A Highly Efficient Bifunctional Catalyst for Alkaline Air-Electrodes Based on a Ag and Co3O4 Hybrid: RRDE and Online DEMS Insights, Electrochimica Acta 151, 332–339, 2015. http://dx.doi.org/10.1016/j.electacta.2014.11.017

Dielectric Materials and Applications: ISyDMA'2016
Materials Research Proceedings **1** (2016) 63-66

Materials Research Forum LLC
doi: http://dx.doi.org/10.21741/2474-395X/1/16

Contribution to the study of power losses due to friction phenomenon between spur gear teeth

Bernard MUSHIRABWOBA*[1], Belfals LAHCEN[1], Brahim NAJJI[2], Abdelilah LASRI[1]

[1]Laboratory of Quality, Security and Maintenance, University Mohammed V Agdal, Rabat, Morocco

[2]Laboratory of Mechanics, Thermics and Materials, Ecole Nationale supérieure des Mines de Rabat, Rabat, Morocco

*Corresponding author: E-mail: mubefils@yahoo.fr

Keywords: Spur Gear, Materials, Friction and Wear Power Loss, Gear Efficiency

Abstract. This paper aims to present a numerical model to calculate the power losses due to the friction phenomenon in spur geared transmissions which are highly used in many domestic and industrial applications. Since the increase of the power to be transmitted at high rotational speeds result in significant power losses and wear phenomenon, the latter must therefore be taken into account during the different phases especially when designing and choosing the materials of the gears since power losses have direct impact on the lifetime of the power transmitters. The present work gives a better understanding of how to numerically evaluate the friction power losses in spur gears and it is a preliminary study to the future work where the impact of friction and wear phenomena on the material lifetime and energy consumption will be discussed.

Introduction

Nowadays, power transmitters which are mechanical systems or mechanisms that are used to accommodate the power according to the needs are highly used in many of our daily activities. While the popularity of the latter is closely linked to better performance they offer, nevertheless, the increase in power to be transmitted at high rotational speeds results in significant power losses. A well designed mechanical system would enable the user to avoid breakage of the mechanism due to thermal expansion and would allow a better design of the cooling systems. In a geared power transmission system, the total power loss can be divided into load-dependent contribution, the losses due to friction phenomenon for example, and into load-independent contribution which includes the power losses related to the lubrication method, to the ventilation on teeth and to the trapping of the air-lubricant between the teeth. In function of the operating conditions, each power loss category controls the overall efficiency of the system. As cited in [1] and in [2], the friction phenomenon related power losses are preponderant at low speeds whereas they are low at high-speed in comparison to the power losses due to the ventilation phenomenon and to the trapping phenomenon which was discussed by the authors in [3]. In order to understand the friction phenomenon resulting from metal-metal in non-lubricated contacts also known as dry friction, many experiments were conducted. In the case of lubricated contacts there are few models to qualitatively and quantitatively identify the friction phenomenon and its direct influence on power loss in geared power transmissions. In this work a numerical method was developed to calculate the friction power losses assuming the contact between teeth as a lubricated one.

Modelling of meshing and frictional power loss calculation

By studying the impact of different geometric and operational parameters on the power losses due to friction and by exploiting the documentation on spur gears cited in [4], a numerical modeling was established; using the latter we can instantaneously know the evolution of different meshing parameters along the line of action. The data summarized in the table 1were used in our study:

Dielectric Materials and Applications: ISyDMA'2016   Materials Research Forum LLC
Materials Research Proceedings **1** (2016) 63-66   doi: http://dx.doi.org/10.21741/2474-395X/1/16

TABLE 1. GEOMETRIC DATA

Parameter	Value
Number of teeth (Pinion)	20
Number of teeth (Wheel)	40
Module (mm)	3
Tooth width (mm)	20
Viscosity (cSt)	60
Pressure angle (degrees)	20
Surface roughness parameter S (μm)	0.07
Effective radius of curvature R (mm)	5
Maximum pressure of Hertz P_h (Gpa)	2
Torque (Nm)	250
Rotational speed (rpm)	1500

Sliding velocity

In [4],the sliding velocity is defined as follows:

$$V_s(x) = 0.1047\left(1 + \frac{1}{i}\right)N(x - x_p) \tag{1}$$

Where i is the gear ratio, N the rotational speed, x the meshing position along the path of contact and x_p the abscissa of the pitch point. Using the data in table 1, we plotted in figure 1below the sliding velocity in the function of the meshing position.

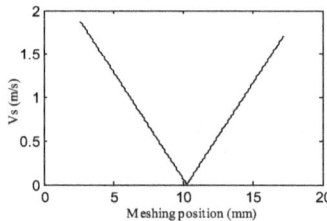

Fig. 1.Variation of the sliding velocity along the line of action

The maximum value of the sliding velocity is found at the starting and ending points of meshing while its minimum value is obtained at the pitch point.

Coefficient of friction

The law of friction in this study is given in [5]as follows:

$$\mu = e^{f(SR,P_h,\upsilon_0)} P_h^{b_2} |SR|^{b_3} V_e^{b_8} \upsilon_0^{b_7} R^{b_6} \tag{2}$$

Where: $f(SR, P_h, \upsilon_0) = b_1 + b_4 |SR| P_h \log_{10}(\upsilon_0) + b_5 e^{-|SR|P_h \log_{10}(\upsilon_0)} + b_9 e^{S}$ (3)

V_e is the entraining velocity in m/s, SR the slide to roll ratio and the constants $b_i$are given as: $b_i = $ -8.916465 ; 1.03303 ; 1.036077 ; -0.354068 ; 2.812084 ; -0.100601 ; 0.752755 ; -0.390958 ; 0.620305 for $i = 1$ to 9. The variation of the coefficient of friction during the meshing phase is shown on figure 2 below:

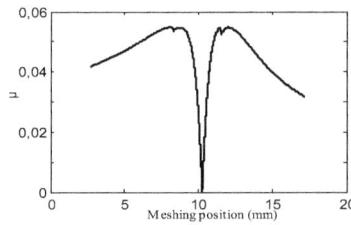

Fig. 2. Variation of the coefficient of friction along the path of contact

The friction coefficient curve shows an increase of the latter from the starting point of meshing and its quick drop near the pitch point is due to the increase of the load in that area.

Sliding power losses
In spur geared transmissions, the sliding power losses are calculated using the basic Coulomb law:

$$P_s(x) = F_N(x)V_s(x)\mu(x) \tag{4}$$

Where: $V_s(x)$ is the sliding velocity, $\mu(x)$ the coefficient of friction and $F_N(x)$ the transmitted normal load given by:

$$F_N(x) = \frac{2C}{dp_1 \times b \times \cos\theta} \tag{5}$$

With; C the torque, dp_1 the pinionpitch diameter, b the tooth width and θ the pressure angle. The sliding power loss is plotted on the figure 3 below

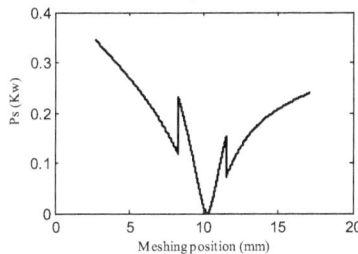

Fig. 3. Variation of the power loss along the line of action

The instantaneous power loss curve shows a minimum value of the sliding power losses which results in the increase of the load and the decrease of the coefficient of friction and the sliding velocity in that area.

Conclusion
Analytical study on instantaneous sliding power loss in spur geared transmissions was carried out. A numerical model that takes into account different gear parameters in the contact was developed. It allows the user to instantaneously know the sliding power losses at every meshing position along the line of action, in order to evaluate the average power loss; a numerical integration will be used in our

Dielectric Materials and Applications: ISyDMA'2016   Materials Research Forum LLC
Materials Research Proceedings 1 (2016) 63-66   doi: http://dx.doi.org/10.21741/2474-395X/1/16

future work by aiming to study the impact of the friction and wear phenomena to gear material lifetime and energy consumption.

References

[1] Diab, Y. ; Ville, F. ; Houjoh, H. ; Sainsot, P. ; Velex, P. ; 2004 ; '*Experimental and numerical investigations on the air-pumping phenomenon in high speed spur and helical gears*', Proc. IMechE vol. 219 Part c:, Journal of Mechanical Engineering Sciences,2005.

[2] Velex P. and Cahouet, V. « Experimental and numerical investigations on the influence of tooth friction in spur and helical gear dynamics » ASME, J. Mech. Des., 2000, Vol. 122, pp 515-522. http://dx.doi.org/10.1115/1.1320821

[3] Abdelilah Lasri, Lahcen Belfals, Brahim Najji and Bernard Mushirabwoba, "*Pressure estimation of the Trapped and Squeezed   oil between Teeth Spaces of Spur Gears*", Applied Mathematical Sciences, Vol. 8, 2014, no. 107, 5317 – 5328, HIKARI Ltd http://dx.doi.org/10.12988/ams.2014.47542.

[4] Gitin M. Maitra: "*Handbook of Gear Design*". Second edition, 1994.

[5] H.Xu, A. Kahraman, N. Anderson, and D. Maddock, "Prediction of mechanical efficiency of parallel-axis gear pairs," ASME, Journal of Mechanical design, Vol, 129, no. 1, p, 58-68, 2007.

Dielectric Materials and Applications: ISyDMA'2016   Materials Research Forum LLC
Materials Research Proceedings 1 (2016) 67-70   doi: http://dx.doi.org/10.21741/2474-395X/1/17

Simulation of electrical properties of InP/InGaAs heterojunction bipolar transistors in microwave

Yamina BERRICHI*[1], Kheireddine GHAFFOUR[2]

[1]Physics Department, Faculty of Sciences, University of Abou-bekr Belkaid, Algeria

[2]Electronics department, Faculty of Technology, University of Abou-bekr Belkaid, Algeria

*Corresponding author: E-mail: amina2010271@yahoo.fr

Keywords: Transistor, HBT, InP/InGaAs, NPN, Heterojunction, Microwave

Abstract. This work is devoted to the simulation of electrical characteristics of the InP/InGaAs double heterojunction bipolar transistor (DHBT) NPN type, the essential aim for extracting the maximum and transition frequencies are important indicators for the use of the component in the microwave. InP/InGaAs DHBTs with surface of $0.25 \times 0.73 \ \mu m^2$, the simulation of this structure has demonstrated a maximum current gain cutoff frequency f_T of 900 GHz, with a simultaneous maximum power gain cutoff frequency f_{MAX} of 500 GHz at the current density J_c of 66 m A/μm. tapical BV_{CE} values exceed 1.9 V.

Introduction

The InP/InGaAs NPN double heterojontion bipolar transistor is very good candidates for many applications because of their advantage: Lower surface recombination, super electron, lower turn-on voltage, higher electron mobility, better thermal dissipation, better microwave performance.

The InP/InGaAs double heterojuntion bipolar transistor (DHBT), is the subject of this work, it is emerging as a very promising high speed device for a wide variety of applications, Discrete InP/InGaAs HBTs capable of operating well above 765 at 25°c and 845 GHz at -55°C have been demonstrated experimentally [1].

In this paper we have presented electronic characteristic of InP/InGaAs HBTs over a wide range of frequencies simulated.

Many parameters will be specifying for the simulation, once the mesh, geometry, doping profiles, method and physical models.

Model of simulation

The modeling of the transport of charge carriers in semiconductor structures is more or less complex according to the degree of approximation to be taken to account properly experimentally observed properties. If quantum effects are taken into account, transport can then be modeled with a classic approach: Fermi-Dirac statistics. And for solving equations semiconductors systems, we chose the method of Newton.

Resultats and discusion

Structure of simulation
The layer structure of the InP/InGaAs DHBT is shown in fig 1.

Dielectric Materials and Applications: ISyDMA'2016   Materials Research Forum LLC
Materials Research Proceedings **1** (2016) 67-70   doi: http://dx.doi.org/10.21741/2474-395X/1/16

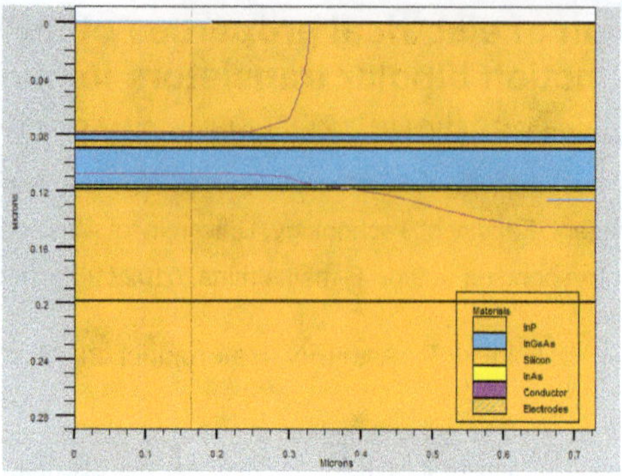

Fig.1. Structure of InP / InGaAs DHBT.

As shown in table I, the HBT consists of 80 nm thick emitter, 30 nm thick base, 26 nm thick collector and 80 nm thick substrate.

TABLE I: PARAMETERS DEFINITION OF MESH AND DOPING

Electrodes	Dopage (cm^{-3})	Peak and junction doping	Thickness (nm)
surface concentration	1×10^{15}		/
Emitter (InP)	7×10^{19}	junc=0.078	80
layer (InGaAs)	Non dopé		5
Layer (AsGa)	Non dopé		5
Base (InGaAs)	3×10^{19} 5×10^{19}	junc=0.11 peak=0.09 char=0.05	26
Layer (InAs)	Non dopé		3
Collector (InP)	3×10^{19}	peak=0.19	80
Substrat (InP)	/	/	10

The Figure 2 shown the collector and base current of InP /InGaAs DHBT, biased in the forward active mode of operation with $V_C = 0.7$ V.

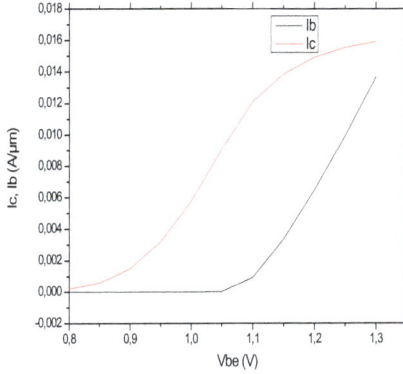

Fig.2. Gummel plot of InP / InGaAs DHBT

A low polarization (0V $<V_{BE}$ <1.05, we can observe the current components of the base (I_B < 1nA) called "non-ideal".

For intermediate polarizations (1.05V $<V_{BE}$ <1.3) the base current and the collector current have ideal behavior.

With high polarization (V_{BE}>1.3), collector current is reduced. Several reasons explain this: the influence of the base resistors and the quasi-saturation phenomenon.

Frequency analysis
The Figure 3 shows the dynamic and unilateral gain of InP / InGaAs DHBT.

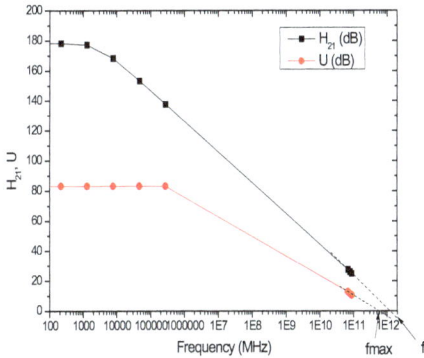

Fig.3. Dynamic and unilateral gain of InP / InGaAs DHBT.

The value of f_T as found by extrapolation (at -20dB/decade) of dynamic gain to 0 dB [2], this extrapolated value shall be denoted by H_{21}, the value of f_{MAX} as found by extrapolation (at -20 dB/ decade) of unilateral gain to 0 dB [2], this extrapolated value shall be denoted by U.

Dielectric Materials and Applications: ISyDMA'2016   Materials Research Forum LLC
Materials Research Proceedings 1 (2016) 67-70   doi: http://dx.doi.org/10.21741/2474-395X/1/16

TABLE II COMPARISON BETWEEN OUR RESULTATS AND OTHER WORK

	ft (GHz)	fmax (GHz)	ref
InP/InGaAs HBT	765 (294K) 845 (218K)	227 (294K) 263 (218K)	[3]
InP/InGaAs HBT	360 (294 K)	800K	[4]
InP/InGaAs HBT	900	500	Our resultants

The HBTs were characterized network analyzer from 100 MHz to 89 GHz. The current gain and unilateral gain for a 0.25×0.73 μm^2 HBT are shown in Fig.3, successively.

A summary of the characteristic of the most recent high-speed bipolar transistor is shown in table II. The f_T and f_{max} product for the InP/InGaAs DHTs devices exceeds 900 GHz and 500 GHz successively, in comparison by the experimentally results of the [3], [4], the values obtained of the maximum current gain cutoff frequency f_t and maximum power gain cutoff frequency f_{MAX} is compared with the experimental results.

On the other hand, we find maximum power gain cutoff frequency f_{MAX} smaller than the maximum current gain cutoff frequency f_t frequency that becomes the high resistivity of the base region (R_B).

Conclusion

This paper presents an extraction of the maximum current gain cutoff frequency (f_T is 900 GHz) and maximum power gain cutoff frequency f_{MAX} (f_{MAX} is 500 GHz) parameters in the microwave. Our resultats prove the quality of the model and the method. Future work will pursue reduction in access resistivity of the base to improve device performance.

References

[1] Snodgrass W, Hafez W, Harff N, Feng M. Pseudomorphic heterojunction bipolar transistor (PHBTs) experimentally demonstrating ft=765 GHz at 25°C increasing to ft=845 GHz at -55°C. Présenté au IEDM (IEEE), 2006.

[2] S. Lee, H.J.Karim, M.Urteaga, S. Krishnan, Y. Wei, M. Dahlstrom and M. Rodwell, transferred-substrate InP- InGaAs- InP double Heterojunction Bipolar Transistors With f_{MAX} of 425GHz, Electronic latters 2001.

[3] Lobisser E, Griffith Z, Jain Z, Thibeault BJ, Rodwell M. 200-nm InGaAs/InP DHBT employing a dual sidewall emitter process demonstracting fmax>800GHz and fT>360GHz. Presented at IEEE Proc. Indium Phosphide and Related Materials, pp. 16-19, 2009.

[4] Snodgrass W, Freng M. Nano-scale type II InP/InGaAs DHBTs to reach THz cutoff frequencies. Presenté au CS MANTECH Conference, 2008.

Dielectric Materials and Applications: ISyDMA'2016   Materials Research Forum LLC
Materials Research Proceedings 1 (2016) 71-74   doi: http://dx.doi.org/10.21741/2474-395X/1/18

Electric field distribution around 400 kV line composite insulators in different connection conditions

M. BOUHAOUCHE*, A. MEKHALDI, M. TEGUAR

Laboratoire de Recherche en Electrotechnique, Ecole Nationale Polytechnique, Algiers, Algeria

*Corresponding author: E-mail: mohamed.bouhaouche@g.enp.edu.dz

Keywords: Composite Insulators, Electric Field Distribution, Suspension, Dead-End, Corona Ring, Conductor, Ground Structure, Impulse Voltage

Abstract. This paper presents simulation results of the electric field distribution along composite insulators used in I, II and V strings suspension and dead-end towers in 400 kV AC power transmission lines in Algeria. A two-dimensional model is built using COMSOL Multiphysics software based on the Finite Element Method. Corona rings are considered on both line and ground ends. Their parameters (ring radius, tube radius and distance from both ends) are kept invariable throughout the analyses. The impact of conductors and ground structure, insulator orientation (suspension and dead-end) and insulators surface state condition (dry and polluted) on the electric field distribution has been analyzed. General conclusions are displayed such as that the dead-end insulators experience higher electric field compared to those used in suspension lines. Finally, the influence of impulse voltage on the electric behavior of insulators is investigated.

Introduction

Reliability of power transmission lines depends on the proper choice of insulator type. For that purpose, composite insulators have been introduced and used to replace the traditional ones made of glass or porcelain [1]. Indeed, composite insulators have lighter weight, higher mechanical strength and better anti-contamination performances due the hydrophobic characteristics of their surface [2]. However, composite insulators still suffer from aging problems, mostly related to electric field distribution, which lead in general to corona phenomena on the surface and flashover.

The evaluation of electric field distribution is therefore helpful to predict corona discharges occurrence on the surface of insulators. Many studies were carried out for the analysis of electric field along composite insulators [3-5]. Numerical techniques are widely used in modern simulation tools to study electric field distribution such Finite Element Method [6].

The electrical stress along composite insulators depends on various factors such as corona rings, tower configuration, voltage magnitude, environmental conditions and insulators orientation, etc [7-9]. In this paper, we evaluate at first the electric field distribution along a 400 kV AC composite insulators used in different orientation configurations (I, II, V and dead end) under dry and clean surface conditions. Then, for each configuration, we took into account the effect of different parameters on the electric field distribution such as grounding structure and conductors, presence of contamination on the surface and impulse voltage.

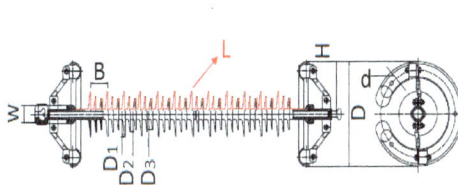

Fig. 1. Composite insulator characteristics.

Dielectric Materials and Applications: ISyDMA'2016   Materials Research Forum LLC
Materials Research Proceedings **1** (2016) 71-74   doi: http://dx.doi.org/10.21741/2474-395X/1/18

Modeling and Materials Properties

In this study, a typical 400 kV SiR composite insulator is modeled by AutoCAD software and then exported to COMSOL Multiphysics for simulations. Figure 1 show the used composite insulator associated to its corona ring made of steel. Their dimensions are given in Table I.

TABLE I DIMENSIONS OF COMPOSITE INSULATOR AND CORONA RING

Shed Number	38/38/38
Shed diameter D1/D2/D3 (mm)	190/131/100
Shed spacing B (mm)	79
Leakage path L (mm)	16300
Width of end fittings W (mm)	72.4
Corona ring diameter D (mm)	370
Corona tube diameter d (mm)	50
Corona ring height H (mm)	110

Relative permittivity of the materials is an important input parameter for the study. Values of relative permittivity of each domain are listed in Table II.

TABLE II MATERIALS OF COMPOSITE INSULATORS

Materials	Relative permittivity (ε_r)
Air background	1
Silicone Rubber	4.3
FRP core	7.2
Forged steel	10^{20}

Results and Discussion

Effect of Insulator Orientation

Under dry and clean surface conditions, we evaluate the electric field distribution along the leakage path "L" of composite insulator. Maximum electric field was higher at the triple junction point (air/silicone rubber/end fittings) at the line end.

TABLE III EFFECT OF INSULATOR ORIENTATION

	Maximum Electric Field (10^5x V/mm)	
	HV end	Ground end
I string	1.76	1.21
II string	1.73	1.19
V sting	2.48	1.36
Dead end	2.96	2.25

From Table III, we can notice that between suspension strings (I, II and V), V string experience higher electrical stress due its diagonal shape which engender a decrease in the effective insulation distance. Moreover, insulators in the dead end configuration experience about 16% higher electrical stress when compared to the critical suspension configuration (V string).

Effect of Grounding Structure and Conductors

To study the effects of the grounded structure and the power line conductor on the electric field distribution near the insulator, we added just below the insulator, a single conductor with a diameter of 31.5 mm and a length equal to 1.5 times the length of the insulator [10]. The insulator will be also suspended from the middle of a 0.8 m x 1 m grounded supporting structure.

Table IV gives a comparison of maximum electric field on both ends (HV and ground) for each insulator configuration (I, II, V and dead end) with and without the grounding structure and conductor.

Dielectric Materials and Applications: ISyDMA'2016   Materials Research Forum LLC
Materials Research Proceedings 1 (2016) 71-74   doi: http://dx.doi.org/10.21741/2474-395X/1/18

TABLE IV EFFECT OF GROUNDING STRUCTURE AND CONDUCTOR

| | Maximum Electric Field (10^5x V/mm) | | | |
| | HV end | | Ground end | |
	With grounded structure and conductor	Without grounded structure and conductor	With grounded structure and conductor	Without grounded structure and conductor
I string	1.54	1.76	1.47	1.21
II string	1.52	1.73	1.35	1.19
V sting	2.67	2.48	1.98	1.36
Dead end	2.81	2.96	2.63	2.25

It is obvious that the presence of the conductor at the line end and the grounded supporting structure reduces the electric field strength at the line end, but increases it near the ground end.

Effect of Surface Conditions
In this case, a uniform pollution layer of 1 mm is distributed on the surface of the insulator. Relative permittivity and electric conductivity of the pollution layer are taken 50 and 0.0071 S/m respectively. Values of maximum electric field in presence of the contamination layer are given in Table V.

TABLE V EFFECT OF SURFACE CONDITIONS

| | Maximum Electric Field (10^5 x V/mm) | | |
	No pollution layer	With pollution layer	% change in electric field
I string	1.76	2.01	12.43
II string	1.73	1.98	12.62
V sting	2.48	2.89	14.18
Dead end	2.96	3.26	9.20

Electric field intensity was higher at the line end (triple junction point) under normal and contaminated surface conditions. Electric field is enhanced between 9 to 14 % in presence of pollution layer.

Effect of Impulse Voltage
We applied a voltage with the below expression:

$$V = 1800 \ (e^{-0.01486.10^{-6}.t} - e^{-2.80145.10^{6}.t}) \ [kV] \qquad (1)$$

For each string configuration, the maximum electric field is higher at both ends of the insulators. From Table VI, we noticed that the dead experience about 13.5% and 26.5% higher electrical stress when compared to the suspension configurations (I, II and V) at both line and ground ends respectively.

TABLE VI EFFECT OF IMPULSE VOLTAGE

| | Maximum Electric Field (10^5x V/mm) | | | |
| | HV end | | Ground end | |
	Alternative voltage	Impulse voltage	Alternative voltage	Alternative voltage
I string	1.76	4.79	1.21	3.12
II string	1.73	4.98	1.19	3.69
V sting	2.48	5.17	1.36	3.98
Dead end	2.96	5.54	2.25	4.24

From Table VI, we can notice also that when the applied voltage is an impulse, maximum electric field is higher about 65% at the line end and 68% at the ground end.

Conclusion

Electric field analysis plays a critical role in the use of composite insulators for different orientation positions and under different service conditions.

Under dry and clean conditions, electric field for composite insulators used in dead end configuration is higher when compared to a suspension configurations. However, under contamination conditions, electric field distribution of composite insulators used in dead end configuration is less sensitive to the pollution distribution.

Hardware such as grounding structure and conductors added to composite insulators change the electric field distribution on both HV and ground ends.

Voltage type influences also the electric field distribution. It was found that electric field is higher in the case of impulse voltage as compared to AC voltage under similar service conditions.

References

[1] J. F. Hall, "History and bibliography of polymeric insulators for outdoor applications", IEEE Transactions on Power Dilevery, Vol. 8, No. 8, pp. 376-385, 1993. http://dx.doi.org/10.1109/61.180359

[2] N. Mavrikakis, K. Siderakis, D. Kourasani, M. Pechynaki and E. Koudoumas, "Hydrophobicity transfer mechanism evaluation of field aged composite insulators", 5th International Conference on Power Engineering, Energy and Electrical Drives, pp. 215-219, 2015. http://dx.doi.org/10.1109/powereng.2015.7266322

[3] J. Wang, Y. Chen, J. Liao and Z. Peng, "Voltage and E-field distribution of UHV composite insulator with connection of porcelain insulators", 11th International Conference on the Propreties and Applications of Dielctric Materials, pp. 628-631, 2015. http://dx.doi.org/10.1109/icpadm.2015.7295350

[4] T. Doshi, R. S. Gorur and J. Hunt, "Electric field computation of composite line insulators up to 1200 kV AC", IEEE Transactions on Dielectrics and Electrical Insulation, Vol. 18, No. 3, pp. 861-867, 2011. http://dx.doi.org/10.1109/TDEI.2011.5931075

[5] M.S.Kalimurugan, S. Arun Sankar and M. Willjuice Iruthayarajan, "Investigation of electric field distribution on AC composite 230kV insulator using corona ring", International Conference on Circuit, Power and Computing Technologies, pp. 142-147, 2014. http://dx.doi.org/10.1109/iccpct.2014.7054773

[6] S. Muthu Kumar and L. Kalaivani, "Electric field distribution analysis of 110 kV composite insulator using finite element modeling", International Conference on Circuit, Power and Computing Technologies, pp. 136-141, 2014.

[7] M. Liang, and K. L. Wong, "Study of electric field distribution on 22 kV insulator under three phase energisation", Conference proceedings of ISEIM, pp. 140-143, 2014.

[8] N. Murugan, G. Sharmila and G. Kannayeram, "Design optimization of high voltage composite insulator using electric field computations", International Conference on Circuit, Power and Computing Technologies, pp. 315-320, 2013. http://dx.doi.org/10.1109/iccpct.2013.6528981

[9] T. Zhao and M. G. Comber, "Calculation of electric field and potential distribution along nonceramic insulators considering the effectsof conductors and transmission towers", IEEE Transactions on Power Dilevery, Vol. 15, No. 1, pp. 313-318, 2000. http://dx.doi.org/10.1109/61.847268

[10] R. Anbarasan and S. Usa, "Electrical field computation of polymeric insulator using reduced dimension modeling", IEEE Transactions on Dielectrics and Electrical Insulation, Vol. 22, No. 2, pp. 739-746, 2015. http://dx.doi.org/10.1109/TDEI.2015.7076770

Dielectric Materials and Applications: ISyDMA'2016   Materials Research Forum LLC
Materials Research Proceedings 1 (2016) 75-78   doi: http://dx.doi.org/10.21741/2474-395X/1/19

Modeling and optimization techniques of boron diffusion parameters in MOS transistor using SILVACO ATHENA and Matlab

N. GUENIFI*, R. MAHAMDI, I. RAHMANI

LEA, Electronics Department, University of Batna 2, Batna (05000) Algeria.

*Correspondingauthor: e-mail : guenifi_2000@yahoo.fr

Keywords: Polysilicon, SiO₂, Boron, Redistribution, Finite Differences Method, Silvaco

Abstract. Silicon oxide (SiO_2) is a good dielectric material in metal-oxide-semiconductor (MOS) structures. The improved SiO_2 quality requires adequate study of doping diffusion in this structure to maintain the absence of the different impurities in the interface Poylsilicon/SiO_2. For this we studied a theoretical model of boron diffusion before and after thermal annealing in a highly doped polysilicon films. The model takes into account the distribution of vacancy mechanism by associating parameters and effects related to high concentrations. Based on the literature the model is solved using the engineering software tool MATLAB, following a well-defined algorithm. The model is validated with the help of simulation results obtained from Silvaco.

Introduction

Obtaining a good silicon oxide in metal-oxide-semiconductor (MOS) structures requires the reduction of doping diffusion in order not to reach the polysilicon/silicon oxide interface by different methods namely the insertion of bi-layer polysilicon (poly-Si) [1] or nitrogen doped silicon (NiDoS) [2] or the heavily doped poly-Si[3,4]. Boron is the most widely used p-type doping in modern microelectronic devices due its high solid solubility in Si and its strong diffusivity. Our aim is mainly to optimize the silicon oxide thickness. In order to avoid solubility problems indeed, we study a theoretical model of boron diffusion before and after thermal annealing in a highly doped poly-Si film. We take into account several mechanisms and effects related to the high concentrations, such as the formation of clusters. The model is solved applying the engineering tool MATLAB. The numerical method used is the finite difference method. The model is validated with the help of simulation results obtained from Silvaco[5].

Modeling of boron diffusion in polysilicium

The simple model (Fick's law) for the low concentrations is no longer suitable for the polycrystalline silicon. In this case, it is necessary to add many terms to yield a model that can provide valuable quantitative results in the field of concentrations employed. The flow of dopant in isotropic medium containing impurities is given by the following expression [3]

$$\frac{\partial c_B}{\partial t} = D_{i,B}\frac{\partial}{\partial x}\left[\frac{1+\beta f}{1+\beta}\left[g + \frac{\alpha C}{\sqrt{(\alpha C)^2 + 4n_i^2}}\right]\frac{\partial C}{\partial x}\right]$$

With $\quad f = \dfrac{2n_i}{(-\alpha C)+\sqrt{(\alpha C)^2 + 4n_i^2}}$, and $\quad D_i = D_{i0}\exp(\frac{-E_a}{kT})$

D_i is the boron intrinsic diffusion coefficient, C is the boron chemical concentration,

α is the boron activation rate and g the term due to the cluster effect which cause the slow down in diffusivity, n_i is the intrinsic concentration, β is the ratio of the diffusivity induced by the charged

Dielectric Materials and Applications: ISyDMA'2016   Materials Research Forum LLC
Materials Research Proceedings **1** (2016) 75-78   doi: http://dx.doi.org/10.21741/2474-395X/1/19

vacancies on the global diffusivity induced by the neutral vacancis, D_{i0} is the diffusion coffecient in polysilicon, Ea activation Energy, k is the Boltzmann constant and T is the temperature.Using theorotical model we calculated the boron concentration starting from an algorithm using an implicit finite difference method [6] with specified boundary conditions and initial conditions by using Matlab.

Résults and comparison of profiles (Silvaco/theoritical model)

A   Boron Profile Before and afterThermal Annealing

The thickness of poly-Si films used in our study was 335 nm, B implanted with energy of 15 KeV and the dose was 4.10^{15} at.cm^{-2}. We have taken concentration equal to 10^{15}at/cm^3. In first evaluation of the simulated B diffusion profiles were adjusted by varying β, α and g. Figure (1) gives the shape boron concentration profile before annealing at 700°C using the theoretical model. by engineering tool MATLAB. Figure (2) shows the model SIMS (Secondary Ion Mass Spectrometry) profile using the famous technological Silvaco. Based on the model of Pearson IV found in [5]. Both plots are similar. Also, we observe that the maximum penetration depth of the boron before thermal annealing is in order of 0.08 μm representing about 51% of the thickness film.

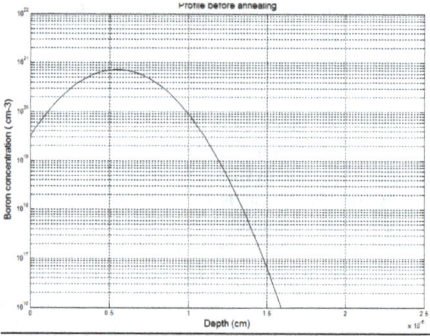

Fig. 1 calculated profile of boron distribution before annealing Dose = 4.10^{15} at.cm^{-2}, E=15 KeV (Matlab)

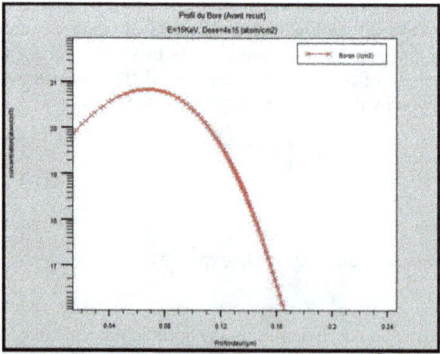

Fig.2- Profile of boron distribution before annealing dose = 4.10^{15} at.cm^{-2}, E=15 KeV, (SILVACO)

Dielectric Materials and Applications: ISyDMA'2016   Materials Research Forum LLC
Materials Research Proceedings 1 (2016) 75-78   doi: http://dx.doi.org/10.21741/2474-395X/1/19

B   *Boron Profile After Thermal Annealing:Effect of Annealing Duration*
Figures (3) and(4) show a superposition of diffusion profiles after annealing. These figures show B concentration (at/cm^3) as a function of Si film depth (μm). For different times of annealing (1min, 5min and 10 min) at 700°C   the boron diffusion depth increases substantially with the thermal annealing time.

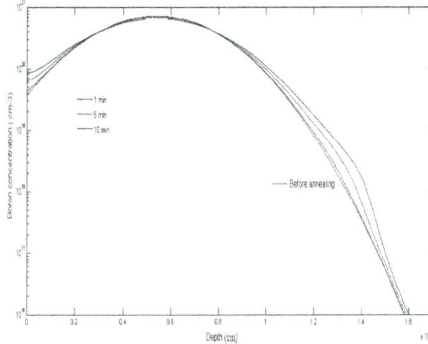

Fig.3-Boronprofilescalculatedbefore and after annealing at different durations t =1 min, t =5 min and t =10 min, annealing temperature T=700°C and dose = 4.10^{15} at.cm^{-2}, E=15 KeV (Matlab)

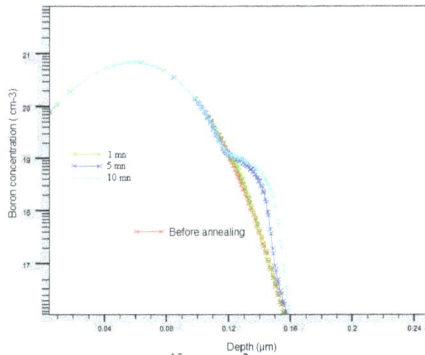

Fig.4.Borondiffusion Profiles (dose = 4.10^{15} at.cm^{-2}.E =15 KeV) after annealing for duration 1min, 5 min and 10 min (SILVACO).

We observe the shoulder phenomena that occurs for the boron solubility limit. This phenomena increases with the annealing duration. In addition, we have observed shoulder when we varied the temperature (not presented here).In all study the results are calculated for the following parameters: E_a=3.75 eV, D_{poly}= 1000D_{mono}, D_{mono} = 7.87 cm^2.s^{-1}, β = 0.11, α = 0.65 and g = 0.1.In this paper, concentration distribution versus depth in three dimensions is reported(Figure 5), taking into account model proposed in reference[3]. The results are compared with those plotted in 2D. A good agreement is observed between both methods (Matlab and Silvaco).

Dielectric Materials and Applications: ISyDMA'2016   Materials Research Forum LLC
Materials Research Proceedings **1** (2016) 75-78   doi: http://dx.doi.org/10.21741/2474-395X/1/19

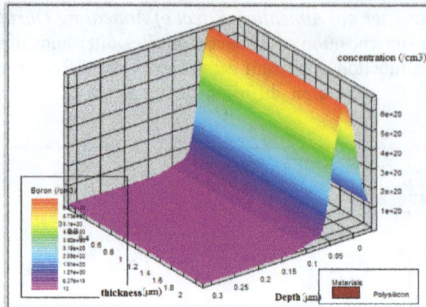

Fig. 5-Boron diffusion profiles as function depth and thickness dose = 4.10^{15} at.cm^{-2}.E =15 KeV, after annealing at 800°C/1mn (Silvaco)

Conclusion

In this work, an algorithm implemented with Matlab has been developed based on theoretical model of boron diffusion in a highly doped poly-Si film. We have used the finite difference method. The model is validated with the help of simulation results obtained from Silvaco. We obtained the profiles for the boron redistribution before and after annealing the proposed models or simulation of the boron diffusion in the poly-Si. The shoulder phenomena are observed in all cases. The results obtained in both simulators are similar. Also, we have obtained a good agreement between our results and those found in literature [3]. This theoretical study shows that technological conditions preserve the quality of the silicon oxide structure studied.

References

[1] R. Mahamdi, L.Saci, F. Mansour, P. Temple-Boyer, E. Scheid and L. Jalabert. "Boron diffusion and activation in polysilicon multilayer films for P+ MOS structure: Characterization and modeling". Microelectronics Journal.Vol.40, N°1, pp.1-4; 2009. http://dx.doi.org/10.1016/j.mejo.2008.08.003

[2] R. Mahamdi, F. Mansour, H. Bouridah, P. Temple-Boyer, E. Scheid, L. Jalabert. "Nitrogen doped silicon films heavily boron implanted for MOS structures: Simulation and characterization". Materials Science in Semiconductor Processing, Vol.N°13, pp.383–388;2010

[3] Ramdane Mahamdi,Farida Mansour, Emmanuel Scheid, Pierre Temple-Boyer and Laurant Jalabert. "Boron Diffusion and Activation During Heat Treatment in Heavily Doped Polysilicon Thin Films for P$^+$ Metal-Oxide-Semiconductor Transistors Gates". Japanese Journal of Applied Physics, Vol. 40, pp. 6723-6727, 2001. http://dx.doi.org/10.1143/JJAP.40.6723

[4] Hachemi Bouridah, Fatiha Bouaziz, Farida Mansour, Ramdane Mahamdi, and Pierre Temple-Boyer. "Study of grains size distribution and electrical activity of heavily boron doped polysilicon thin films". Materials Science in Semiconductor Processing; Vol.14, Issues 3–4, pp. 261–265, 2011. http://dx.doi.org/10.1016/j.mssp.2011.04.006

[5] SILVACO International. ATHENA User's Manual. Vol. 1–2; 2010.

[6] J. Cranck, "The Mathematics of Diffusion", 2nd ed, Clarendon Press, Oxford, 1975.

Dielectric Materials and Applications: ISyDMA'2016
Materials Research Proceedings 1 (2016) 79-84

Materials Research Forum LLC
doi: http://dx.doi.org/10.21741/2474-395X/1/20

Study and design of nanostructures on lithium niobate for dielectric material sensors: application in electrical insulation systems

Abdelhamid Mohammed El Amin LECHLECH[*1], Djamel KALAIDJI[2], Khadidja HENAOUI[2]

[1]Electrical and Electronic Engineering Department Aboubakr Belkaid University, Unit Research in Materials and Renewable Energy Tlemcen, Algeria

[2]Physics Department, Aboubakr Belkaid University, Unit Research in Materials and Renewable Energy Tlemcen, Algeria

*Correspondingauthor: e-mail: hamid.ing.ebm@gmail.com

Keywords: Nanotechnology, Microelectronics, E/O Effect, Geometric Shapes, Crystal

Abstract. The measure, in industry, of voltage and electric field is essential for the design of insulation systems and this factor despite the performance of advanced numerical computation that allows only very difficult to estimate the electric field distribution disturbed by the presence of space charges. Indeed, the space charges distribution (an electron cloud located in a space which can be a vacuum, a gas, a liquid or a solid which can interfere with the movement of the electron flux flowing) depends poorly known physical factors such as mobility of the charged particles and also of the ionization rate of gas molecules. Therefore, it is necessary to measure precisely the intensity of the electrical field for more efficiency and reliabilities operations on high voltage systems. The development of nanotechnology, microelectronics and integrated optics (realization of all optical components on a single substrate) led to the birth of the electrical field optical sensors. The electric field sensors using the E/O effect can be divided into two categories: low-field sensors and high-field sensors. In this context, we studied the optical and electro-optical properties of this material; Then, geometric shapes and configurations (for crystal) are studied in order to choose which is most appropriate for a given application.

Introduction

Nanosciences and nanotechnologies can be defined as the set of studies and manufacturing processes and handling structures (electronic, chemical, etc. ...), devices and hardware systems to the nanoscale. In this context, Nanoscience is the study of phenomena and manipulation of matter at the atomic, molecular and macromolecular scale where properties differ significantly from those that prevail on a larger scale. Nanotechnology, for their part, concern the design, characterization, production and application of structures, devices and systems by controlling shape and size at nanometer scale.

The purpose of this work was to give, in general, the basic elements for the design of optoelectronic sensors, that is to say consisting of optical and electronic elements.

After studying the main applications of optoelectronics, the different steps in the design, as well as large families of sensors, we described the base quantities in radiometry and their relationships. Then we are interested in the spread of radiation, the role of optics, to arrive at the calculation of the flow, or optical signal received by the detector, which converts it to an electrical signal.

The combination of "electronic" terms and "optical" is the origin of various vocables: it is said of a material that is electro-optical if its optical properties (transmission, refractive index, birefringence) are modifiable in the action of an electric field.

Dielectric Materials and Applications: ISyDMA'2016   Materials Research Forum LLC
Materials Research Proceedings **1** (2016) 79-84   doi: http://dx.doi.org/10.21741/2474-395X/1/20

Study of the E/O effect: The Pockels effect

The Pockels effect is the appearance of birefringence in an environment created by a static or variable electric field. Birefringence appears is proportional to the electric field, in contrast to the Kerr effect which is proportional to the square of the field. The Pockels effect appears only in crystals without symmetry with respect to an axis, such as lithium niobate or gallium arsenide.

In an anisotropic dielectric environment (like crystalline), optical properties (refractive index) seen by an electromagnetic wave propagating them, depend on the direction of propagation of this one. For a propagation direction given (fixed), there are two (2) refractive indices. These indices are associated with electromagnetic waves having polarization states that can spread without alteration.

A wave propagating in any direction can be decomposed into two independent plane wave with wave propagation linear modes. The two principal components of the wave will each have a refractive index whose difference Δn is called natural birefringence of the environment in the direction considered.

Birefringence means that there are two refractive indices or two directions of propagation for each incident ray.

An electromagnetic wave is represented by the data of its vectors, electric field $\vec{E}$ and magnetic field $\vec{H}$. The influence of these fields on the material described by the introduction of two vectors: the vector of the electric displacement $\vec{D}$ and the vector of magnetic induction $\vec{B}$. [1] [2]

In a non-magnetic material environment (crystalline), vector fields are connected by Maxwell's equations are written:

$$\vec{rot}\,E = \frac{-\partial \vec{B}}{\partial t} \quad , \quad \mathrm{div}\,\vec{B} = 0 \quad , \quad \vec{rot}\,H = \frac{\partial \vec{D}}{\partial t} \quad , \quad \mathrm{div}\,\vec{D} = 0$$

In addition to Maxwell's equations, the electric field E and magnetic field B satisfy:
• The constitutive relationship (not taking into account the optical activity)

$$D = \varepsilon_0 . [\varepsilon] . E$$

• A rule of energy conservation, verified by electromagnetic energy density

$$\begin{cases} w = w_e + w_m \\ w_e = {}^{1}\!/_{2}.E.D \\ w_m = {}^{1}\!/_{2}.B.H \\ S = E * H \dots vecteur\ de\ POYNTING \end{cases} \dots \frac{dw}{dt} + \mathrm{div}\,S = 0$$

Then, we will discuss the relative permittivity tensor ε dependent early 9 parameters:

$$\varepsilon = \begin{bmatrix} \varepsilon_{11} & \varepsilon_{12} & \varepsilon_{13} \\ \varepsilon_{12} & \varepsilon_{22} & \varepsilon_{23} \\ \varepsilon_{13} & \varepsilon_{23} & \varepsilon_{33} \end{bmatrix}$$

The tensor is symmetric (depends 6 parameters) may be diagonalized. This gives a matrix dependent of 3 parameters:

$$\begin{bmatrix} \varepsilon_x & 0 & 0 \\ 0 & \varepsilon_y & 0 \\ 0 & 0 & \varepsilon_z \end{bmatrix}$$

The search for such a marker is facilitated by the symmetry elements of the crystal structure of the material. Three (3) coefficients (ε_x, ε_y, ε_z) are necessary for the characterization of an anisotropic medium, and their relative values allow a classification of optical media.

Once the tensor diagonalized and by analogy with the isotropic media, we can introduce the refractive indices n_x, n_y, n_z related to the permittivity by the relationship $\varepsilon_i = n_i^2$ (i=x,y,z). These indices are called "principal refractive indices" of the relevant optical medium. Their behavior when the medium is subjected to an electric field, is essential for the development of an optical electric field sensor.

So it is important to characterize the relationships between these indices to the electric field is what leads us to define the Index ellipsoid.

Index ellipsoid

In the absence of electric field, the index ellipsoid is described by the following general equation: [3]

Dielectric Materials and Applications: ISyDMA'2016   Materials Research Forum LLC
Materials Research Proceedings 1 (2016) 79-84   doi: http://dx.doi.org/10.21741/2474-395X/1/20

$$\sum_{ij} \eta_{ij} x_i . x_j = 1$$

$\eta_{ij} = \frac{1}{\varepsilon_{ij}}$: tensor of electrical impermeability.

In own space, the equation becomes:

$$\frac{x^2}{n_{11}^2} + \frac{y^2}{n_{22}^2} + \frac{z^2}{n_{33}^2} = 1$$

Using the contracted notation of Voigt for transforming the pairs of indices i, j single index p:

(i, j)	(1, 1)	(2, 2)	(3, 3)	(2, 3) (3, 2)	(1, 3) (3, 1)	(1, 2) (2, 1)
p	1	2	3	4	5	6

$$\frac{x^2}{n_1^2} + \frac{y^2}{n_2^2} + \frac{z^2}{n_3^2} = 1 \ldots \quad \Longrightarrow \quad \frac{x^2}{n_x^2} + \frac{y^2}{n_y^2} + \frac{z^2}{n_z^2} = 1$$

Here, the x, y, z represent the principal axes of the crystal and n_x, n_y, n_z the corresponding refractive indices.

The index ellipsoid (indicator) allows simple geometric representation of the optical properties of anisotropic crystals, and a determination of the refractive indices in a direction of propagation of a well defined light wave.

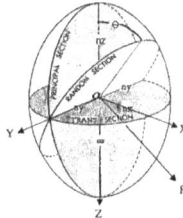

Fig. 1. Index Ellipsoid

The previous figure shows the index ellipsoid in general form. The principal axes X, Y, Z have the lengths: $2n_x$, $2n_y$, $2n_z$.

Random section : wave plane through the origin and perpendicular to a light beam propagating along an arbitrary direction (op). It cuts the ellipsoid along an ellipse.

In the case of uniaxial crystals, we have an ellipsoid of revolution about the axis OZ which is the optical axis of the crystal. The general equation of the ellipsoid becomes:

$$\frac{x^2 + y^2}{n_o^2} + \frac{z^2}{n_e^2} = 1$$

n_o: ordinary refractive index.

n_e: extraordinary refractive index that is in the direction of the optical axis.

Any material submitted to the action of an electromagnetic wave is the seat of an induced polarization which fully determines the response of a medium to excitation radiation. for example, high light intensities emitted by the lasers produce electric fields whose magnitude approaching that of the electric field ensuring the cohesion of the electrons: this is the source of the optical nonlinearity of the dielectric environment.

In general, when an electromagnetic field is applied to environment, it interferes with the dielectric properties. In an isotropic environment, a birefringence is induced; in an anisotropic environment, the birefringence is changed (or altered). [4]

The electro-optic effect results in a variation of the dielectric constant of the dielectric and a rotation axis of a crystal, under the action of an electric field. The electric field induces a deformation of the crystal of the index ellipsoid.

Dielectric Materials and Applications: ISyDMA'2016   Materials Research Forum LLC
Materials Research Proceedings **1** (2016) 79-84   doi: http://dx.doi.org/10.21741/2474-395X/1/20

Materials with a linear electro-optic effect the most used are those with E/O coefficients as large as possible and the other characteristics are consistent with the use made of it; hence the use of Lithium Niobate.

LITHIUM NIOBATE

Lithium niobate is a ferroelectric material as the oxygen octahedron. It has a single transition structural phase in the temperature $T_C \approx 1200$ ° C. Beyond T_C in its paraelectric phase, it is centrosymmetric. Below TC and at ambient temperature, which is ferroelectric with rhombohedral symmetry (trigonal) $C_{3V} = 3m$. The polar axis is of order 3. It is made for it to be mono-domain.

It is a negative uniaxial material $n_e < n_o$. It has a large value for the spontaneous polarization $P_S = 70$ uC / cm^2. This sharp polarization leads to strongly LiNbO3 piezoelectric, pyroelectric and of course electro-optics. Moreover, it has the advantage, of the experimental point of view, of having no optical rotation (rotation of the polarization of the light beam at the output of the crystal). [5]

Influence of geometric factors

The arrangement of the lithium niobate crystal in any electric field distribution has a major influence on his own intense field distribution. At first, we must determine the influence of the applied electric field distribution relative to the position of the crystal. In a second step, the geometric shape of the sample should be chosen to enable a more uniform distribution of the internal field as possible for accurate measurement of the electric field.

The configuration with an electric field in the Z direction is the most interesting since it offers the greatest effect I / O in addition to having a larger electric field strength in the crystal due to its low relative permittivity. However, its dependence on thermo-optical effect unfortunately forced us to put aside.

Regarding configurations with an electric field in the X or Y directions, they both offer the same possibilities with the exception of the rotation of π / 4 of the index ellipsoid that differentiates them. These are the most interesting configurations.

Finally, we will opt for the configuration with any field in the X-Y plane as it uses the features of the previous two configurations and will facilitate the practical use of the sensor in any electric field distribution. The following figure shows this configuration. So the spread of the light beam will be along the optical axis of the crystal. This will have as major advantage of eliminating the natural birefringence of the crystal and to minimize the thermo-optic effect. Thus the section of the crystal parallel to the wave plane and on which is applied the electric field will be in the X-Y plane.

➢ First configuration

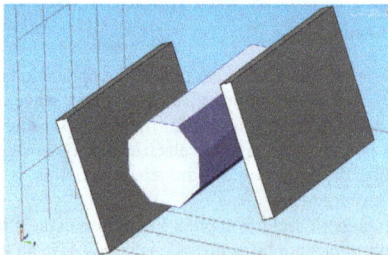

Fig. 2. Optical electric field sensor.

The crystal of lithium niobate is octagonal and is oriented so that the electromagnetic wave propagates along the optical axis which is the axis Z. The two electrodes generate an electric field along the X-Y plane (the two electrodes form an angle α with the Y axis).

Dielectric Materials and Applications: ISyDMA'2016   Materials Research Forum LLC
Materials Research Proceedings **1** (2016) 79-84   doi: http://dx.doi.org/10.21741/2474-395X/1/20

➢ Second configuration

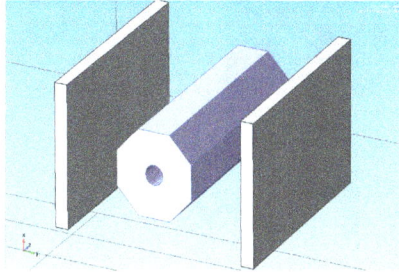

Fig. 3. Optical sensor waveguide / Bragg grating

In this configuration, the crystal contains either a single waveguide is a waveguide with a Bragg grating:
- For single waveguides, the dielectric medium surrounded by a second medium of lower refractive index form a trap for the light: it is the principle of light confinement. Experimentally, the procedure is done by photo-inscription longitudinal propagation. In this case can produce a beam along its propagation through the crystal spatial soliton (solitary wave that propagates without deformation in a medium), ie a beam that creates its own light guide.
- The Bragg grating is a periodic structure formed by a change in refractive index of the waveguide. This structure behaves substantially as a mirror for a spectral band around a characteristic wavelength $\lambda_\beta = 2 \times n_e \times \Lambda$ (2 $\times$ effective index $\times$ pitch of modulation). It remains transparent to the other wavelength different λ_β. In this case, the sensor is selective in terms of wavelength. Also, when a Bragg grating is subjected to a temperature change, it expands or contracts, thereby changing its pitch (this principle can be used for measuring the temperature).

➢ Third configuration

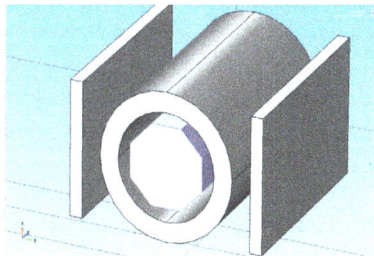

Fig. 4. Adding a cladding for the optical sensor

The addition of a protection ring (Teflon) around the crystal can improve the signal or rather the propagation of the electromagnetic wave throughout the crystal. This is the principle of the pipe. But this ring being Teflon blocks the electric field distribution induced by the electrodes, in the crystal. To remedy this problem, one can simply provide a perforated sleeve at the surfaces. This improves the signal without removing the influence of the famous electrodes.

Dielectric Materials and Applications: ISyDMA'2016   Materials Research Forum LLC
Materials Research Proceedings **1** (2016) 79-84   doi: http://dx.doi.org/10.21741/2474-395X/1/20

➢ Fourth configuration

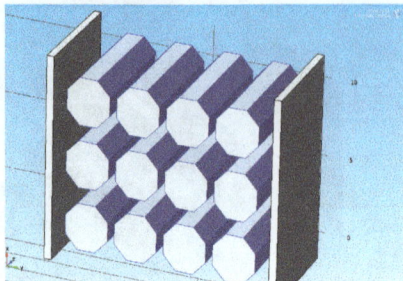

Fig. 5. Stack crystals

The previous figure shows a stack of crystal lithium niobate. The idea is to sweep a range of wavelengths, since each crystal is disposed, relative to the two electrodes, in a different way compared to the other crystals. In this configuration, several factors must be taken into account in particular the influence of the electric field on each crystal will cause, through the interaction between neighboring crystal, disruption of the electric field within the same crystal .

Conclusion

Currently, optical technology plays an important role in the field of nanotechnology. The growing control of the manufacture of optical components now enables the realization of simple sensors such use, precise and stable. The work focused on the theoretical study of an optical sensor based on electro-optical effect induced by the lithium niobate.

Then we presented a method for producing an optoelectronic sensor, we focused on the physical phenomenon that is the detection of an electromagnetic wave and we have identified several radiometric quantities.

Beyond these values of the electromagnetic wave and other factors involved in the embodiment of the sensor. In this context, we studied the electro-optic effect or Pockels induced anisotropic birefringent material effect, which has attracted the interest of choosing the crystal of lithium niobate. A material whose properties at the nanoscale, prove fascinating.

References

[1] Physique des semiconducteurs et des composants électroniques. Henry Mathieu, Hervé Fanet : 6e édition. – Paris 2009

[2] M. S. Birman and M. Z. Solomyak, L2-theory of the Maxwell operator in arbitrary domains, Russian Mat. Surveys, (1987), pp. 75-96.

[3] Optique anisotrope F. Treussart, 2008

[4] Roger Grousson : Propriétés photoréfractives du niobate de lithium et leurs applications au traitement d'image Thèse PhD, université Pierre et Marie Curie- PARIS 6, 2012

[5] E. F. Weller. Lithium niobate - a new type of ferroelectrics : growth, structure and properties. Elsevier, Amesterdam, 1967

Dielectric Materials and Applications: ISyDMA'2016
Materials Research Proceedings 1 (2016) 85-88

Materials Research Forum LLC
doi: http://dx.doi.org/10.21741/2474-395X/1/21

Electronic and optical properties of polymer MEH-PPV and their applications in hybrid optoelectronic devices

Esmaà KHENNOUS*[1], Ihab Eddine YAHIAOUI[2], Hamza ABID[3]

[1]Applied Materials Laboratory, Research Center (CFTE), University of Sidi Bel-Abbes, Sidi Bel-Abbes, Algeria

[2]Modeling and Simulation in Materials Science Laboratory, Physics Department, University of Sidi Bel-Abbes, Sidi Bel-Abbes, Algeria

[3]Applied Materials Laboratory, Research Center (CFTE), University of Sidi Bel-Abbes, Sidi Bel-Abbes, Algeria

*Corresponding author: E-mail: khennous.esmaa@gmail.com

Keywords: MEH-PPV, DFT, FP-LAPW, Electronic Properties, Optical Properties, Organic/inorganic Hybrid

Abstract. Conjugated polymers are a novel class of materials has been extensively used in optoelectronic devices. Well-known PPV derivative Poly [2-methoxy-5-(2-ethylhexyl-1,4-phenylenevinylene] (MEH-PPV) have a wide applications such as organic light emitting diodes OLED, organic emitting transistor OLET ,photovoltaic and organic solar cells. This document focuses on the electronic and optical properties of MEH-PPV and their applications in different optoelectronic devices based on hybrid polymer–semiconductor materials. In this work we studied theoretically the electronic and optical properties of the polymer MEH-PPV using the FP-LAPW calculation method implemented in the wien2k code. The results were determined in the context of the theory of density functional (DFT) made in the approximation of GGA+mBJ. We find our calculated results in good agreement with the available experimental data.

Introduction

The Conjugated polymers have been extensively used in optoelectronic devices as active components. It's found from the literature that the Poly [2-methoxy-5-(2-ethylhexyl-1,4-phenylenevinylene] (MEH-PPV) is a π-conjugated polymer has been considered as one of the most potential conducting polymers for various optoelectronic applications .

The goal of this work is to show the electronic and optical properties of the MEH-PPV structure and their different applications. For the electronic properties, our studies are based on the full-potential linear augmented plane-wave (FP-LAPW) method within density functional theory (DFT) as implemented in the Wien2K code [1]. The calculations are based on the XC (exchange-correlation): modified Becke-Johnson (mBJ) [2]. The mBJ not only gives better band gaps closer to experimental but also the correct band dispersion.

In the next section, we describe the computational method adaptor for the theoretical calculation and discuss our results. In section III, we summarize some different applications of MEH-PPV in hybrid optoelectronic devices found in literature. Finally, summary of our work will be given in section IV.

Theoretical studies

A. Computational details

The unit cell of MEH-PPV: Poly [2-methoxy-5-(2-ethylhexyl-1,4-phenylenevinylene] ($C_{17}H_{24}O_2$) is divided into non overlapping muffin-tin (MT) spheres around the atomic sites, and an interstitial region. Values R_{MT} taken in this study were chosen equal to 1.21, 0.71 and 1.34 (a.u) for Carbon C, Hydrogen H and Oxygen O respectively. A plane-wave expansion with $R_{MT} K_{MAX}$ equal to 7, the

Dielectric Materials and Applications: ISyDMA'2016   Materials Research Forum LLC
Materials Research Proceedings 1 (2016) 85-88   doi: http://dx.doi.org/10.21741/2474-395X/1/21

charge density is Fourier expanded up to $G_{MAX}=12$ and the l-pansion of the non-spherical potential is performed up to $l_{max}=10$. A mesh of 35 special k-points is taken in the irreducible wedge of the Brillion zone (IBZ) for structural optimization. Self-consistent calculations are considered to have converged when the total energy of the system is stable within 0.1 mRyd. The atomic positions of $C_{17}H_{24}O_2$ (MEH-PPV) are optimized at equilibrium state. The parameters characterizing the geometry of the unit cell are: $a=7.120(A°)$, $b=16.049(A°)$, $c=6.469(A°)$ and $\alpha=\beta=\gamma=90°$.

B. Results and discussion

We have calculated the electronic band structure for MEH-PPV using the generalized gradient approximation within modified Becke-Johnson correction (GGA+mBJ) approaches as shown in Fig. 1.(a) The value obtained of band gap energy is 2.40eV .Our result is in good agreement with experimental band gap (2.45 eV [3]). This energy corresponds to the energy of visible photons with absorption wavelength of 517 nm. The following equation could calculate the wavelength from Eg :

$$Eg\ (eV) = 1240/\lambda\ (nm) \tag{1}$$

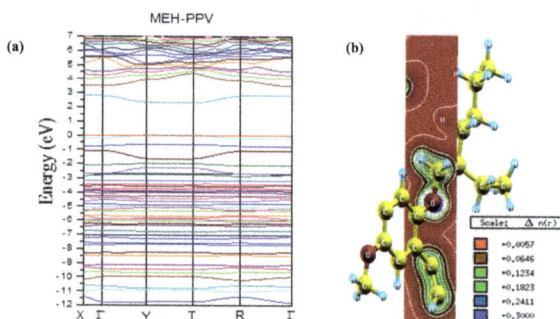

Fig. 1.(a) The calculated band structure using mBJ witch give the energy gap values of about 2.4 eV. (b) Electronic space charge density distribution contour in the (110) plane One can be observed from Δ_n (r) the scale of the cherge density colors where blue has the maximum charge accumulation site.

We can see that the direct gap character of this polymer is very important for the optical transitions. The proper understanding of the optical properties of a material is important for its applications in photonic devices and optical communications. The theoretical and experimental studies from literature demonstrate that the ordinary part of absorption coefficient of MEH-PPV film shows a strong absorption peak in the wavelength interval (400–600 nm) and a maximum peak at ($\lambda_{max}\sim$ 500nm) [4-5].

The electronic charge density contours in the crystallographic plane (110) are illustrated in Fig. 1. (b) The bonds between the atoms can be classified according to their Pauling electro negativity values. The bounds between C-C are covalent whereas the bonds C-O and C-H are ionic due to the large electro negativity between these atoms: 2.55 for Carbone, 2.20 for Hydrogen and 3.04 for Oxygen. It's clear from Δ_n (r) the scale of the charge density colors where blue color (+3.0000) corresponds to the maximum charge accumulating site. The O atom is more electronegative than C and H atoms, as one can clearly see that the charge accumulates more near O along the bonds and the charge around O uniformly distributed.

Dielectric Materials and Applications: ISyDMA'2016
Materials Research Proceedings 1 (2016) 85-88

Materials Research Forum LLC
doi: http://dx.doi.org/10.21741/2474-395X/1/21

MEH-PPV hybrid hetrojonction

MEH-PPV is a photoactive polymer that forms excitons on exposure to light and a p-type semiconductor with a relatively low conductivity due to the low hole and electron mobilities, when compared to inorganic semiconductor materials that is known as an electron donor and a hole transport material [6]. To overcome this problem, physicists and scientists have always been interested in the possibility of combining organic and inorganic compounds, specifically π - conjugates polymers and semiconductors nanocrystals. Table1 shows some performance parameters of optoelectronic devices based on hybrid polymer–semiconductor structure. The addition of an inorganic acceptor material to form organic–inorganic hybrid devices should theoretically improve the performance of OPV and OLED, due to additional advantages such an enhanced absorption and improved charge transport characteristics. However, to date, the efficiency of hybrid solar cells have been very low, when compared to their all-organic counterparts.

TABLE I SELECTED PERFORMANCE PARAMETERS OF HYBRID POLYMER–SEMICONDUCTOR DEVICES.

Components	Basic Structure	$\lambda_{EL.max}$ [nm]	Power Efficiency	Devices	Ref.
MEH-PPV: 20 wt% CdSe (ZnS)	Quantum dots	590	0.20 [lm/W][a]	OLED	[7]
/MEH-PPV-POSS: x wt% CdS0.75Se0.25	Quantum dots	588	0.43 [lm/W][a]	OLED	[7]
MEH-PPV:CdSe	Nanocrystals	-	0.85%	PV cell	[8]
MEH-PPV:ZnO	Ordered nanord arrayof ZnO	-	0.61%	PV cell	[9]
MEH-PPV-CdS	CdS nanowire	-	1.62%	Solar cell	[3]

[a] [lm/W]: lumens per watt (unit of luminous efficacy).

Conclusion

In summary, DFT calculation for the electronic crystal structure of the polymer MEH-PPV was performed using the FP-LAPW method within in Wien2k code. The chemical bonding is investigated through the electronic charge density space distribution contours in the (110) crystallographic plane. Finally, we presented some performance parameters of optoelectronic devices based on hybrid polymer–semiconductor structure reported in literature.

References

[1] P. Blaha, k. Schwarz, G.K.H. Madsen, D. Kvasnicka, J. Luitz, in: Karlhein Schwarz (Ed.), Wien2k An Augmented Plane Wave Plus Local Orbital Program for Calculating the Crystal Properties, Vienna University of Technology, Austria,2001.

[2] F. Tran, P. Blaha, "Accurate Band Gaps of Semiconductors and Insulators with a Semilocal Exchange-Correlation Potential," Phys. Rev. Lett. **102**, 226401 – Published 3 June 2009.

[3] J-C. Lee, W. Lee, S-H Han, T.G. Kim, and Y-M Sung, "Synthesis of hybrid solar cells using CdS nanowire array grown on conductive glass substrates," Electrochemistry Communications 11 (2009) 231–234. http://dx.doi.org/10.1016/j.elecom.2008.11.021

[4] Y.W. Jung, J.S. Byun, Y.H. Cha, Y.D. Kim," Ellipsometric study on the optical property of UV exposed MEH-PPV polymer film," Synthetic Metals160 (2010) 651–654. http://dx.doi.org/10.1016/j.synthmet.2009.12.021

[5] W. Xiaoyang,X. Yanmei,Z. Chunping et al. "Calculation of Optical Parameter of MEH-PPV Film," Acta Photonica Sinica, 2005, 34(5): 746-749.

[6] W.U. Huynh, J.J. Dittmer, A.P. Alivisatos, " Hybrid nanorod—polymer solar cells," Science 295 (2002) 2427–2430. http://dx.doi.org/10.1126/science.1069156

Dielectric Materials and Applications: ISyDMA'2016   Materials Research Forum LLC
Materials Research Proceedings 1 (2016) 85-88   doi: http://dx.doi.org/10.21741/2474-395X/1/21

[7] G. Saygili, G. Ünal, S. Özcelik, C. Varlikli," Highly efficient orange–red electroluminescence from a single layer MEH-PPV POSS:CdS0.75Se0.25 hybrid PLED," Materials Science and Engineering, B 177 (2012) 921–928 . http://dx.doi.org/10.1016/j.mseb.2012.04.011

[8] L.Han, D. Qin, X. Jiang, Y. Liu, L. Wang, J. Chen, Y.Cao, "Synthesis of high quality zinc-blende CdSe nanocrystals and their application in hybrid solar cells," Nanotechnology 2006, 17, 4736–4742. http://dx.doi.org/10.1088/0957-4484/17/18/035

[9] H-W. Choi, K-S Lee, T. L. Alford, "Optimization of antireflective zinc oxide nanorod arrays on seedless substrate for bulk-heterojunction organic solar cells," Appl. Phys. Lett. 2012, 101, doi: 10.1063/1.4757997. http://dx.doi.org/10.1063/1.4757997

Dielectric Materials and Applications: ISyDMA'2016
Materials Research Proceedings 1 (2016) 89-91

Materials Research Forum LLC
doi: http://dx.doi.org/10.21741/2474-395X/1/22

The electrical conductivity behavior of an organic insulator caused by presence of dead ends under high electric field

Amina BENALLOU*[1], Baghdad HADRI[1], Juan MARTINEZ-VEGA[2]

[1]Laboratoire d'électromagnétisme et optique guidée LEGO, Université Abdelhamid Ibn Badis, Mostaganem, Mostaganem, Algeria

[2]Laboratoire d'électromagnétisme et optique guidée, Université de Toulouse; UPS, LAPLACE F-31062, CNRS; Toulouse, France.

*Corresponding author: amina.benallou@gmail.com

Keywords: Polymers, Electrical Conductivity, Organic Insulator, Percolation, Dead End

Abstract. In this paper, an analytical model based on the percolation theory and a numerical model based on Monte Carlo method was developed and used to study the electrical conductivity behavior of an organic insulator under high electric field. The percolation model considers that the real network of organic insulator is consisted of the traps randomly distributed and connected to each other by conductance G_{ij}. The phenomenon of charges trapping and detrapping depends strongly on certain parameters such as the intensity of the applied electrical field, the dead ends length of traps, and the random distribution of the energies of traps constituting the dead end. Using gaussian energy distribution negative electrical conductivity law as exponential function of the electric field was found. The simulation results were compared with an analytical model, they are in agreement, proving the accuracy of this model.

Introduction

The transport phenomenon in the organic insulators (polymers) was treated by using the percolation theory [1,2]. When a percolation network such as the organic insulator reaches percolation threshold that is the critical concentration of traps by which a cluster of infinite traps appears. This cluster consists of finite set of clusters of traps connected to each other them and connecting the two electrodes. This infinite cluster (cluster percolating) contains additional finite clusters of particular traps that are connected to it only on one side .They are called dead ends.

The percolation model could be apply in order to determine the electrical conductivity behavior as the function of electric field. The dead ends of traps are considered to be not participate to the process of conduction in the organic materials when a weak electric field is applied [3, 4]. In a high electric field, the dead end effect is often neglected during the electrical conduction modeling. In this paper, it will be developed an analytical model based on the percolation theory. It could be explain the fluctuations caused by the dead ends of traps on the behavior of electrical conductivity, in high electric field region. This model will be validated by a numerical simulation.

Percolation approach

The charge carriers are trapped by the dead ends during a rather long time of stay. However, when the organic insulator material is subjected to the high electrical field effect, trapping time increases until some localized charge carriers are detrapped and trap again on other traps belonging to dead end and so on. This phenomenon of trapping and detrapping depends strongly on certain parameters such as the intensity of the applied electrical field, the length of the dead ends of traps, and the random distribution of the energies of traps constituting the dead end [5].

Dielectric Materials and Applications: ISyDMA'2016   Materials Research Forum LLC
Materials Research Proceedings 1 (2016) 89-91   doi: http://dx.doi.org/10.21741/2474-395X/1/22

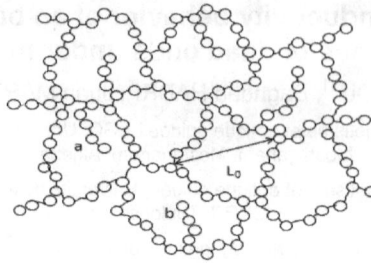

Fig.1. Schematic representation of a portion of infinite cluster constituted of a set of finite clusters of length L_0 containing dead ends (a) and (b).

After calculated the total number of finite clusters trap in a dead end in order to determine the electrical conductivity versus the electric field for Gaussian distribution.

The probability of finding a dead end of traps with length x_0 for Gaussian distribution is determined as follow:

$$p(x_0, \delta) \cong \exp\left(-\frac{\sqrt{\pi} x_0 \delta}{L_0}\right) \tag{1}$$

Where p is the probability of finding a dead end of traps with length x_0 ,
L_0: correlation length

$\delta = \frac{eEx_0}{kT}$ Dimensionless electric field

Electron average residence time in the dead end is determined as

$$\prec t \succ \cong t_0 \exp\left[\frac{eEL_0}{4\sqrt{\pi}kT} - \frac{eEL_0}{8\sqrt{\pi}kT}\right] \tag{2}$$

t_0 is residence time without electric field E, k is Boltzmann constant and T is temperature
Electrical conductivity is determined as

$$\sigma(E, W) \cong \sigma_0 \exp\left[-\frac{eEL_0}{8\sqrt{\pi}kT}\right] \tag{3}$$

Where σ_0 Electrical conductivity without electric field.

Simulation
Previously it has been developed and showed that, the dead ends of traps effect on the behavior of the electrical conductivity in a high electrical field. When the carriers of charges were trapping by the dead end of traps, the behavior of the electrical conductivity were decaying exponentially. In other words the dead ends of traps decrease a part of the electrical conductivity according to the distribution of the energies of the traps. The simulation is to demonstrate and validate the effect of fluctuations in dead ends of traps on the electrical conductivity.

We note that the electrical conductivity curve of the simulated organic insulating shows fluctuations for values of electric field. To check if the dead end of traps affects the electrical conductivity, it should be deleting to see if the compensation of the fluctuation will take place. We have cutting some dead ends. Our simulated results shown in figure 2 that. The simulation results show firstly that electrical conductivity curve (black) $\sigma_1=f(E)$ is completely different to $\sigma_2=f(E)$

electrical conductivity when a few dead end are cutting (red). So the dead ends are responsible for the observed fluctuations on the electrical conductivity curve.

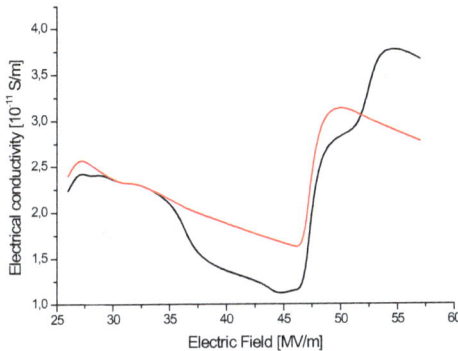

Fig.2. Comparison between electrical conductivity versus electric field. $\sigma_1=f(E)$ global electrical conductivity (black) . $\sigma_2=f(E)$ electrical conductivity when a few dead end are cutting (red).

Conclusion

From the research that has been carried out, we can conclude that the electrical conductivity as function of electric field in an organic insulator was studied by using percolation approach. It was shown that the electrical conductivity due to dead ends of traps were decaying. By employing the simple Monte Carlo algorithm in order to simulate the electrical conductivity behavior versus the electric field with the presence of dead ends of traps, the numerical results were showed fluctuations in electrical conductivity curve. These fluctuations could be explained by the effect of dead ends .

References

[1] Wu Kai, Xie Hengkun and Ge Jingpang ,percolation and Dielectric Breakdown,Electrical Engineering Department of Xi'an Jiaotong University, Xi'an. 710049, P.R.China,1994.

[2] Nir Tessler, Yevgeni Preezant, Noam Rappaport, and Yohai Roichman, «Charge Transport in Disordered Organic Materials and Its Relevance to Thin-Film Devices: A Tutorial Review», Adv. Mater. 2009, 21, pp2741–2761. http://dx.doi.org/10.1002/adma.200803541

[3] D.I.Aladashvili., Z.A.Adamiya., K.G.Lavadovskii., E.Levin., B.I.Shkloskii., «Nonohmic hopping conductivity of weakly compensated semiconductors »,*Sov Phys semicond* 24 (2),234-249 Fev1990

[4] B.Hadri., «Etude de la conductivité électrique en régime non linéaire dans les matériaux inhomogènes », Thèse de Magister (1994), p57, Université d'Oran Es-Sénia, Algérie.

[5] B. Hadri, J. Martinez-Vega, «Investigation of a negative differential conductance in insulating polymers for high electric fields by percolation approach », RS-RIGE Volume 11, pp410,419

Dielectric Materials and Applications: ISyDMA'2016   Materials Research Forum LLC
Materials Research Proceedings 1 (2016) 92-95   doi: http://dx.doi.org/10.21741/2474-395X/1/23

Dynamic piezoelectric response of cellular micro-structured PDMS piezo-electret material as a function of polymeric reticulation ratio

Achraf KACHROUDI[1,2,3,*], Skandar BASROUR[1,2], Libor RUFER[1,2], Fathi JOMNI[3]

[1] Univ. Grenoble Alpes, TIMA Laboratory, F-38031 Grenoble, France.

[2] CNRS, TIMA Laboratory, F-38031 Grenoble, France.

[3] Université de Tunis El Manar, Faculté des Sciences de Tunis, Laboratoire Matériaux, organisation et propriétés (LMOP), 2092, Tunis, Tunisie.

*Corresponding author: E-mail: achraf.kachroudi@imag.fr

Keywords: Piezo-Electret, Micro-Structured Cellular PDMS

Abstract. This paper reports the design and the micro-fabrication by a low-cost process of a new micro-structured PDMS material referred as piezo-electret material. This material presents a very low Young's modulus compared to the conventional piezoelectric materials that facilitates its integration in wearable devices applications. Studies of the inverse piezoelectricity through the dielectric resonance spectroscopy allow determining the piezoelectric parameters of the micro-structured material. The polymer crosslinking effect on these properties is also studied.

Introduction

Piezo-electret is a micro-structured polymer that exhibits piezoelectric effect. This effect arises from their anisotropic structures, in fact, they are polymer matrix containing micro-cavities trapping air. The ionization of the air generates opposite charges that are implanted in the inner micro-cavities surfaces, each micro-cavity is considered as macro-dipole. The macroscopic dipole moment is determined by the quantity of the charges and the distance between the separated charges. So, piezo-electret is sensitive to any external mechanical or electrical stress. In fact, an external mechanical load can compress piezo-electret. Therefore, the distance between opposite charges decreases and compensating charge distribution are generated in the conductive electrodes. But, in the case of external applied ac-voltage, piezo-electret vibrates periodically. Due to these piezoelectric like effect of piezo-electrets and their low Young's modulus, there are utilized as functional materials in electromechanical micro-sensors [4-6]. In this paper, a novel low-cost micro-fabrication molding process is reported. Piezoelectric parameters of the obtained piezo-elecret material are studied through the inverse piezoelectricity and under the effect of the polymeric reticulation.

Fabrication processes

Sample preparation

The material used in this work are in form of a sandwich of two solid layers separated by a micro-structured layer containing cylindrical micro-cavities (Fig. 1). The PDMS used in this work is the kit Sylgard 184 from Dow Corning. According to our previous work [1, 2], the structures geometric parameters are fixed to optimize their piezoelectric response. The two solid layers are *55µm* thicknesses and the micro-structured layer is of *40µm* thickness. The micro-cavities are of *100µm* diameter separated by a pitch of *150µm* [1]. The different layers are bonded by an oxygen plasma treatment. Structures with proportions of 1:5, 1:10, 1:15 and 1:20 between the prepolymer and the crosslinking agent are prepared in order to study the effect of the polymer reticulation on the piezoelectric response of the structures.

Dielectric Materials and Applications: ISyDMA'2016   Materials Research Forum LLC
Materials Research Proceedings 1 (2016) 92-95   doi: http://dx.doi.org/10.21741/2474-395X/1/23

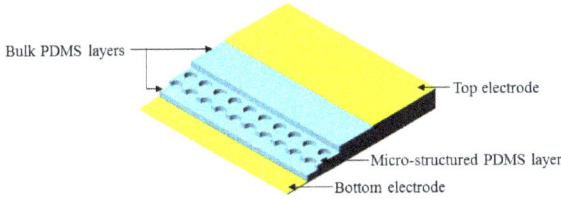

Fig. 1. Schematic illustration of the micro-structured PDMS material with top and bottom gold electrodes.

Electrical charging process

A quasi-static triangular voltage across the samples electrodes with amplitudes between *1kV* and *4kV* is applied with a frequency of *0.5Hz* for *15* minutes. The electric voltage is estimated by Paschen law to trigger electric breakdown in the air micro-cavities and generate macro-dipoles. The electrical charging process is composed of a function generator [Keithley, 3390, 50MHz] coupled with high voltage amplifier [Spellman, SL60].

IR charachterizations of the samples

Transmittance FTIR is performed on the bulk samples prepared with different weight ratios (1:5, 1:10, 1:15 and 1:20). The IR recorded data in Fig. 2 allow to analyze the effect of the mixing ratios between the two components of the kit on the resulting elastomeric network. Silicon hydride functionalities are characteristic groups of the curing agent. The polymerization reaction is achieved by the hydrosilylation of the pre-polymer vinyl-terminations with the cross-linker Si-H groups. According to Fig. 2 and the zoom reported as inset of the figure, the most pronounced intensity variation increasing the mixing ratio is attributed to the silicon-hydride group which has direct influence onto the resulting elastomeric network.

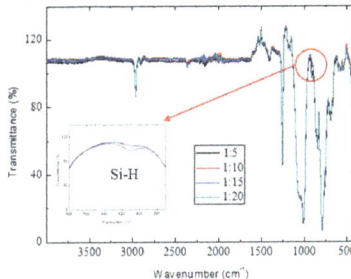

Fig. 2. FTIR transmittance spectra for PDMS at different mixing ratio.

This peaks height variation indicates lower cross-linking density when the proportion of the curing agent is reduced on the mixture. This can strongly influence the mechanical properties of the final samples, it means that samples are increasingly flexible for mixing ratio ranging from 1:5 to 1:20.

Piezoelectric studies through dielectric spectroscopy

After the electrical charging process of each structure, the dielectric resonance spectroscopy is used to study the samples inverse piezoelectricity. In fact, samples deformed upon an ac-voltage periodically. If the frequency matches with a mechanical vibration mode, a resonance phenomenon appears. Fig. 3 shows the dielectric losses spectra.

Dielectric Materials and Applications: ISyDMA'2016   Materials Research Forum LLC
Materials Research Proceedings 1 (2016) 92-95   doi: http://dx.doi.org/10.21741/2474-395X/1/23

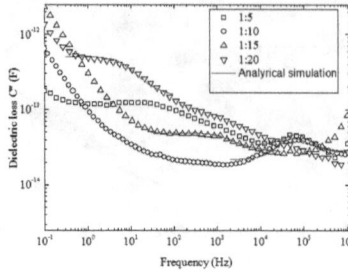

Fig. 3. Dielectric losses of the poled structures for different PDMS mixing ratio.

The obtained data are fitted by a least square means through the imaginary part of Equation1 around the resonance[2, 3].

$$C'' = \text{Im}(C_{TE}^{*} = \frac{\varepsilon_{33}A}{h} \frac{1}{1 - k_{33}^{2} \frac{\tan(\pi f / 2 f_{r})}{(\pi f / 2 f_{r})}}) \tag{1}$$

A and h represent the electrodes area and the thickness of the sample respectively. ε_{33}, k_{33} and f_{r} are respectively the complex permittivity, the electromechanical coupling factor in the thickness direction and the anti-resonance frequency. These parameters are related through Equations 2 and 3.

$$f_{r} = \frac{1}{2h}\sqrt{\frac{c_{33}}{\rho}} \tag{2}$$

$$k_{33}^{2} = \frac{d_{33}^{2}c_{33}}{\varepsilon_{0}\varepsilon_{33}} \tag{3}$$

Where c_{33} and ρ are the thickness stiffness and the mass density of the samples respectively.

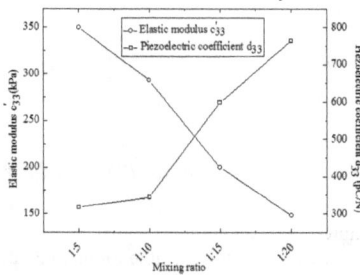

Fig. 4. Elastic modulus and longitudinal piezoelectric longitudinal coefficient as function of the mixing ratio.

From Fig. 4, one can see that the piezoelectric coefficient is highly dependent on the material elastic modulus. In fact, d_{33} increases when the elastic modulus decreases. This effect is achieved due to the shift of the resonance frequency to the low frequencies by reducing the crosslinking agent proportion during the samples preparation. This reduces the polymeric reticulation of the PDMS and decreases its elastic modulus. The higher piezoelectric coefficient of *750pC/N* is obtained in the case of the 1:20 mixing ratio samples.

Conclusion

This paper reports the important effect of the polymeric reticulation on the piezoelectric properties of a cellular piezo-electret based on the PDMS material. The piezoelectric coefficient is greatly dependent on the mechanical properties studied through the elastic modulus which depends also on the polymeric reticulation of the material. Longitudinal piezoelectric coefficient is enhanced to be *750pC/N* for the 1:20 mixing ratio compared to *310pC/N* for the 1:5 mixing ration.

References

[1] A. Kachroudi, S. Basrour, L. Rufer, A. Sylvestre, and F. Jomni, *Smart Mater. Struct.*, vol. 24, no. 12, p. 125013, 2015. http://dx.doi.org/10.1088/0964-1726/24/12/125013

[2] A. Kachroudi, S. Basrour, L. Rufer, and F. Jomni, *J. Phys. Conf. Ser.*, vol. 660, no. 1, p. 012040, 2015. http://dx.doi.org/10.1088/1742-6596/660/1/012040

[3] P. Fang, F. Wang, W. Wirges, R. Gerhard, and H. C. Basso, *Appl. Phys. A*, vol. 103, no. 2, pp. 455–461, Aug. 2010. http://dx.doi.org/10.1007/s00339-010-6008-2

[4] J. J. Wang, J. M. Hsieh, R. W. Tsai and Y. C. Su, *16th International solid-state Sensors, Actuators and Microsystems Conference* DOI: 10.1109/TRANSDUCERS.2011.5969868, June 2011. http://dx.doi.org/10.1109/TRANSDUCERS.2011.5969868

[5] J. J. Wang, T-H. Hsu, C. N. Yeh, J-W. Tsai and Y. C. Su, *Journal of Micromechanics and Microengineering*, vol. 22, no. 1, 015013, 2012. http://dx.doi.org/10.1088/0960-1317/22/1/015013

[6] J. Shi, D. Zhu, S. P. Beeby, *Journal of physics: conference series* vol. 557, 012104, 2014.

Dielectric Materials and Applications: ISyDMA'2016   Materials Research Forum LLC
Materials Research Proceedings 1 (2016) 96-99   doi: http://dx.doi.org/10.21741/2474-395X/1/24

A non-destructive planar biosensor for dielectric materials characterization

Nabila AOUABDIA[1*], Nour Eddine BELHADJ-TAHAR[2], ALQUIE Georges[2]

[1]Electronic Department, Mentouri Brothers University Microsystems and Instrumentation Laboratory – LMI, Constantine; Algeria

[2]Sorbonne Universités, UPMC Univ Paris 06, UR2, L2E, F-75005, Paris, France

*Corresponding author: E-mail: n_aouabdia@yahoo.fr

Keywords: Rectangular Patch Resonator (RPR), Characterization Biological Materials, HFSS Simulation, Electromagnetic Applicator, Non-Destructive Control, Dielectric Properties

Abstract. In-body implanted antennas are surrounded by materials (muscle, fat tissue, skin, etc.), which have special electromagnetic parameters. The effects of these near-field mediums on the implanted antenna are unknown. Performance patch resonator, as the resonant frequency and the quality factor depends on the dielectric parameters of the various materials involved in their structures. In applications of microwave system for the dielectric substrate and superstrate are made of materials with low losses for the best operation. When used as sensors, some of the dielectric layers may be made with an unknown material; changes of parameters of the resonator, mainly the change of the increase in the frequency and quality factor are closer to the complex permittivity of the unknown material. In the particular application, a patch sensor can be used to evaluate the specific permittivity layers by comparing the measured parameters of the patch with a reference structure and those obtained with the unknown material. This work aims to study a planar biosensor for characterizing biological materials in order to derive the dielectric parameters. The medical applications are to detect abnormalities body using the structures raised as nondestructive applicators. To focus on this issue, it is necessary to make simulations with HFSS on a rectangular planar resonator by a coaxial fed, to use this device as an applicator to characterize various homogeneous and heterogeneous materials such as muscle, skin and fat.

Introduction

This work presents a novel planar electromagnetic sensor operating at microwave frequencies for real-time evaluation of the dielectric properties of biological materials, which was designed with nondestructive control biomedical applications. Dielectric permittivity measurements are compared with results obtained from resonant sensor using measurements of resonant frequency and Q-factor [1], [3], [4]. In this context, two approaches were defined. In a first part, a conception, a modeling, a simulation with commercial software HFSS, a realization and measurements were treated for validate a Rectangular Patch Resonators (RPR) prototypes [1]. The second approach of our work which is developed is focused on the characterization of biological materials in vitro using the RPR prototypes proposed as an applicator in the non-destructive control and the medical domain to find the abnormalities of these tissues such as: eczema, psoriasis, cancer, etc [1]. The exactness of the obtained results is estimated using prototypes operating near 6 GHz, taking into account only the fundamental mode resonant frequency. Our contribution to this work is by empirical development equations that helped us through the empty effective permittivity and biological sample to extract the complex permittivity of the sample under test and through the quality factor of having dissipation factor. Once these dielectric parameters calculated, we performed them with HFSS simulations and finally we compared these simulations with our measurements to validate our calculations of the complex permittivity and loss factor of each sample [1].

Dielectric Materials and Applications: ISyDMA'2016   Materials Research Forum LLC
Materials Research Proceedings **1** (2016) 96-99   doi: http://dx.doi.org/10.21741/2474-395X/1/24

Theoritical Approach

Materials and Methods

After modeled, developed and validated several prototypes RPR fed by a coaxial cable **Figure1** whose dimensions are summarized in **Table1** *[2]*.

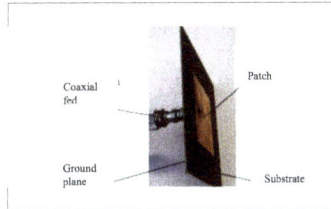

Figure1. RPR fed by coaxial cable

Table1. Dimensions of RPR [1].

Name	Material	Substrate thickness (mm)	W_{sub} (mm)	L_{sub} (mm)	W_{patch} (mm)	L_{patch} (mm)
RPR$_1$	Duroid	0.84	52	41	25.08	15.438

Figure2. The VNA-RPR interface [1].

Figure2 shows the workspace with the dielectric measurements was performed at room temperature 25°C using the following equipment *[1]*.

Mathematical approach

The empirically approach we have treated is a mathematical expansion that we have developed. Starting from the resonance and the gap on the resonant frequency, the effective permittivity of the calculations is then derived and is given by the following formula *[2]*:

$$\varepsilon_{eff} = \varepsilon_{0eff}\left[1 + \frac{2\Delta f}{f_1}\right] \tag{1}$$

with:

$$\varepsilon_{0eff} = \frac{\varepsilon_{sub}+1}{2} + \frac{\varepsilon_{sub}-1}{2\sqrt{1+12\frac{h}{W}}} \tag{2}$$

and f_1 is the resonant frequency of the sample fundamental mode, Δf is the frequency difference between the empty RPR and RPR with the sample superstrate *[2]*.

When W/h >> 1 and the metal thickness are insignificant, the effective permittivity can be calculated as suggested in *[2]*:

$$\varepsilon_{eff} = \frac{\varepsilon_{sub} + \varepsilon_{sup}}{2} + \frac{\varepsilon_{sub} - \varepsilon_{sup}}{2} \cdot \frac{1}{\sqrt{1+12h/W}} \tag{3}$$

ε_{sub}, ε_{sup} are the substrate and the superstrate permittivities respectively.

Then, the superstrate permittivity is deduced by *[2]*:

$$\varepsilon_{sup} = \frac{2\varepsilon_{eff} - \varepsilon_{sub}(1+A)}{1-A} \tag{4}$$

with : $A = \frac{1}{\sqrt{1+12h/W}}$

The loss factor can be calculated using the quality factor and is given by the following formula:

$$tg\,\delta = \frac{1}{RC\omega_0} = \frac{L\omega_0}{R} = \frac{1}{Q} = \frac{\varepsilon''}{\varepsilon'} = \frac{\Delta f}{f_0} \tag{5}$$

Using the previous formulas, we calculated the dielectric parameters of above samples via prototypes operating near 6 GHz at 25°C temperature and are included in **Table2** *[2]*.

Then the new data are introduced in the HFSS simulator and compared the results with measurements for each sample *[2]*.

Result and Discussion

The results of HFSS simulations and measurements for each sample are given by the figures as follows *[2]*:

> **Fat, d$_2$ = 10 mm**

Figure3 shows a qualitative agreement between measurement and HFSS with the presence of a trivial shift in the frequency and the S_{11} parameter, mainly due to dispersion caused by the deformations present in the rough sample surface *[2]*.

We can say that this result is in good agreement and the calculated parameters coincide with the dielectric parameters of the sample *[5]*, with an average relative error on $\Delta f_r/f_r$ 3.4% for the first mode and 4% for the second mode 2.1% for the third mode *[2]*. Note that the result is almost the same with a significant improvement as literature results.

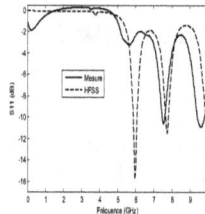

Figure3. The S11 parameter versus Fat frequency for calculated dielectric parameters [2].

> **Butter d$_2$=10 mm**

A glance at **Figure4** shows a qualitative agreement between measurement and HFSS with the presence of a trivial shift in the frequency and the S_{11} parameter, mainly due to dispersion caused by the presence of a slight water quantity and the effect of temperature variations on the sample.

We can say that this result is in good agreement and the calculated parameters coincide with the dielectric parameters of the sample *[5]*, with an average relative error on $\Delta f_r/f_r$ 2.6% for the first mode, 3.4% for the second mode 15.3% for the third mode *[2]*.

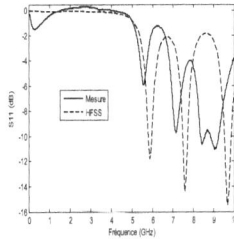

Figure4. The S11 parameter versus butter frequency for calculated dielectric parameters [2].

Table2. Calculated dielectric Parameters of material samples [2].

biological Materials	Calculated Permittivity (ε)	calculated Loss Factor ($tg\,\delta$)
Fat	1.824	0.051
Butter	2.265	0.075

Conclusion

Modeling and experimentation of a RPR covered with a dielectric superstrate are investigated. The RPR criteria are established theoretically and experimentally *[2]*. In this work, we have carefully developed an empirically electromagnetic (EM) sensor method for characterizing biological materials whose dielectric parameters are unknown. In the first time, we made the RPR prototypes, which served as an applicator in this complex analysis. We subsequently treated the problem using our contribution, by a mathematical development that has enabled us through the emptiness effective permittivity and biological sample to extract the complex permittivity of the sample under test and across the quality factor of having dissipation factor. Once these dielectric parameters calculated, we performed HFSS simulations with the last ones and finally we compared these simulations with our measurements to validate our calculations of the complex permittivity and loss factor of each sample. In all cases, we obtained good agreements between our measurements, HFSS simulations and the ones found in the literature.

References

[1] N. Aouabdia, "Etude d'un Capteur à Base de Résonateur Planaire pour Applications au Contrôle Non Destructif ". PhD co-supervised thesis, Laboratoire Microsystème et Instrumentation (LMI)-Université des Frères Mentouri Constantine 1 (UFMC1) & Laboratoire d'Electronique et Electromagnétisme (L2E)-Université Pierre & Marie Curie (UPMC), July 2012.

[2] N. Aouabdia, N. Belhadj-Tahar, G. Alquié, F. Benabdelaziz, "Theoretical and Experimental Evaluation of Superstrate Effect on Rectangular Patch Resonator Parameters". Progress In Electromagnetics Research B, Vol. 32, pp129-147, 2011. http://dx.doi.org/10.2528/PIERB11052610

[3] Preece, A.W., Johnson, R.H., Craig, A.A., Green, J.L., Clarke, R.N., & Gregory, A.P. *(1994)*. "Dielectric Measurement of Reference Liquids and Tissue Equivalent Materials with Non-Invasive Sensors", *IEEE MTT-S Digest, 1061-1064*. http://dx.doi.org/10.1109/mwsym.1994.335172

[4] Lin, C.C., Kuo, M.T., & Chang, H.C. *(2010)*. "Review: Raman Spectroscopy–A Novel Tool for Non-Invasive Analysis of Ocular Surface Fluid", *Journal of Medical and Biological Engineering, 30(6), 343-354*. http://dx.doi.org/10.5405/jmbe.846

[5] http://niremf.ifac.cnr.it/tissprop/

Dielectric Materials and Applications: ISyDMA'2016   Materials Research Forum LLC
Materials Research Proceedings 1 (2016) 100-103   doi: http://dx.doi.org/10.21741/2474-395X/1/25

Elaboration and characterization of polyaniline based polymer composites: a comparative study

D. MEZDOUR

Laboratoire d'étude des matériaux (LEM), Université Mohamed Seddik ben Yahia, Jijel, Algeria

d_mezdour@mail.univ-jijel.dz

Keywords: Composite, Polymer, Permittivity, PANI

Abstract. This study presents a comparison of dielectric properties of composite materials consisting of insulating polymers (Polyamides 6 and 12) containing polyaniline (PANI) as a conductive phase. Samples were obtained by in situ polymerization of the aniline monomer in presence of the insulating matrices. The results obtained by dielectric relaxation spectroscopy showed that it was possible to increase the permittivity of the samples by increasing PANI content. The PANI also induces relaxation phenomena, clearly visible at low temperature. These relaxations seem to be of the same nature in polyamides (PA12 and PA6) films because appearing at nearly the same frequency. It was also shown that dielectric relaxation phenomena are thermally activated.

Introduction

In response to the need for the power ground decoupling to secure the integrity of high speed signals and to reduce the electromagnetic interferences [1], conductive polymers with high dielectric constant are developed by the electronics industry. Several works were dedicated to the elaboration of PANI composites and to some mixtures of PANI/Polymer with high dielectric constants [2-4]. Because of their flexibility and good compatibility with organic printed circuit boards, polymer nanocomposites with high dielectric constants are promising candidates as dielectrics in the embedded passive-components technology [3]. On the other hand, fabrication of thin film capacitors on multi-chip modules is essential to reduce circuit's size in electronic systems of very high speed. To produce such films from intrinsically conducting polymers (ICPs) is complicated because of the difficulty of processing. The present work deals with the preparation and characterization of composite films from an extreme dilution of an ICP with not coloured polymers.

Experimental procedure

Formation of the layered PA6/PANI films

PA6/PANI films were prepared by oxidative polymerization process of PANI in the polyamide 6 matrix. The previously weighed film samples of PA6 were placed in the aniline solution until the desired aniline film content is reached. Swelled polymer films were then subjected to the chemical aniline polymerization in oxidant solution. Only one surface was concerned by the process. The obtained green transparent films were washed by distilled water and placed in soxlet apparatus for 24 h to extract with n-hexane by-products and unreacted aniline. Finally, films were dried under dynamic vacuum for 24 h.

Elaboration of polyamide12/PANI films

PA12/PANI films were synthesized by dissolving PA12/PANI powders of different concentrations in m-cresol. The solutions were deposited on glass slides and left drying. The synthesis of PA12/PANI powders is described elsewhere [5].

Results and discussion

Dielectric characterization in the low frequency (10^{-1}-10^6 Hz) range was monitored by a Novocontrol broad band dielectric spectrometer. Temperature was varied between -160 °C and 30 °C. Gold circular electrodes are positioned onto both flat sides of the samples. In the case of PA6 films,

Dielectric Materials and Applications: ISyDMA'2016   Materials Research Forum LLC
Materials Research Proceedings 1 (2016) 100-103   doi: http://dx.doi.org/10.21741/2474-395X/1/25

samples were sandwiched between two electrodes making the contact with the insulating side of the films.

Dielectric permitivity of films

The existence of conductive pathways of PANI through the matrix of PA12 suggests the existence of a polarization due to the accumulation of conductive charges in the conductive clusters besides the dipoles of the matrix and the PANI (N-H). This phenomenon known as the Maxwell-Wagner-Sillars (MWS) interfacial polarization [6] is a consequence of the difference between the dielectric constant and the conductivity of the PANI clusters and those of the polymer matrix and can lead to relaxation phenomena. Fig. 1 represents the frequency dependence of the permittivity of PA films for various concentrations of polyaniline at $T = -100$ °C. Unlike the pure PA12, with composites containing a conductive phase, the quantity of accumulated charges is going to increase because of the appearance of a polarization in the interfaces PA12/PANI. The dielectric constant, measured at 1 Hz increases from 4.3 to nearly 4600 by increasing the quantity of polyaniline from 0.1 to 5 % only (Fig. 1.a). At this concentration the conductivity is of about 10^{-10} S/cm, a value compatible with dielectric applications. A similar behavior was obtained for PA6/PANI films but at a less magnitude. At 1 Hz, ε' does not exceed a value of 5 (Fig. 1.b).

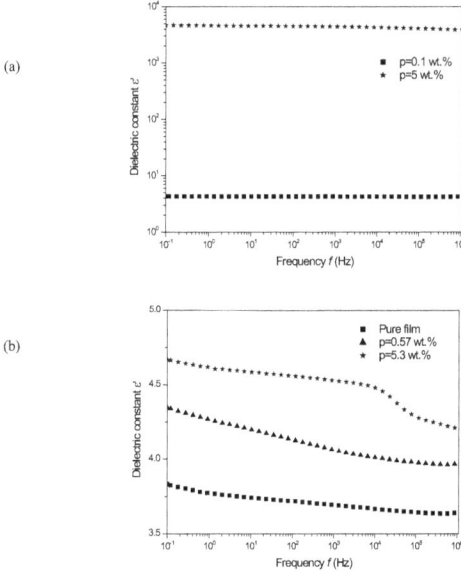

Fig. 1. Evolution of the relative permittivity ε' versus frequency for films of containing various concentrations in weigh of polyaniline at T=-100°C for (a) PA12/PANI films, (b) PA6/PANI films.

Dielectric loss

The evolution of ε'' versus frequency for both categories of films (PA12 and PA6) was recorded at different temperatures as shown in Fig. 2 at 5 and 5.3 % of PANI respectively. PA6 films exhibit a broader peak at low frequencies (1 Hz-10^2 Hz) and a well defined one between 10^4 and 10^5 Hz, denoting the existence of two relaxation processes. It appears that the first peak located at low frequencies shifts toward high frequencies with increasing temperature T expressing a thermally

Dielectric Materials and Applications: ISyDMA'2016   Materials Research Forum LLC
Materials Research Proceedings 1 (2016) 100-103   doi: http://dx.doi.org/10.21741/2474-395X/1/25

activated process for both categories of films but in a large extent for PA6 films. The frequency range didn't allow observing the complete second peak for PA12 films. Since in the general case of the relaxations of interface polarization involving a conductive phase and an insulating phase, the frequency of relaxation is proportional to the conductivity of the conductive material [7], the additional relaxation peak observed at high frequencies is connected to the presence of carriers in the PANI containing layer [8].

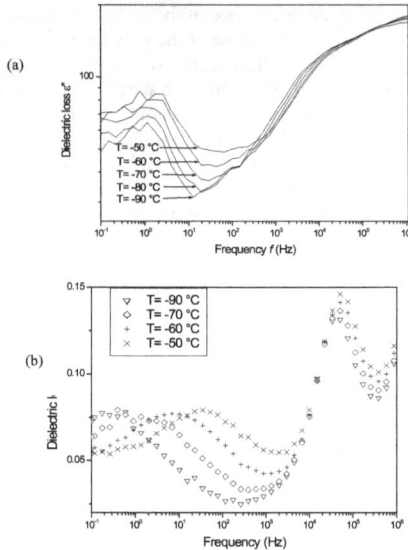

Fig.2. Evolution of dielectric loss ε" versus frequency at various temperatures for (a) a PA12 film containing 5 wt. % of PANI and (b) a PA6 film prepared with 5.3 wt. % of aniline.

Conclusion

It has been shown in this study that, compared to PA6 films, the values of ε' obtained for PA12 films are interesting and reveal that a concentration threshold of PANI exists for the dielectric behavior of these composites. Dielectric constant values of about 4600 were obtained allowing their use as high-k dielectrics in electronic components. Relaxation phenomena were also observed in both categories of films and were linked to the interfacial polarization. These phenomena are thermally activated and reveal the existence of some carriers motion in the PANI clusters at relatively high concentrations.

References

[1] R. Popielarz, C.K. Chiang, R. Nozaki, and J. Obrzut, "Dielectric properties of polymer/ferroelectric ceramic composites from 100 Hz to 10 GHz, " Macromolecules, vol. 34, pp. 5910-5915, July 2001. http://dx.doi.org/10.1021/ma001576b

[2] C.-H. Ho, C.-D. Liu, C.-H. Hsieh, K.-H. Hsieh, and S.-N. Lee, "High dielectric constant polyaniline/poly(acrylic acid) composites prepared by in situ polymerization," Synth. Met., vol. 158, pp. 630-637, June 2008. http://dx.doi.org/10.1016/j.synthmet.2008.04.014

[3] Y. Shen, Y. Lin, and C.-W. Nan, "Interfacial effect on dielectric properties of polymer nanocomposites filled with core/shell-structured particles," Adv. Funct. Mater., vol. 17, pp. 2405-2410, September 2007. http://dx.doi.org/10.1002/adfm.200700200

[4] A. Fattoum, M. Arous, F. Gmati, W. Dhaoui, and A. Belhadj Mohamed, "Influence of dopant on dielectric properties of polyaniline weakly doped with dichloro and trichloroacetic acids," J. Phys. D: Appl. Phys., vol. 40, pp. 4347-4354, June 2007. http://dx.doi.org/10.1088/0022-3727/40/14/033

[5] D. Mezdour, M. Tabellout, S. Sahli, and K. Fatyeyeva, "Electrical Properties Investigation in PA12/PANI Composites," Macromol. Symp., vol. 290, pp. 175-184, April 2010. http://dx.doi.org/10.1002/masy.201050402

[6] C.C. Ku, and R. Liepins, Electrical Properties of Polymers, Munich, New York: Hanser Publishers, 1987.

[7] B.K.P. Scaife, Principles of Dielectrics, Oxford: Oxford University Clarendon Press, 1989.

[8] A.A. Pud, M. Tabellout, A. Kassiba, A.A. Korzhenko, S.P. Rogalsky, G.S. Shapoval, F. Houzé, O. Schneegans, and J.R. Emery, "The poly(ethylene terephthalate)/ polyaniline composite: AFM, DRS and EPR investigations of some doping effects, " J. Mater. Sci., vol. 36(14), pp. 3355-3363, July 2001. http://dx.doi.org/10.1023/A:1017983206220

Dielectric Materials and Applications: ISyDMA'2016   Materials Research Forum LLC
Materials Research Proceedings **1** (2016) 104-107   doi: http://dx.doi.org/10.21741/2474-395X/1/26

Structural and composition properties of ZnO thin films elaborated by spray pyrolysis

Salah BOULMELH[1], Lynda SACI[*1], Farida MANSOUR[1], Ramdane MAHAMADI[2]

[1]LEMEAMED laboratory, Departement of Electronics,University des frères Mentouri, 25000 Constantine, Algeria

[2]LEA laboratory, Departement of Electronics, University Batna 2, 05000, Algeria

*Corresponding author: E-mail: lynda.saci@lec-umc.org

Keywords: Spray Pyrolysis, Undoped ZnO, XRD, Raman, Substrate Temperature

Abstract. Undoped Zinc oxide (ZnO) thin films were de-posited onto microscope glass substrates in the temperature range of 300 °C – 450 °C using spray pyrolysis technique. The study of the composition and structural properties of these films were investigated by means X-ray diffraction and Raman spectroscopy versus the increasing substrate temperature (Ts). The XRD results presented that the films prepared at a substrate temperature greater than 300 °C exhibit the hexagonal wurtzite with a preferential orientation along the (100) direction. Furthermore, the lattice parameters showed that the presence of low micro-stresses and the crystallinity of all samples were improved with thermal annealing. The Raman spectroscopy results showed, the presence of the E_2^{low}, E_2^{high} located around 98 cm^{-1} and 437.5 cm^{-1} respectively, and the deconvolution of the band located between 520cm^{-1} and 620cm^{-1} were formed of an $E_1(LO)$, $A_1(LO)$ located about 581cm^{-1} and 551cm^{-1} , these phonons modes were interpreted by ZnO wurtzite phase. In addition, the shift of the position of E1 (LO), A1 (LO) peaks indicates the presence of micro-stress in the deposited layers. The results obtained are in good agreement with those found by XRD. Finally, the obtained results showed that the deposited films can be used as components in micro technology to know biosensors.

Introduction

Undoped ZnO thin films are semiconductors materials who have a dielectric character (a large resistivity around the megohms). Moreover, having the direct band gap energy of approximately 3.37 eV. It is a promising material due to its lumicence properties and its high chemical stability. In addition, its abundance in nature made its use more economical. ZnO can be used for many applications such as acoustooptic devices [1], photovoltaic [2], solar cells [3] and gas sensors, [4]. Several techniques have been used to prepare ZnO thin films. For instance, RF magnetron sputtering [5], chemical vapor deposition (CVD) [6], sol–gel method [7], thermal evaporation and spray pyrolysis [8]. In this work, we studied undoped ZnO thin films prepared by spray pyrolysis technique in order to investigate the influence of the glass substrate temperature on the composition and structural properties of films. In order to extract the optimum conditions and applied as anti-reflection layer in solar cells or sensitive layer in bio-sensors.

Experimental procedures

After cleaning the glass substrates with methanol, the un-doped ZnO films were deposited using a Spray Pyrolysis technique. A homogeneous solution was prepared by dissolving 0.2 M of zinc acetate dehydrated $(Zn(C_2H_3O_2)_2 \; 2H_2O)$ precursor diluted in deionized water. The deposited temperature substrate was 300 °C - 450 °C, for 15 *minutes* duration. The crystalline structure was studied by X-ray diffraction measurements and the films composition was characterized by means of Raman spectrometer.

Dielectric Materials and Applications: ISyDMA'2016   Materials Research Forum LLC
Materials Research Proceedings 1 (2016) 104-107   doi: http://dx.doi.org/10.21741/2474-395X/1/26

Results and discussion

Structural properties

The crystal structure and orientation of ZnO films were investigated by XRD. The fig.1 shows the superposition of X-ray spectrums. The peaks appearing at 2θ =31.86° , 34.45°, 36.30° and 56.86° can be well indexed to the polycrystalline hexagonal wurtzite structure, in a good agreement with the X-ray diffraction pattern of hexagonal ZnO reported in the JCPDS 361451 card. We can note that after 300°C the peak belonged to the (100) plane is the most intense versus temperature substrate increasing, indicating the preferential orientation of ZnO film (Fig. 1).

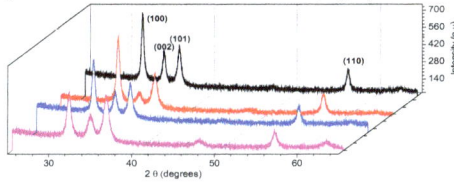

Fig. 1. Superposition of spectrums X-ray diffraction of ZnO thins films.

The table I regroup the calculated of the lattice constants, the stress, the crystallite size (D), the width at half maximum and the intensitys. The calculated values of a and c are around 3.230 °A 5.206 °A respectively.

TABLE I LATTICE CONSTANTS, STRESS VALUES AND CRYSTALLITE SIZE.

T_S	Lattice constants (A)		Stress (GPa)	D (nm)	FHWM (rad)	Intesity (a.u)
	a	c				
300	3.2400	5.2030	0.1790	18.75	0.4408	493
350	3.2392	5.2024	0.2058	21.43	0.3857	616.6
400	3.2404	5.2008	0.2205	25	0.3306	641.5
450	3.2416	5.2024	0.2774	30	0.2755	710

The shift between the lattice constants and there ported values on the JCPDS this also corresponds to a shift in 2θ from 31,8676° to 31,8510°. If the diffraction peaks shift to lower angles, such a shift may be attributed to a tensile stress in the film [9], [10]. From the table1 we note an increasing of the stresses via temperature substrate growth. Moreover, the crystallite size and the crystallinity of the ZnO particles increase with increasing temperatures substrate as observed from the decrease in the full width at half maximum (FWHM) value and increase in the intensity of the XRD patterns, respectively. Furthermore, no peaks corresponding to any other crystalline material is found in the recorded diffract grams. The crystallite size (D) is calculated for the grown films using the Scherers formula [11]. The variation of the grain size for undoped ZnO films versus substrate temperature (see table1). The grain size increases with substrate temperature (300 °C- 450 °C) from 18.75nm to 30.00nm. This improves the crystalline quality of ZnO films.

Composition properties

The Fig. 2 shows the Raman spectra of the ZnO films deposited at 350 °C. The detection of two vibration mode E_2^{low}, E_2^{high} where, the first one, is attributed to the lattice vibrations of zinc atoms. The second is related to the lattice vibration of the oxygen atoms and the presence of the wurtzite phase of ZnO and indicates good crystallinity [12]. This result is in good argument with XRD. The peak corresponds to the A_1 (TO) symmetry mode of ZnO [13]. Moreover, the strong and broad band located between 500 and 650 cm^{-1} that varies with the substrate temperature has been deconvolved. The deconvolution (Fig. 3) shows that this band is formed of glass substrate peak located around 560

cm^{-1} and A_1 (LO), E_1 (LO) (see table II). According to [14], they found that slightly lower E_1 (LO) and A_1 (LO) mode frequencies for the film. They suggest the existence of vacancy point defects within the wurtzite lattice. Moreover, B. Cao and al, say that the presence of the two modes E_1 (LO) and A_1 (LO) in the Raman spectra, for the film, confirms the random orientation of ZnO nanoparticles in conformity with the XRD results. As the reaction temperature is increased, the separation between A_1 (LO) and E_1 (LO) mode is also increased and indicates the better crystallinity of the samples [15].

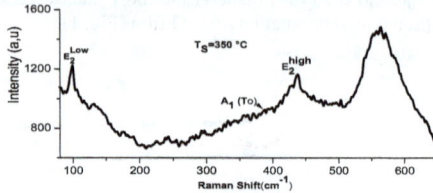

Fig. 2. Raman spectra of ZnO film.

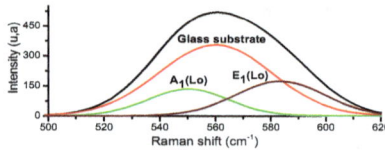

Fig. 3. Deconvolution Raman spectra of ZnO film.

TABLE II THE PEAK ASSIGNEMENT FROM THE RAMAN SPECTRA OF ZNO FILMS DEPOSITED AT DIFFERENT SUBSTRATE TEMPERATURES.

Raman shift (cm^{-1})					
T_S (°C)	E_2^{low}	A_1(TO)	E_2^{high}	A_1(LO)	E_1(Lo)
300	97.5	384	437	551.23	580.93
350	98	387	437.5	550.2	583.45
400	98.5	385	438	549.83	584.35
450	97.5	385.5	437.5	543.33	586.2

Conclusion

The X-ray diffraction analysis shows, the presence of the ZnO wurtzite for ZnO films for Ts > 300 °C with preferential orientation (100) and the improvement of the crystalline films. These results are in good concordance with Raman spectroscopy by the presence of E_2^{low}, E_2^{high} peaks. Moreover, the existence of two peaks, E_1 (LO) and A_1 (LO) suggest the existence of vacancy point defects within the wurtzite lattice and confirms the random orientation of ZnO this is in conformity with the XRD results. Finally, this study showed that the films deposited at Ts=350 °C have dielectric character (values of the measured resistivity are around the megohms), which can be used as sensitive layer in bio-sensors or antireflective layer of solar cells.

References

[1] GC. R. Gorla, N. W. Emanetoglu, S. Liang, W. E. Mayo, and Y. Lu, "Structural, optical, and surface acoustic wave properties of epitaxial ZnO films grown on (0112) sapphire by metalorganic chemical vapor deposition," *Journal of Applied Physics*, vol. 85, no. 5, p. 2595, March.1999.

[2] A. Belaidi, T. Dittrich, D. Kieven, J. Tornow, K. Schwarzburg, M. Kunst, N. Allsop, M. C. Lux-Steiner, and S. Gavrilov, "ZnO-nanorod arrays for solar cells with extremely thin sulfidic absorber,"

Journal of Solar Energy Materials and Solar Cells, vol. 93, no. 67, pp. 1033–1036, 2009. http://dx.doi.org/10.1016/j.solmat.2008.11.035

[3] N. Jabena Begum, K. Ravichandran, "Effect of source material on the transparent conducting properties of sprayed ZnO: Al thin films for solar cell applications," *Journal of Physics and Chemistry of Solids*, vol. 74, pp. 841–848, 2013. http://dx.doi.org/10.1016/j.jpcs.2013.01.029

[4] Chien-Yuan Lu, Sheng-Po Chang,Shoou-Jinn Chang and I-Cherng Chen. "ZnO Nanowire-Based Oxygen Gas Sensor" *IEEE Sensors Journal*, vol. 9(4), pp.485 – 489, 2009. http://dx.doi.org/10.1109/JSEN.2009.2014425

[5] A. Ismail, M.J. Abdullah, "The structural and optical properties of ZnO thin films prepared at different RF sputtering power"*Journal of King Saud University – Science*, Vol 25, pp.209–215, 2013.

[6] Pai-Chun Chang, Zhiyong Fan, Dawei Wang, Wei-Yu Tseng, Wen-An Chiou, Juan Hong, and Jia G. Lu, "ZnO Nanowires Synthesized by Vapor Trapping CVD Method"*Chemistry Materials*. Vol.16, pp.5133-5137, 2004. http://dx.doi.org/10.1021/cm049182c

[7] Chien-Yie Tsay , Hua-Chi Cheng , Yen-Ting Tung, Wei-Hsing Tuan , Chung-Kwei Lin., "Effect of Sn-doped on microstructural and optical properties of ZnO thin films deposited by sol–gel method," *Thin Solid Films* .vo. l517, pp. 1032–1036, 2008.

[8] N. Jabena Begum, R. Mohan, K. Ravichandran, "Effect of solvent volume on the physical properties of aluminium doped nanocrystalline zinc oxide thin films deposited using a simplified spray pyrolysis technique," *Superlattices and Microstructures*, vol. 53, pp. 89-98, 2013. http://dx.doi.org/10.1016/j.spmi.2012.09.012

[9] C. H. Lee and L. Y. Lin, "Characteristics of spray pyrolytic ZnO thin films," *Applied Surface Science*, vol. 92, pp. 163–166, 1996. http://dx.doi.org/10.1016/0169-4332(95)00223-5

[10] B. J. Lokhande, P. S. Patil, and M. D. Uplane, "Deposition of highly oriented ZnO films by spray pyrolysis and their structural, optical and electrical characterization," *Materials Letters*, vol. 57, no. 3, pp. 573– 579, 2002. http://dx.doi.org/10.1016/S0167-577X(02)00832-7

[11] A. L. Patterson, "The Scherrer Formula for X-Ray Particle Size Determination," *Physical Review*, vol. 56, no. 10, pp. 978–982, Nov. 1939. http://dx.doi.org/10.1103/PhysRev.56.978

[12] R. Ayouchi, F. Martin, D. Leinen, and J. R. Ramos-Barrado, "Growth of pure ZnO thin films prepared by chemical spray pyrolysis on silicon," *Journal of Crystal Growth*, vol. 247, no. 3, pp. 497–504, 2003. http://dx.doi.org/10.1016/S0022-0248(02)01917-6

[13] X. Q. Wei, B. Y. Man, M. Liu, C. S. Xue, H. Z. Zhuang, and C. Yang, "Blue luminescent centers and microstructural evaluation by XPS and Raman in ZnO thin films annealed in vacuum, N_2 and O_2," *Physica B: Condensed Matter*, vol. 388, no. 12, pp. 145–152, Jan. 2007. http://dx.doi.org/10.1016/j.physb.2006.05.346

[14] N. Ashkenov, B. N. Mbenkum, C. Bundesmann, V. Riede, M. Lorenz, D. Spemann, E. M. Kaidashev, A. Kasic, M. Schubert, M. Grundmann, G. Wagner, H. Neumann, V. Darakchieva, H. Arwin, and B. Monemar, "Infrared dielectric functions and phonon modes of high-quality ZnO films," *Journal of Applied Physics*, vol. 93, no. 1, pp. 126–133, Jan. 2003. http://dx.doi.org/10.1063/1.1526935

[15] B. Cao, W. Cai, H. Zeng, and G. Duan, "Morphology evolution and photoluminescence properties of ZnO films electrochemically deposited on conductive glass substrates," *Journal of Applied Physics*, vol. 99, no. 7, p. 073516, Apr. 2006. http://dx.doi.org/10.1063/1.2188132

Dielectric Materials and Applications: ISyDMA'2016   Materials Research Forum LLC
Materials Research Proceedings 1 (2016) 108-109   doi: http://dx.doi.org/10.21741/2474-395X/1/27

Theoretical investigations, spectroscopy and spin orbit couplings for heavy metal complexes MX (M=Hg, and X=H, O, S)

N. EZARFI*, A. TOUIMI BENJELLOUN, S. SABOR, M.BENZAKOUR, M. MCHARFI, A. DAOUDI

Equipe de Chimie Informatique et Modélisation (ECIM), Laboratoire d'Ingénierie des Matériaux, de Modélisation et d'Environnement (LIMME), Faculté des Sciences Dhar El Mahraz (FSDM), Université Sidi Mohammed Ben Abdallah (USMBA), B.P. 1796, Fès-Atlas, Fès, Morocco

*Author for correspondence: zarfi25@hotmail.com

Keywords: Mercury Oxide, Mercury Sulphide, CCSD (T), Spectroscopic Properties

Abstract. Heavy metal oxides and sulphides MX (M=heavy metal and X=O, S) have been widely used in the industry such as catalysts. Recently, because of their involvement in the cycle of atmospheric chemistry, several studies in gas phase are performed [1-5].. Electronic structures, spectroscopic properties as well as ionization energies for these species are often inaccessible. The main subject of the present work is reportin g dipole moments (μ), harmonic frequencies (ωe), equilibrium distances (Re) as well as single ionization energies (IE1) for low-lying electronic states of MX (M=Hg and X=H,S, O). These calculations were done using CCSD(T) [6-7] levels of theory with a large basis sets. The scalar relativistic effects have been accurately included by making use of relativistic effective potentials.

Introduction

Different species of mercury have different physical/chemical properties and thus behave quite differently in air pollution control equipment and in the atmosphere. The emissions of mercury and mercury compounds into the atmosphere are of special significance. Mercury metal is widely distributed in nature at very low concentrations. In recent decades, the spectroscopic properties of HgX (X=O, S and H) represent an active research field in physical chemistry. High resolution spectroscopic studies of these species by conventional spectroscopic technique are lacking. Presently, we perform state-of-art ab initio computations dealing with the potential energy curves and wavefunctions (PECs) of the lowest electronic states of HgX with X=O, S and H. All electronic computations were performed using the GAUSSIAN package. The standard aug-cc-pVTZ basis sets have been employed for the O, S and H atoms. The Hg basis sets corresponded to aug-cc-pVTZ (-PP) recently realised by Figgen et al. in connection with the fully relativistic ECP60MDF pseudo-potential that leaves 20 electrons ($5s^2 5p^6 5d^{10} 6s^2$) to be explicitly correlated in the ab initio calculations.

Results and discussion

A good agreement is found for molecules between the calculated values and the experimental ones and theoretical results . Table gathers the results obtained for HgH, HgS and HgO molecules as calculated with GAUSSIEN .

For Hg-H, the CCSD(T)/AVTZ calculations give a bond length (Re 1.75Å) and a dipolar moment (0.46 Debye), in excellent agreement with experiment the deviations amounting only 0.01 Å and 0.01 Debye.

For HgS and HgO, we compared our results with a study done with MRCI + Q which generates the spin-orbit coupling.

Dielectric Materials and Applications: ISyDMA'2016   Materials Research Forum LLC
Materials Research Proceedings 1 (2016) 108-109   doi: http://dx.doi.org/10.21741/2474-395X/1/27

For the HgO molecule, it has been found that the $^3\Pi$ state is lower than the $^1\Sigma^+$ state , their ionization potential is 9.52 eV .

For the HgS molecule, it has been found that the $^1\Sigma^+$ state is lower than the $^1\Sigma^+$ state , their ionization potential is 9.26 eV and 9.58 eV.

Second, the fact that the oxygen atom is smaller than sulfur results in a shorter bond length (for HgO, the equilibrium distances of the $^1\Sigma^+$ and $^3\Pi$ states are 1.91 Å and 2.22 Å, respectively , to be compared with 2.24 Å and 2.57 Å r, respectively, for HgS).

Table : spectroscopic constants of HgH,HgO and HgS with CCSD(T)/AVTZ R_e (distance Hg-X) , ω_e (frequency) , μ (dipolar moment)

		R_e(Å)	ω_e(cm^{-1})	PI(eV)	μ(Debye)
HgH	CCSD(T)/AVTZ	1.75	1346.04	8.03	0.46
	Exp[8]	1.74	1203,2	-	0.47
HgO $^1\Sigma^+$	CCSD(T)/AVTZ	1.90	609.7	9.52	5.14
	MRCI+Q 5Z [4]	1.91	598.80	-	-
HgO $^3\Pi$	CCSD(T)/AVTZ	2.16	290.9	9.52	1.42
	MRCI+Q 5Z [4]	2.22	287.8	-	-
HgS $^1\Sigma^+$	CCSD(T)/AVTZ	2.26	352.24	9.26	5.9
	MRCI+Q 5Z [5]	2.24	363.0	-	-
HgS $^3\Pi$	CCSD(T)/AVTZ	2.55	167.71	9.58	0.46
	MRCI+Q 5Z [5]	2.57	147.8	-	-

References

[1] S. Sabor, A. Touimi Benjelloun, M. Mogren Al Mogren and M. Hochlaf, Phys. Chem. Chem. Phys. 16, 21356 (2014). http://dx.doi.org/10.1039/C4CP03136A

[2] A. Touimi Benjelloun, A. Daoudi, and H. Chermette, J. Chem. Phys. 121, 15 (2004).

[3] A. Touimi Benjelloun, A. Daoudi, and H. Chermette, J. Mol. Phys, 103, 2-3, (2005). http://dx.doi.org/10.1080/00268970512331317282

[4] B. C. Shepler and K. A. Peterson, J. Phys. Chem. A 107, 1783 (2003). http://dx.doi.org/10.1021/jp027512f

[5] C. Cressiot , M.Guitou , A. Mitrushchenkov Molecular Physics, Vol. 105, No. 9, 10 May 1207–1216 (2007).

[6]- Pople, J. A; Head-Gordon M.; and Raghavachari, K. J. Chem. Phys. 87, 5968 (1987). http://dx.doi.org/10.1063/1.453520

[7] - Pople, G. Berthier , J. A and Nesbet, R. K. J. Chem. Phys, 22, 571 (1954)

[8] Nikolai S. Mosyagin, Anatoly V. Titov GRECP/RCC, St.-Petersburg district 188300, Russia (9 juin 2001).

Dielectric Materials and Applications: ISyDMA'2016
Materials Research Proceedings 1 (2016) 110-113

Materials Research Forum LLC
doi: http://dx.doi.org/10.21741/2474-395X/1/28

Characterization of streamer patterns in insulating vegetable oil subjected under impulse voltages by fractal analysis

Aissa YOUSFI*, Boubakeur ZEGNINI

Laboratoire d'étude et de développement des Matériaux semi-conducteurs et diélectriques-LeDMaScD, Université Amar Telidji de Laghouat, BP 37 G route de Ghardaïa Laghouat 03000 - Algeria

*Corresponding author: E-mail: aissayousfip@yahoo.fr

Keywords: Fractal Dimension, Box Counting Method, Pattern, Insulating Vegetable Oil, Power Transformer, Impulse Voltage

Abstract. This paper presents the variation of fractal dimension with final stopping lengths and different applied voltage. This investigation deals mainly with the fractal properties of 2D (Dimension) patterns of real propagating streamers obtained in insulating vegetable oil. The streamer patterns are generated experimentally under impulse voltages. Simulation program is used to calculate the fractal dimension for experimentally obtained sequence of initiation and propagation of streamers in vegetable oil by using box counting method. It is found that fractal dimension is a function of magnitude of the voltage, polarity of the voltage. Fractal analysis is a good and suitable tool for the analysis of streamers initiation and propagation in insulating vegetable oil used in transformers.

Introduction

The increasing crisis of petroleum oil that is currently leading to uncertainty in its sustainable supply has forced the researchers to find suitable alternate sources. In this context, the vegetable oils are natural products and have renewable sources with plenty of supply that can replace the mineral based oils for applications in transformers [1]. Researchers and industries are performing investigations on vegetable oils for providing them as insulating oils in transformers and pollution free environment. Vegetable oil on the other hand is environmental friendly, biodegradable, renewable, cheap, highly available and safer alternative insulating and cooling medium for transformers. Therefore it is of significant importance to study electrical performance of transformer liquids in divergent fields in terms of streamer characteristics.

Many investigators [2], [3] used high speed photography to record the shape and the growth dynamics of the pre-breakdown streamers in liquids. It shows that these streamer parameters are associated with various test conditions such as voltage polarities, additives and liquid type. However, these parameters can only indicate one-dimensional information of streamer patterns. Two-dimensional information such as surface complexity and its dynamic change with time and have not yet been well studied. Fractal analysis was increasingly used as an index to quantitatively describe the complexity of a pre-breakdown phenomenon, i.e. electrical treeing in solid insulating materials and creeping discharge patterns over solid insulators immersed in insulating oils [4-6]. In this paper the fractal index of streamers was estimated by using the box-counting method. Variations of fractal dimension of streamers with different polarity and applied voltages were evaluated.

Experimental descriptions

The experiment set up for detection of initiation and propagation of streamers in the vegetable oils is similar to that we used in [7]. The impulse generator was used to produce standard lightning impulse voltages which were measured through an RC voltage divider and Digital Impulse Measuring System. A test cell comprising of point-plane electrode system was designed and used. Tungsten and high carbon steel needles with tip radius (rp) of 10 μm, electrode gap 20mm were used. The plane electrode was made of brass having a diameter of 50 mm with its edges rounded. A resistor was

Dielectric Materials and Applications: ISyDMA'2016   Materials Research Forum LLC
Materials Research Proceedings 1 (2016) 110-113   doi: http://dx.doi.org/10.21741/2474-395X/1/28

connected at the output of the shock generator to limit the breakdown current and injected energy into the test cell. A charge-coupled device (CCD) camera is used to record the propagation of streamers in vegetable oil tree during the experiment. The shadowgraph system consisting of a CCD Camera, which could be used with or without a flash, was employed.

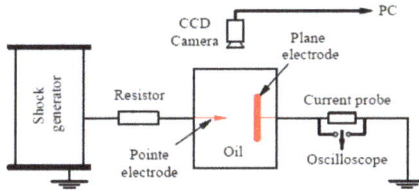

Fig. 1. Experimental arrangement.

The most important physico chemical characteristics properties of tested vegetable oil are summarized in Table 1.

TABLE I Proprieties of tested vegetable oil

Unit	Measured characteristics		
	Property	Temperature	Value
Kg/dm³	Density	23°C	0.98
mm/s	Viscosity	30°C	28
		60°C	12
mg/KOH/g	Acidity		0.1315
ppm	Water content		125
kVrms	AC Breakdown Voltage using IEC (60Hz)		43
	Dielectric Dissipation Factor (tan δ)	23°C	0.0027
		80°C	0.019
	Permittivity (εr)	23°C	3.07
		80°C	2.85

Experimental Results

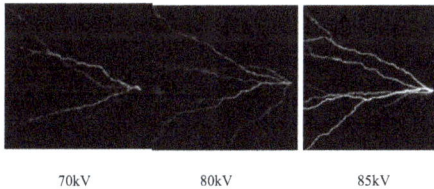

70kV   80kV   85kV

Fig. 2. Capture of propagated streamers in vegetable oil under positive polarity lightning impulses;

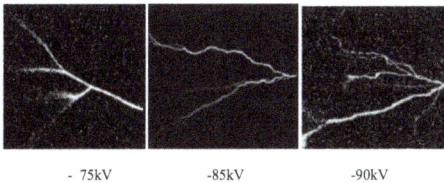

- 75kV   -85kV   -90kV

Fig. 3. Capture of propagated streamers in vegetable oil under negative polarity lightning impulses.

Dielectric Materials and Applications: ISyDMA'2016   Materials Research Forum LLC
Materials Research Proceedings 1 (2016) 110-113   doi: http://dx.doi.org/10.21741/2474-395X/1/28

The streamer currents are measured through a non-inductive resistor connected in series with the test cell, and thanks to a high time resolution oscilloscope. The most important properties of tested vegetable oil are the dielectric ones beside the usual physico-chemical characteristics. It was more difficult to capture the initiation event of streamers in vegetable oil under positive polarity lightning impulses. The reason is that the streamers once initiated, propagate very swiftly with long branches. Figures 2 and 3 show propagated streamer shapes captured with increasing voltage.

Fractal analysis by using the box-counting method

The evaluation of fractal nature of streamer patterns is achieved by using fractal dimension D_f .Literally there are several methods for the estimation of fractal dimension [4].However, only the box-counting method is considered in this work, since this method is one of the most popular methods for estimating the structure of self-similar patterns and is easy to achieve by MATLAB programming in order to calculate the fractal dimension for experimentally obtained streamers by using box counting method . The idea of the box algorithm is to subdivide the domain into boxes, and count the number of boxes that contain observations of the data. Firstly, image of the streamer pattern was converted to binary image by using an image processing program. Secondly, streamer pattern in the binary image is covered by square boxes of size r (r= 256, 128, ..., 1). When a box of size r covers a streamer patterns, the following relationship is shown as:

$$N_r \approx r^{-Df} \tag{1}$$

Where is N_r the number of covering boxes measured by scale rand D_f is the fractal dimension. Equation (1) can be rewritten

as: $$D_f = - \lim (\log N_r / \log r \ ) \tag{2}$$

$$r \rightarrow \infty$$

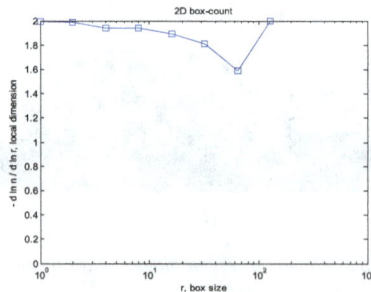

Fig.4. The results of 2D box count.

In figure 4 the local slope shows that the image the streamer patterns under positive polarity lightning impulses (+70kV) is indeed approximately fractal, with a fractal dimension Df = 1.91 +/- 0.015 for scales 30 <r < 105.The effect of the increase of the applied voltage (+85 kV) can be seen clearly in the simulated fractal index of streamer Df = 1.94 +/- 0.021 for scales 20 <r < 90.The increasing of fractal dimension is due to the propagation of main channel of the streamer pattern and extension of micro-channels with the increase of the applied voltages.

Conclusion

In this paper the fractal index of streamers was estimated by using the box-counting method. 2D (Dimension) patterns of real propagating streamers obtained in insulating vegetable oil. The streamers patterns are generated experimentally under impulse voltages. Under positive polarity lightning impulses, which streamers propagate very fast the streamers propagate much faster than under negative polarity, while their more luminous filaments indicated that they were more conducting than their negative counterparts. As soon as the voltage was increased, large size filaments appeared and propagated toward the plane electrode much faster compared to streamers under negative polarity.

References

[1] Essam A. Al-Ammar, Optical observation of streamer propagation and breakdown in seed based insulating oil under impulse voltages *international Journal of physical science, Vol 9 (13)pp 292-301, 16 July 2014*.(2014)

[2] P. Rain and O. Lesaint, "Prebreakdown phenomena in mineral oil understep and ac voltage in large-gap divergent fields", *IEEE Trans. Dielectr. Electr. Insul., Vol. 1*, pp. 692-701, (1994). http://dx.doi.org/10.1109/94.311712

[3] S. Otsuki, K. Yamazawa, K. Sugiura, and H. Yamashita, "A study of three-dimensional measurement of pre-breakdown streamer in dielectric liquids", IEEE Conf. Electr. Insul.and Dielectr. Phenom., Vol. 2, pp. 644-647, (1997). http://dx.doi.org/10.1109/ceidp.1997.641157

[4] K. Kudo, "Fractal analysis of electrical trees," *IEEE Transactions on Dielectrics and Electrical Insulation*, vol. 5, pp. 713-727 (1998). http://dx.doi.org/10.1109/94.729694

[5] Kebbabi and A. Beroual, "Fractal analysis of creeping discharge patterns propagating at solid/liquid interfaces-influence of the nature and geometry of solid insulators," *Annual Report Conference on in Electrical Insulation and Dielectric Phenomena (CEIDP '05)*, pp. 132-135.(2005).

[6] W. Lu, X. F. Wang and Q. Liu, Fractal Index of Streamer Patterns in Insulating Liquids under Lightning Impulse Voltages, *2014 IEEE International Conference on Liquid Dielectrics*, Bled, Slovenia, June 30 - July 3, 2014 ,pp.1-4.(2014). http://dx.doi.org/10.1109/icdl.2014.6893158

[7]]Viet-Hung Dang, A. Beroual and C. Perrier, Investigations on Streamers Phenomena in Mineral, Synthetic and Natural Ester Oils under Lightning Impulse Voltage, *IEEE Transactions on Dielectrics and Electrical Insulation* Vol. 19, No. 5; October 2012, pp1521-1527 (2012).

[8] K.J Anoop ,K.Kanchana, Characterization of electrical trees in Bakelite insulator by fractal analysis, *Int. Trans. Electr. Energ. Syst.* (2014)

Dielectric Materials and Applications: ISyDMA'2016   Materials Research Forum LLC
Materials Research Proceedings 1 (2016) 114-118   doi: http://dx.doi.org/10.21741/2474-395X/1/29

Prediction of the AC electrical conductivity of carbon black filled polymer composites

S. A. ELHAD KASSIM*[1,2], M. E. ACHOUR[1], L. C. COSTA[3]

[1]LASTID Laboratory, Physics Department, Faculty of Sciences, Ibn Tofail University, BP 133, 14000 Kenitra, Morocco

[2]LEMA Laboratory, Faculty of Sciences and Technique,University of Comoros, Corniche, Moroni, Comoros

[3]I3N and Physics Department,University of Aveiro, 3810-193 Aveiro, Portugal

*Corresponding author: E-mail: elhadkassims@yahoo.fr

Keywords: Carbon Black, AC Electrical Conductivity, Percolation, Nanocomposites

Abstract. Since the discovery of conductive polymer composite in the 70s; different mixing laws developed from the effective medium theory have been proposed to describe the physical behavior of these materials. Two models based by percolation theory have been applied to predict the AC electrical conductivity of carbon black/epoxy composites. McLachlan and Mamunya models show satisfactory results considering the parameters related to these models as adjustable parameter. These adjusted parameters do not vary significantly as a function of frequency.

Introduction

Polymer composites filled with carbon nanoparticles are of interest for many fields of engineering. This interest comes from the fact that the electrical characteristics of such composites are close to the properties of metals, whereas the mechanical properties and processing methods are typical for plastics [1]. Conductive polymer composites have gained a great significance due to their practical applications in electrical and electronic equipments as capacitors, transistor, sensors, electrostatic discharge materials, electromagnetic interference shielding materials, etc [2-4]. Extrinsically conductive composites are prepared by the inclusion of different inorganic and organic conducting particulate like metal particles, carbon blacks, carbon nanotubes, graphene into the insulating polymer matrix. Compared to metals, these composites are light in weight and flexible in nature. In our present paper, we study the electrical properties of cabon black (CB)/epoxy composites.

The experimental AC conductivity values were compared with those obtained by generalized effective medium model McLachlan and Mamunya model. These models are usually applied to predict the DC conductivity [5-6]. In this work, these two models are used to describe the behavior of AC conductivity.

Experimental

Materials

The samples are composed of spherical carbon black (CB) particles randomly dispersed in an insulating polymer matrix. The polymer used as host matrix for fabrication of the composites is diglycidylic ether of biosphere F (DGEBF) epoxy. The carbon black used in this work is type Raven 2000, produced by Cabot Corporation, USA and polymer, from the Ciba Geigy Company, Switzerland. The essential feature of Raven 2000 (carbon black particles) and polymer matrix epoxy are summarized in the Table 1. A full description of this procedure can be found in Ref. [7].

Table 1 *Summary of the essential features of carbon black particles*

Component type	Particle size (nm)	Electrical conductivity, $(\Omega.m)^{-1}$	Density $(Kg.m^{-3})$
Carbon black	9,6	656	1817
Epoxy	--	$1,4 \times 10^{-16}$	1180

Electrical measurements

The dielectric measurements of the series of samples Raven 2000/DGEBF were performed in Paul Pascal Research Center of the Bordeaux I University, France. The equipment used is an impedance analyzer HP 4194A, functioning between 100 Hz to 15 MHz. The samples were cut into cylindrical form of 3 mm thick and 12 mm diameter. The circular surfaces were polished and covered with a thin silver layer to serve as the contact with the electrodes.

Results and discussion

The AC electrical conductivity as a function of the volume fraction of the filler is shown in Figure 1 for different frequencies, at constant temperature of 300 K. At a certain concentration, the percolation threshold, the conductivity value increases considerably as in the DC, for low frequencies. The value of the threshold concentration of the sample series is 3.60%.

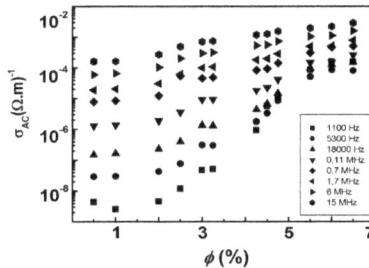

Fig.1. AC conductivity as a function of concentration of filler, at different frequencies

Two models were applied to describe the electrical behavior of the AC conductivity, McLachlan model and Mamunya models.

McLachlan model

The variation of the AC conductivity of the composite, as a function of the carbon black particles content, can be critically examined by the GEM equation since it takes into account the intrinsic conductivities, geometries and orientation of the filler and polymer particles relatively to an applied electric field. McLachlan [8, 9] postulated the generalized effective medium (GEM) Eq. (1) for electrical conductivity of binary composites systems, as:

$$(1-\phi)\frac{(\sigma_m^{1/t}-\sigma_{AC}^{1/t})}{(\sigma_m^{1/t}+A\sigma_{AC}^{1/t})} + \phi\frac{(\sigma_c^{1/t}-\sigma_{AC}^{1/t})}{(\sigma_c^{1/t}+A\sigma_{AC}^{1/t})} = 0 \qquad (1)$$

Where σ_c and σ_m are the conductivities of the carbon black and matrix polymer respectively, and A,

Dielectric Materials and Applications: ISyDMA'2016   Materials Research Forum LLC
Materials Research Proceedings 1 (2016) 114-118   doi: http://dx.doi.org/10.21741/2474-395X/1/29

$$A = \frac{1-\phi_c}{\phi_c} \tag{2}$$

McLachlan model describes the behavior of the conductivity as a function of the volume fraction at different frequencies. This model shows satisfactory results between the calculated and the experimental values at different frequencies by considering the exponent t as an adjustable parameter. Figure 2 presents the comparison between the values of the conductivity calculated and experimental values at a frequency of 15 MHz. The solid curve is a calculation based on the McLachlan equation, Eq. (1), and full square is the experimental data. At different frequencies, this model also shows good agreements with t, an adjustable parameter. The various values t obtained after modeling a function of the frequency are shown in Figure 3. These values are almost identical as a function of frequency.

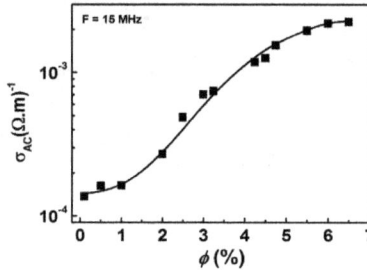

Fig.2. AC conductivity as a function of CB volume fraction for Raven 2000/DGEBF (full square) at frequency of 15 MHz. The solid curve is a calculation based on the McLachlan model

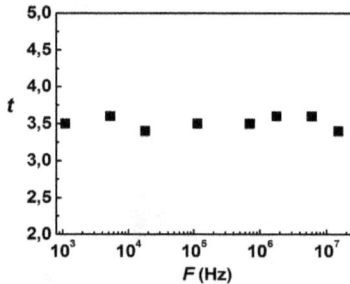

Fig.3. Exponent t of McLachlan as a function of frequency for Raven 2000/GDEBF

Mamunya model

Mamuny [1, 10-11] have studied the conductivity of composites in different polymers to evaluate the influence of the polymer–filler interactions on the conductivity. The proposed model takes into account the surface energy of the polymer as well as the filler and his aspect ratio. The Mamunya equation can be written as follows:

$$Log\sigma_{AC} = Log\sigma_{\phi_c} + \left(Log\sigma_{Max} - Log\sigma_{\phi_c}\right) \left(\frac{\phi-\phi_c}{\phi_{Max}-\phi_c}\right)^k \tag{3}$$

Where ϕ_{Max} is the maximum packing fraction, σ_{ϕ_c}, the electrical conductivity at the critical volume fraction, σ_{Max} the conductivity at the highest volume fraction, $\phi = \phi_{Max}$. k is a parameter that depends on the filler fraction, ϕ, percolation threshold, ϕ_c, and interfacial tension, γ_{Pf},

$$k = \frac{R\phi_c}{(\phi-\phi_c)^n} \tag{4}$$

$$R = D + G\gamma_{Pf} \tag{5}$$

R and n are additional parameters, and D and G are constants.

The variations of conductivity σ_{AC} values calculated from the Mamunya model and experimental values at 15 MHz are shown in Figure 4. This Mamunya model, with R and n adjustable parameters, gives good agreement between the experimental results and calculated values, above the percolation threshold. The solid curve is a calculation based on the Mamunya equation, Eq. (3) and full square is the experimental data. These satisfactory results are observed at different frequencies for volume fractions above the threshold concentration. For concentrations below the threshold, the model does not fit accurately the experimental results. These parameters R and n obtained after modeling are presented in Figure 5. These values do not vary significantly as a function of frequency.

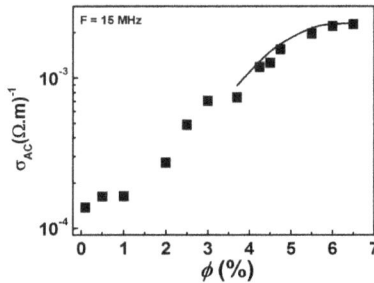

Fig.4. AC conductivity as a function of CB volume fraction for Raven 2000/GDEBF at frequency of 15 MHz. The solid curve is a calculation based on the Mamunya model above the percolation threshold

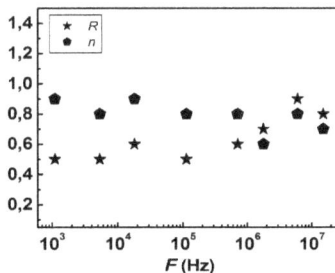

Fig.5. Variation of parameters R and n of McLachlan as a function of frequency for Raven 2000/DGEBF

Dielectric Materials and Applications: ISyDMA'2016   Materials Research Forum LLC
Materials Research Proceedings 1 (2016) 114-118   doi: http://dx.doi.org/10.21741/2474-395X/1/29

Conclusion

In this paper, two models were applied to predict the AC electrical conductivity of carbon black/epoxy for different frequencies. The McLachlan and Mamunya models give good agreement between the calculated and the measured values of the AC conductivity, considering the exponent t of McLachlan equation and the parameters R and n Mamunya as adjustable parameters. These adjustable parameters obtained after modeling are comparable regardless of the chosen frequency. These models are not influenced by the variation of the frequency.

Acknowledgment

The authors acknowledge the support from National Center for Scientific and Technical Research (CNRST) and FCT-CNRST bilateral cooperation and FEDER by funds through the COMPETE 2020 Programme and National Funds through FCT - Portuguese Foundation for Science and Technology under the project.

References

[1] E.P. Mamunya, V.V. Davydenko, P. Pissis and E.V. Lebedev, "Electrical and thermal conductivity of polymers filled with metal powders" Eur. Polym. J. vol. 38, pp. 1887-1897, 2002. http://dx.doi.org/10.1016/S0014-3057(02)00064-2

[2] M. Rahaman, T.K. Chaki and D. Khastgir, "Modeling of DC conductivity for ethylene vinyl acetate (EVA)/polyaniline conductive composites prepared through insitu polymerization of aniline in EVA matrix" compos. Sci. Technol. vol. 72, pp. 1575-1580, 2012.

[3] S. Shang, L. Liang, X. Yang and Y. Wei, "Polymethylmethacrylate-carbon nanotubes composites prepared by microemulsion polymerization for gas sensor" Compos. Sci. Technol. vol. 69, pp. 1156-1165, 2009. http://dx.doi.org/10.1016/j.compscitech.2009.02.013

[4] L. Jiongxin, K.S. Moon, B. K. Kim and C. P. Wong, "High dielectric constant polyaniline/epoxy composites via insitu polymerization for embedded capacitor applications". Polym. Vol. 48, pp. 1510-1516, 2007. http://dx.doi.org/10.1016/j.polymer.2007.01.057

[5] S. A. Elhad Kassim, M. E. Achour, L.C. Costa and F. Lahjomri, "Modelling the DC electrical conductivity of polymer/carbon black composites" J. Electrost. Vol. 72, pp. 187-191, 2014. http://dx.doi.org/10.1016/j.elstat.2014.02.002

[6] S. A. Elhad Kassim, M. E. Achour, L.C. Costa and F. Lahjomri, "Prediction of the DC electrical conductivity of carbon black filled polymer composites" Polym. Bull. Vol. 72, pp. 2561-2571, 2015. http://dx.doi.org/10.1007/s00289-015-1421-5

[7] M.E. Achour, Electromagnetic properties of carbon black filled epoxy polymer composites, in: C. Brosseau (Ed.), Prospects in Filled Polymers Engineering: Mesostructure, Elasticity Network, and Macroscopic Properties, Transworld Research Network, Kerala, 2008, pp. 129-174.

[8] D.S. McLachlan, "Analytical Functions for the dc and ac Conductivity of Conductor-Insulator Composites" J. Electroceram. Vol. 5, pp. 93-110, 2000. http://dx.doi.org/10.1023/A:1009954017351

[9] D.S. McLachlan, "A quantitative analysis of the volume fraction dependence of the resistivity of cermets using a general effective media equation" J. Appl. Phys. Vol. 68 pp. 347114-347119, 1990. http://dx.doi.org/10.1063/1.347114

[10] E.P. Mamunya, V.F. Shumskii, E.V. Lebedev, "Rheological properties and electric conductivity of carbon black-filled polyethylene and polypropylene" Polym. Sci. vol. 36, pp. 835-839, 1994.

[11] E.P. Mamunya, V.V. Davidenko, E.V. Lebedev, "Effect of polymer-filler interface interactions on percolation conductivity of thermoplastics filled with carbon black" Compos. Interface, vol. 4, pp. 169-176, 1997. http://dx.doi.org/10.1163/156855497X00145

Dielectric Materials and Applications: ISyDMA'2016
Materials Research Proceedings 1 (2016) 119-121

Materials Research Forum LLC
doi: http://dx.doi.org/10.21741/2474-395X/1/30

Nanoscaled chalcogenide films for optical applications

Plamen PETKOV*[1], Ani STOILOVA[1], Tamara PETKOVA[2]

Departments of Physics, TFTLab, University of Chemical Technology and Metallurgy, 1756 Sofia, Bulgaria

Solid State Department, Institute of Electrochemistry and Energy Systems BAS, 1113 Sofia, Bulgaria

*Corresponding author: E-mail: p.petkov@uctm.edu

Keywords: Chalcogenides, Thin Films, Optical Properties, Electronic Structure

Abstract. Chalcogenide glasses from the $[Ge(Te/Se)_5]_{1-x}In_x$ system with 5,10,15,20,25 mol % In have been investigated. The composition trends of the glassy properties are discussed with the view of structural transformation in the main matrix. The optical spectra of the films have been studied to understand the role of indium on the film behavior. An optical characterization method, based on the transmission and the reflection spectra at normal incidence of uniform, thin films, has been used to obtain the thicknesses and optical constants corresponding to the as-deposited and annealed samples. The dispersion of the refractive index is discussed in terms of the single-oscillator Wemple-Di Domenico (WDD) model. The variations in the refractive index, the band gap, and the oscillation energy of the films after annealing are discussed with respect to rearrangement of the main structure units.

Introduction

Germanium chalcogenides have traditionally been used for mid-IR fibers with transmission windows up to 15 μm. In addition, germanium doped chalcogenide glasses and films exhibit photoconductivity and band gap reduction with increasing doping, suggesting potential application in low-cost mid-IR detectors. These behaviors make further investigations of the optical properties and the band structure of interest [1-5].

For many of above mentioned applications, thin films are necessary. The practical application of thin amorphous chalcogenide films, especially of the vitreous Ge-containing condensates, is closely connected with their transparency in the visible and near IR spectral regions and with the possibility to create optical media with defined values of the refractive index, dispersion, and extinction coefficients [6-7]. The relatively low energy of the chemical bonds in the Ge-based chalcogenide glasses offers the opportunity for photostructural transformations and a number of other light-induced effects, all of which are usually accompanied by considerable changes in the chalcogenide's optical constants.

We present the preparation of films from Ge–Te(Se)–In glasses by thermal evaporation. We establish a correlation between the film composition and the optical properties of the glasses as determined from chemical and physical properties.

Experimental

Glassy alloys were prepared by quenching technique. The respective amounts of the elements with purity of 6N were mixed up and sealed in evacuated quartz ampoules and were heated with a rate of 2 K/min to 1200 K. The melt was rapidly quenched in ice water.

Thin films were deposited from the relevant bulk glasses by thermal vacuum evaporation using a Leybold LB 370 set-up under standard conditions. The evaporation temperature ranged from 650 to 820 K was controlled by Ni-Ni/Cr thermocouple.

Dielectric Materials and Applications: ISyDMA'2016   Materials Research Forum LLC
Materials Research Proceedings **1** (2016) 119-121   doi: http://dx.doi.org/10.21741/2474-395X/1/30

Results and discussion
In an ideal glass the value of the mechanical constraint, Nco, is equal to 3, which defines the maximum in the mechanical strength of the network. In the studied glasses the value Nco=3 is reached in samples with 10 mol. % indium indicating structure transition in this composition. The variations in the density and molar volume show an almost linear increase with indium introduction with small deviation from the trend in samples with 10 mol. % In where the threshold is expected as predicted from mechanical constrain value.

The as prepared films are red coloured, homogeneous on the surface and in the depth as established from the SEM cross-section patterns. The top-view SEM pictures reveal the uniform and smooth surfaces of the films under this study. Neither defects nor traces of initial nucleation are visible. The chemical composition of the thin films is close to the composition of the used targets. The amorphous nature of the films is evidenced by TEM and XRD studies. The electron diffractogramms prove also the amorphous character of the layers by absence of sharp rings on the patterns.

The variations in the evaporation and condensation energy values can be viewed in terms of compositional and structural changes in the films. During the evaporation process of binary Ge-Se(Te) glasses the vapor phase most probably consists of Ge-Se(Te) and Se(Te)-Se(Te)n units[8-10]. The increasing of the indium content into glassy matrix constituted from selenium (tellurium) and germanium atoms probably leads to gradual structural change: the degree of cross-linking and packing increases because of grown average bond energy in the system [12].

The transmission spectra recorded within the spectral range 400– 2500 nm show interference maxima and minima approaching the transmission of the substrate. A red shift of in the film transmission is observed with increasing the indium content. This shift proves that below the optical absorption edge, the light can generate mobile carriers in the films. The transmission spectra recorded after annealing exhibit red shift in the optical absorption edge due most probably to structure transformation and increase in the optical density of the film.

Spectral dependencies of the refractive index are defined from the optical transmission spectra of as-deposited, and annealed thin films using the modified Swanepoel method [13]. The unusual growth in the index values can be described as due to the mechanism if the film preparation. The indium atoms most probably are commonly located around film/substrate interface due to migration. After annealing the values of the refractive index become higher which suggest reorganization in the film structure.

The absorption coefficient α is determined from the envelope of the interference maxima and minima of transmission spectra using a modified Swanepoel equation. In the strong absorption region ($\alpha \geq 10^4$ cm^{-1}), which involves optical transitions between the valence and conduction bands, the absorption coefficient is calculated using the 'non-direct transition' model proposed by Tauc [14].

The films absorption grows with the indium content as confirmed by the red shift in the α coefficient. The optical gap Eg$_{opt}$ for non-direct transition can be obtained from intercept $(\alpha h)^{1/2}$ vs. (hv) plot with the energy axis at $(\alpha h)^{1/2} = 0$ and can also be derived from the spectral distribution of the absorption coefficient (α) calculated as energy at $\alpha = 2 \times 10^4$ cm^{-1} (also known as Eg^{04} method). The optical band gap calculated by both approaches reveals reduction of the values with indium content.

According to the single-effective oscillator model proposed by WDD [11] the optical data can describe with two values: energy of the effective dispersion E_o and oscillator strength E_d. The energy spectrum of electron states in the range $hv > E_g$ can be studied from the reflectivity spectrum of light in the fundamental absorption band where hv is the energy of quantum and Eg is the optical gap. Of special interest is the study of the optical features of non-crystalline semiconductors near the absorption edge. It is known that the absorption edge of non-crystalline materials is sensitive to the composition and to the material structure as well as to external factors such as electric and magnetic fields, optical, heat, electronic and other radiations. Under the influence of the mentioned factors optical parameters of non-crystalline semiconductors suffer reversible and irreversible changes. The reduction of Eg with indium content can be accounted for in terms of differences in the structure of the films prepared by vacuum thermal evaporation. The decrease is consistent with the formation of new "broken" bonds due to the indium atoms. These defective bonds can act as carriers' traps resulting in the band gap narrowing. The optical band gap values of the annealed films show lower values which suggest the increase in the number of defective bonds after temperature treatment.

The decrease in E_d and E_0 may also be attributed to the formation of an increased number of defect centres and the more disorder with the increase in In content. This is in accordance with the Mott and Davis model [15] which states that the width of the localized states near the mobility edges depends on the degree of disorder and the defects present in the amorphous structure. The unsaturated bonds are responsible for the formation of some defects in the films. Such defects produce localized states in the amorphous solids. The change in the properties after the threshold at N=2.4 is due to changes in the local structure arrangement. The threshold does not affect the optical band gap but changes the trend of the refractive index values. The decrease of E_0 could be attributed to the dissolving of In atoms with larger atomic radius than Se atoms forming Se–In bonds with the longest bonding distances and increasing of the number of scattering center. The oscillator strength or the dispersion energy measures the average strength of the interband optical transitions and follows the trend of the refractive index values.

After temperature treatment the dispersion in the average energy gap values becomes smaller due to higher optical density of the film after annealing.

Acknowledgement
The authors are very grateful for the financial support of this study under Bulgarian Science Fund – Grant DFNI-T 02/26-12.12.14 - "New hydride photoreactive crystal – liquid crystal – graphene structures"

References
[1] M.Vlcek, S.Schroeter, A.Fiserova, J Mat. Sci: Mat. Elect. **20** 290 (2009)

[2] M.Shurgalin, V.Fuflyigin, E.Anderson, J. Phys. D:. **38** 4037 (2005). http://dx.doi.org/10.1088/0022-3727/38/22/006

[3] A. Ganjoo, K. Shimakawa, Recent Res. Dev. Appl. Phys., **2** 129 (1999)

[4] E. Marquez, T. Wagner, J.Gonzalez-Leal, R. Prieto-Alcon, R. Jimenez-Garay, P. Ewen, J. Non-Cryst. Solids 274 62 (2000). http://dx.doi.org/10.1016/S0022-3093(00)00184-8

[5] V. Lyubin, M. Klebanov, B. Sfez, J. Non-Cryst. Solids **359** 183 (2004)

[6] V.M. Lyubin, M. Klebanov, B. Sfez , Mat. Lett. **58** 1706 (2004). http://dx.doi.org/10.1016/j.matlet.2003.11.029

[7] M.A. Majeed Kan, M. Zulfequar, M. Husain, J. Opt. Mat. **22** 21 (2003). http://dx.doi.org/10.1016/S0925-3467(02)00234-3

[8] T. Petkova, Y. Nedeva, P. Petkov, J.Optoel..Adv. Mater. **3** 855 (2001)

[9] P. Petkov, S. Parvanov, Y. Nedeva, Phys. Chem. Glas. **41** 377 (2000)

[10] P. Petkov, C. Vodenicharov, C. Kanasirski, Phys. Stat. Sol. (a) **168**, 447 (1998). http://dx.doi.org/10.1002/(SICI)1521-396X(199808)168:2<447::AID-PSSA447>3.0.CO;2-7

[11] S. H. Wemple, M. DiDomenico, Phys. Rev. B, **3** 1338 (1971). http://dx.doi.org/10.1103/PhysRevB.3.1338

[12] A. Stoilova, P. Petkova, Y. Nedeva, AIP Conf. Proc. **1203** 398 (2010). http://dx.doi.org/10.1063/1.3322475

[13] R. Swanepoel, J. Phys. E: Sci. Instrum. **17** 896 (1984)

[14] J.Tauc, Amorphous and Liquid Semiconductors, Plenum: New York,1979

[15] N. Mott, E.Davis, Electronic processes in non-crystalline materials, Clarendon Press, London, 1970

Dielectric Materials and Applications: ISyDMA'2016
Materials Research Proceedings **1** (2016) 122-126

Materials Research Forum LLC
doi: http://dx.doi.org/10.21741/2474-395X/1/31

Study of corrosion by alternating currents on buried and cathodically protected pipelines near high-voltage power lines

Fatiha BABAGHAYOU*, Boubakeur ZEGNINI, Tahar SEGHIER

Laboratory for the study and development of Semiconductors and dielectric materials, Amar Telidji University Laghouat, Algeria

*Corresponding author: E-mail: babag.fatiha@gmail.com

Keywords: Corrosion, Cathodic Protection, Pourbaix diagram, Reference Potential, Immunity Zone

Abstract. The Buried pipelines of gas and oil, are already protected from corrosion, first by an insulation coating and second by a cathodic protection (CP). The latter reduces the amount of corrosion in order to avoid eventual damage. But, the presence of alternating-current´s interferences due to the high voltage power lines nearby, could cause serious corrosion damage on metallic structures, even in CP conditions. Our laboratory's research aims to explain the basic mechanisms of corrosion (electrical and electrochemical) by the alternating current and how to minimize its impact on pipelines. We started an experimental study by modeling the studied phenomenon. The results show that the induced AC current causes the corrosion of the sample that was protected by the CP. Then we continued our study by a numerical simulation of the process. Which gives an E_{off} (Reference potential measured close to the iron sample) exceeding the immunity zone, a strong oxidation of iron, significant reduction of oxygen, a high liberation of Hydrogen and a clear sample deformation. Lastly, we set up a monitoring and correction program to bring the E_{off}, in the presence of AC, in the corrosion immunity zone according to standards, this program was able to minimize the reactions of the electrochemical process and prevents sample deformation.

Introduction

Since 1986 some examples of corrosion by alternating current AC (50hz) on gas canalization were found [1]. Thorough studies have been carried out in this field, by many researchers [2]. Our study aims first to explaining the basic mechanisms of corrosion (electrical and electrochemical) by the alternating current, second conducting electrochemical tests on a laboratory sample and simultaneously doing a simulation study in order to digitally represent the CP phenomenon and AC corrosion and how to minimize its impact on pipelines [7]-[9].

Theorical Study of Corrosion by AC

Rigorous theoretical research showed us the main AC corrosion mechanism of pipelines API type, it illustrated to us multiple points such as the different types of interference [2][4], the corrosion by AC [1][2], the explanation of electrochemical corrosion[3][4], the Pourbaix diagram of corrosion and immunity areas [4]-[6], cathodic protection [2]-[4], the measurement parameters of corrosion [2] and the principal models of the corrosion mechanism by AC [1].

Carrying Out Electrochimical Tests on Laboratory Samples

We conducted electrochemical tests on a laboratory model, it presents the application of DC (CP) and the induced AC on an Iron sample of pipeline type API 5L X52. Below the reactions of the electrochemical process:

Anodic Process

$$Fe \rightarrow Fe^{2+} + 2e^{-} \tag{1}$$

Cathodic Process

$$4e^- + O_2 + 2H_2O \rightarrow 4OH^- \tag{2}$$

$$2e^- + 2H_2O \rightarrow 2OH^- + H_2 \tag{3}$$

$$2e^- + 2H^+ \rightarrow H_2 \tag{4}$$

Fig.1. Test Banch

Fig.2.Experimental results

We can summarize the laboratory investigation conducted by the following steps:

1. Checking the corrosion process by the application of the procedure "free potential" (in the absence of CP and AC) : The E_{off} measured decreases and stabilizes at -500 mV, the device also shows us that the oxidation of the sample took place.
2. Finding the immunity zone for which the sample does not undergo oxidation: By applying the LP analysis method (Linear Polarization Log i (E_{off}) [10] on two different samples, we chose the immunity zone between -0.96V and -0.67 V.
3. Finding, in the absence of the induced AC, the adequate potential of CP (DC): Based on the results of step 2, we can also choose the adequate CP in the middle of the restraining zone (CP≈-0.8V).
4. According to tests results, for CP= -0.8V and -0.96V< E_{off} < -0.67V, the sample does not undergo any corrosion, it is in the immunity zone, these results are consistent with the standards.
5. Varying the induced current (AC) and monitoring the potential $E_{off,}$, to highlight the cases of exceeding the restraining zone :

For CP = -0.8V and AC = 100 mV, the sample was in reduction state and E_{off} was in immunity area.
For CP = -0.8V and AC = 200 mV or AC = 400 mV, the sample was in oxidation state and E_{off} exceeded the immunity area.

Numerical Simulation of Corrosion by AC

We did a simulation study in order to, digitally, represent the CP phenomenon , AC corrosion, and optimizing AC corrosion behavior by changing the CP in case of AC interferences, by a monitoring program, which returns process status to the immunity zone $-0.85v<E_{off} <-1.15v$ [7] (E_{off} is the reference potential sensed by the reference electrode), by referring to Pourbaix diagram [5] and PC criteria suggested by Hosokawa [6]. Numerical simulations are based on:

A. The parameterization of current distribution in the cell:
Electrolyte:

$$\nabla.i_l = Q_l, \quad i_l = -\sigma_l \nabla \Phi_l \tag{5}$$

Electrode:

$$\nabla.i_s = Q_s, \quad i_s = -\sigma_s \nabla \Phi_s \tag{6}$$

Electrode-Electrolyte-Interface:

$$\eta = Q_s - Q_l - E_{eq} \tag{7}$$

Dielectric Materials and Applications: ISyDMA'2016   Materials Research Forum LLC
Materials Research Proceedings 1 (2016) 122-126   doi: http://dx.doi.org/10.21741/2474-395X/1/31

where Φ_l: is the electrolyte potential, Φ_s: electric potential, σ_l and σ_s : electrolyte conductivity, Q_l and Q_s : general current source term, E_{eq} : equilibrium potential.
    where $\Phi_s = E_{CP} + E_{AC} = E_{CP} + A \cdot sin(2\pi.f.t)$, ($A$: the Amplitude).

B. *The transport of chemical species in order to configuring the chemical species movement (Fe, O_2 and H_2):*

$$(\sigma c_i / \sigma t) + \nabla.(-D_i \nabla c_i) = R_i \qquad (8)$$

$$N_i = -D_i \nabla ci \qquad (9)$$

where for the specie i, c_i: the concentration (mol/m^3), D_i: the diffusion coefficient (m^2/s), N_i: the flux vector and R_i: the reaction rate expression ($mol/(m3.s)$).

C. *The math ODE and DAO to support the AC monitoring program and bring the situation to the state of immunity.*

D. *The deformation of the sample by corrosion on the basis of the kinetic reactions to the sample and the speed v(m/s) of iron dissolution at the anode:*

$$i_{cat} = -i_{0,cat} \cdot 10^{\eta/Acat} \quad and \quad i_{tafel} = -i_{0,an} \cdot 10^{\eta/Aan} \qquad (10)$$

where i_0 : is the exchange current density, η : the over-potential of the reaction and A: the Tafel slope.

$$v = i_{an}/2F.M/\rho \quad where \quad i_{an} = (i_{tafel}.i_{lim})/(i_{tafel} + i_{lim}) \qquad (11)$$

where M: average molar mass, ρ: the density of the iron, $F(C/mol)$: Faraday constant and i_{lim}: limiting current density.

Numerical Simulation Results

A *The potential E_{off} in the presence of AC and CP:*

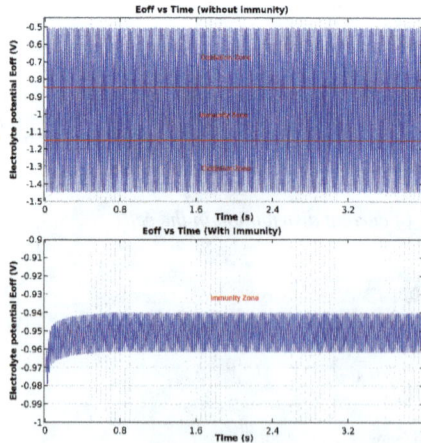

Fig.3. E_{off} in the presence of CP and AC

B   *The Concentration of iron ions Fe^{2+} near to the sample:*

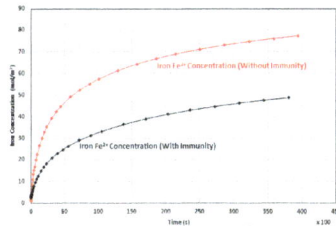

Fig.4. Concentration of iron ions Fe^{2+} near the sample.

Discussion of Résults

Fig.3 presents the E_{off} in the presence of CP and an induced current AC, without then with the immunity program. This representation allowed us to discuss, following the curves of the corrosion of the sample, based on the standards [2]-[6]. First, it shows that E_{off} exceeded the immunity area where there's a significant risk of corrosion. Then for the same value of AC and introducing the resolution of the immunity program, we were able to bring the E_{off} in immunity zone. Fig.4 shows the increase in the Iron ions Fe^{2+} concentration near the sample, thus demonstrating that a reaction of oxidation of iron (1), so its corrosion took place. And after the implementation of the immunity program, we decrease this concentration.

Conclusion

In the practical study, we were able to record data such as: first the free potential that means without CP and without AC, second the proper CP in the absence of AC and finally the application of AC to a model subjected to a CP. Regarding the numerical simulation, the results indicated that the induced AC current causes the corrosion of the sample which was protected by the CP, because we found an E_{off} exceeding the immunity zone and a strong oxidation of iron. We have also simulated the oxygen and the hydrogen concentration in the electrolyte, and the sample deformation. Then, our monitoring program has allowed us to bring the E_{off}, in the presence of AC in corrosion immunity zone according to standards, to minimize the reactions of the electrochemical process and it equally prevents the sample's deformation. To sum up, it is important to establish an execution of immunity program, in addition to the CP system in places where pipelines are near power lines and causing interference AC.

References

[1]   Lucio Di Biase, Ferdinand Stalder,"Corrosion par courant alternatif sur les canalisations cathodiquement protégées", CEOCOR - 2001.

[2]   Maud BARBALAT, "Apport des techniques électrochimiques pour l'amélioration de l'estimation de l'efficacité de la protection cathodique des canalisations enterrées", ÉCOLE DOCTORALE Gay Lussac, Sciences pour l'environnement, 2012 ;

[3]   Andrea Brenna, "A proposal of AC corrosion mechanism of carbon steel in cathodic protection condition", Politecnico di Milano 2009-2011 ;

[4]   Rob Wakelin, Bob Gummow, and Tom Lewis, "CP 3–Cathodic Protection Technologist - COURSE MANUAL", NACE, July 2008 ;

[5]   Nguyen-Thuy LE, "Protection cathodique V.1.1", INERIS, Jan 2008 ;

[6]   C.GABRIELLI, H.TAKENOUTI, "Méthodes électrochimiques appliquées à la corrosion - Techniques stationnaires", Techniques de l'Ingénieur, Juin 2010 ;

Dielectric Materials and Applications: ISyDMA'2016   Materials Research Forum LLC
Materials Research Proceedings 1 (2016) 122-126   doi: http://dx.doi.org/10.21741/2474-395X/1/31

[7]   R.A. Gummow, "A/C Interference Guideline Final Report", Canada Energy Pipeline Association, June (2014) ;

[8]   B.D. Yadav, "Due to Interference from High Voltage Transmission Line", Pipeline Technology Journal p. 30-34 (2016) ;

[9]   C. Brelsford, "Mitigating AC Corrosion on Cathodically Protected Pipeline", Pipeline & Gas Journal, October (2015) ;

[10]   Ziouche, N. Zoubir, A. Hamraras and H. Beldjouhar, "Comportement Electrochimique d'un Acier Inoxydable Duplex (LDX2101) dans Différentes Concentrations de HCL", Recherche et développement, Bulletin d'information N°6 Décembre (2014).

Dielectric Materials and Applications: ISyDMA'2016
Materials Research Proceedings 1 (2016) 127-130

Materials Research Forum LLC
doi: http://dx.doi.org/10.21741/2474-395X/1/32

Image-based CFD modelling of cerebral blood flow

Sekhane DJALAL[*,1], Karim MANSOUR[1,2]

[1]LEMEAMED. Université des Frères Mentouri Constantine 1, Route d'Ain El bey, Constantine
Algeria

[2]Faculté de Médecine de Constantine. Université Constantine 3 –Chalet des pins – Constantine
– Algeria

Corresponding author: E-mail: djalalsekhane@gmail.com

Keywords: MRI, CFD, Wall Shear Stress, Pressure

Abstract. Detailed knowledge of the hemodynamics in the cerebral blood tree is valuable not only for studying the mechanisms of alimentation in the cortex, but also to investigate the role of hemodynamics in the development of cerebrovascular pathologies. The aim of this work is to present preliminary data processing pipeline focussing on the extraction of a 3D patient specific smooth geometric model from magnetic resonance imaging (MRI) and estimate hemodynamic factors using computational fluid dynamics (CFD).

Introduction

Understand the blood flow in cerebral arterial tree have a major importance to understand the mechanisms of the transport of blood in the cortex. However, using Computational Fluid Dynamics (CFD) combined with medical imaging methods has increased with the big development known in these domains [1]. The use of this method seem to be necessary due to the limitation of the current imaging modalities to provide hemodynamic information and its capability to provide effective and safe information of blood flow inside the vessels[9,10]. In this work, a numerical simulation based on 3D Navies-stokes laws used to investigate the different hemodynamic parameters. The study concentrates on the Circle of Willis (CoW), the base of cerebral vasculature network.

Material and methods

A. Vascular model :

Patient specific images of the circle of Willis were obtained from Magnetic Resonance Angiography (MRA), using 3D Time Of Flight (TOF) sequence with the following parameters: TE=3.1ms,TR=25ms and slice thickness=1.4mm.

Using 3D slicer software, a three dimensional geometry of the COW was created (fig1).

B. Mathematical and blood model:

The blood dynamics was modelled with the Navies-Stokes equation. In the case of a laminar flow, Navies-Stokes equation can be written as:

$$\rho\left(\frac{\partial u}{\partial t} + u.\nabla u\right) = -\nabla p + \nabla\left(\mu(\nabla u + (\nabla u)^T) - \frac{3}{2}\mu(\nabla.u)I\right) + F$$

Where ρ is the density, u the velocity and μ the dynamic viscosity.The blood is considered newtonian and incompressible with a density ρ=1060kg/m^3 and dynamic viscosity μ= 4mPa.s [2].

C. Numerical method:

To solve Navies-Stokes equation, we used COMSOL multiphysics, which uses finite elements method for spatial discretisation. The simulation was performed for three cycles and the typical flow rate is 4.6ml/s [3].

The vessels walls was assumed rigidand no slip boundary condition was applied.

Dielectric Materials and Applications: ISyDMA'2016   Materials Research Forum LLC
Materials Research Proceedings **1** (2016) 127-130   doi: http://dx.doi.org/10.21741/2474-395X/1/32

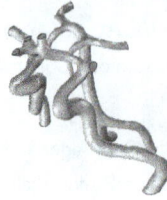

Fig1: Blood Vessels of the

Results

Fig 2 shows the pressure distribution at the systole time on a complete circle of Willis of a young health person, the pressure take values from 103 to 106 mmHg distributed from high pressure in the internal carotids and vertebral arteries to low localised in anterior, middle and posterior cerebral arteries.

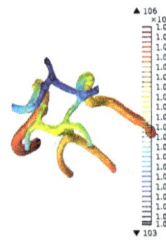

Fig 2: Pressure distribution inside the Circle of Willis

Fig 3 shows the wall shear stress (WSS) values resulting from the blood flow inside the different compartments of the circle of Willis, and obtained from the product of dynamic viscosity and the shear rate solution.

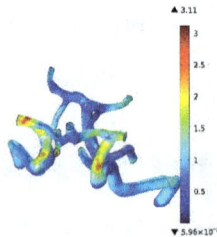

Fig 3: Wall shear stress map at the diastolic time.

Fig 4 shows the streamlines in the circle of Willis, it gives a cartography of the flow speed inside the different segments.

Dielectric Materials and Applications: ISyDMA'2016   Materials Research Forum LLC
Materials Research Proceedings 1 (2016) 127-130   doi: http://dx.doi.org/10.21741/2474-395X/1/32

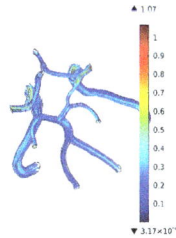

Fig 4: Stream lines inside the different segment of the circle of Willis

Discussion

In this study, we present results from the simulation of blood flow inside a complete circle of Willis, using CFD combined with patient specific images from a young person. Several hemodynamic parameters were used to investigate the effect of the blood flow on the major vessels of the brain. It is known that the WSS is involved in many pathophysiological processes related to vascular diseases [4], like aneurysms. It's suggested that high values of WSS may cause the initiation of an aneurysm, while low values facilitate the growth in the creation [5]. In our case, the values of WSS were low and the possibility of creation of an aneurysm is also low. However the highest values of WSS were localised near to bifurcation and at this location, creation of an aneurysm is eventual but still far. The pressure is also involved in many vascular diseases, like atherosclerosis defined as a deposition of plates on the vessel wall [6] and more aneurysms [4]. Where high values of pressure leads to cholesterol deposition on the carotid arteries [7]. Our results shows normal values of pressure [103-106 mmHg], which keeps low probability for a vascular disease caused by the pressure.

Conclussion

In this study CFD based on patient specific images was performed on a healthy young person to investigate the hemodynamic factors. The hemodynamic factors investigated have shown no abnormal values on the different compartments, no risk for cerebro-vascular disease has been found.

References

[1] Juan Cebral, Image-Based CFD Modelling of Cerebral Aneurysms, Computational Biomechanics for Medicine: Deformation and Flow VIII, 140p. , 2012.

[2] Yufeng Hua et al.Influence of Parent Artery Segmentation and Boundary Conditions on HemodynamicCharacteristics of Intracranial Aneurysms.Yonsei Medical journal. 2015. http://dx.doi.org/10.3349/ymj.2015.56.5.1328

[3] J. Xiang, A. Siddiqui, and H. Meng, "The effect of inlet waveforms on computationalhemodynamics of patient-specific intracranial aneurysms," Journal of biomechanics, vol. 47, no. 16, pp. 3882-3890, 2014. http://dx.doi.org/10.1016/j.jbiomech.2014.09.034

[4] M. Cibis, W. V. Potters, F. J. Gijsen et al., "Wall shear stress calculations based on 3D cine phase contrast MRI and computational fluid dynamics: a comparison study in healthy carotid arteries," NMR in Biomedicine, vol. 27, no. 7, pp. 826-834, 2014. http://dx.doi.org/10.1002/nbm.3126

[5] M. Shojima, M. Oshima, K. Takagi *et al.*, "Magnitude and role of wall shear stress on cerebral aneurysm computational fluid dynamic study of 20middle cerebral artery aneurysms," *Stroke,* vol. 35, no. 11, pp. 2500-2505,2004. http://dx.doi.org/10.1161/01.STR.0000144648.89172.0f

Dielectric Materials and Applications: ISyDMA'2016   Materials Research Forum LLC
Materials Research Proceedings **1** (2016) 127-130   doi: http://dx.doi.org/10.21741/2474-395X/1/32

[6] N. Westerhof, N. Stergiopulos, and M. I. Noble, Snapshots of hemodynamics: an aid for clinical research and graduate education: Springer Science & Business Media, 2010. http://dx.doi.org/10.1007/978-1-4419-6363-5

[7] P. Sobieszczyk, and J. Beckman, "Carotid artery disease," Circulation, vol. 114, no. 7, pp. e244-e247, 2006. http://dx.doi.org/10.1161/CIRCULATIONAHA.105.542860

[8] T. David, S. Moore. "Modeling perfusion in the cerebral vasculature". Medical Enginnering& Physics 30, (2008) 1227-45. http://dx.doi.org/10.1016/j.medengphy.2008.09.008

[9] V. C. Rispoli, J. F. Nielsen, K. S. Nayak *et al.*, "Computational fluid dynamics simulations of blood flow regularized by 3D phase contrast MRI," *Biomedical engineering online,* vol. 14, no. 1, pp. 110, 2015. http://dx.doi.org/10.1186/s12938-015-0104-7

[10] Y. Hua, J. H. Oh, and Y. B. Kim, "Influence of Parent Artery Segmentation and Boundary Conditions on Hemodynamic Characteristics of Intracranial Aneurysms," *Yonsei medical journal,* vol. 56, no. 5, pp. 1328-1337, 2015. http://dx.doi.org/10.3349/ymj.2015.56.5.1328

Dielectric Materials and Applications: ISyDMA'2016   Materials Research Forum LLC
Materials Research Proceedings 1 (2016) 131-134   doi: http://dx.doi.org/10.21741/2474-395X/1/33

Effect of TiO$_2$ nanoparticles on physical properties of Tripropyleneglycol Diacrylate /liquid crystal (TPGDA/ E7) systems

Salima BOUADJELA*[1], Fatima Zohra ABDOUNE[1], Maschke ULRICH[2], Lahcene MECHERNENE[1]

[1]Département de Physique, Faculté des Sciences Université Abou bakrBelkaïd, Tlemcen, Algeria

[2]Unité de Matériaux et deTransformations UMET (UMRCNRS N°8207),Bâtiment C6, Université des Sciences et Technologies de Lille, F-59655, Villeneuve d'Ascq Cedex, France

Corresponding author: E-mail: slimane_hou@yahoo.fr

Keywords: TPGDA, E7, PDLC, FTIR, PIPS, TiO$_2$ Nanoparticles

Abstract. In this paper, TiO$_2$ nanoparticle doped polymer dispersed liquid crystal (PDLC) lenses were made from a mixture of prepolymer, E7 liquid crystal and TiO$_2$ nanoparticles by a polymerization induced phase separation (PIPS) process for smart electronic glasses with auto-shading and auto-focusing functions. Infrared spectroscopy investigation made it possible to obtain the monomer conversion rates, in order to determine the effect of the presence of nanoparticles on the kinetics of polymerization and phase separation.

Introduction

Polymer/liquid crystal systems are the subject of intensive investigations in many laboratories around the world. This interest is motivated by their potential use in many fields of high technology involving electronic equipments, display systems, commutable windows, etc [1,2]. The preparation of these films is often based on the polymerization induced phase separation (PIPS) process by UV light. A proper control of the phase separation phenomena of the polymer/LC composite system is necessary to obtain different morphologies, depending essentially on the polymerization conditions [3].

Nanoparticles (NP) have attracted great interest in recent years because of unique mechanical, optical and magnetical properties [4]. Recently, the merging of nanomaterials and nanotechnology into electro-optical (EO) device technology such as LCDs may attract researchers who are interested in inaugurating a new kind of combination of different fields [5-7]. Among the rest, metal NP have received much attention from the view point of optical, magnetic, and biological properties which depend not only on size and structure, but also on covering materials that play the role of stabilizers as well [8].

In this work we are particularly interested in an experimental study of the effect of the insertion of TiO$_2$ nanoparticles of starting materials (monomers) on properties of elaborated polymer/LC systems.

Experimental

A. Materials

The LC is an eutectic mixture of nematic LCs, commercially known as E7, consisting four cyanoparaphenylene derivatives (5CB, 7CB, 8OCB, 5CT) [9]. It exhibits a single nematic isotropic transition temperature T_{NI} (LC) = 61°C.

The monomer is Tripropylene glycol diacrylate (TPGDA). 2wt% (of the acrylate mixture) of a conventional photo initiator (LucirinTPO, BASF) was added to the initial mixtures. The TiO$_2$ NP was obtained from Sigma-Aldrich and used as received. It exhibit sizes <30 nm.

Dielectric Materials and Applications: ISyDMA'2016   Materials Research Forum LLC
Materials Research Proceedings 1 (2016) 131-134   doi: http://dx.doi.org/10.21741/2474-395X/1/33

B. Sample preparation

Two parallel UV curing lamps, type TL08 (Philips) with an output power of 18_W each, were employed as static low-power UV light source fixed inside a wooden box and separated from each other by around 10 cm. They have a maximum output at a wavelength of X = 365nm, and a dose rate of 0.5 mJ/cm^2.s. The exposure times ranged from a few seconds up to several minutes depending on monomer reactivity. This UV equipment was chosen deliberately to lead to relatively slow photopolymerization and phase separation reactions. Exposure to UV irradiation was conducted at room temperature.

The analysis is performed using a Fourier Transform spectrometer (Agilent cary 640 FTIR) which sends onto the sample infrared radiation and measures the wavelengths at which the material absorbs.

The morphology was studied using a polarizing optical microscope Olympus BX-41 connected to a digital camera and a computer that can record images with high resolution.

Results and discussion

Fourier Transform infrared spectroscopy (FTIR spectroscopy) is a versatile method used to investigate the extent of curing using carbon-carbon double-bond consumption. The difunctional acrylate monomers lead to formation of cross-linked polymer networks. It is evident that high monomer conversions should be reached to minimise the undesired effects of unreacted monomer molecules. Polymerization/cross-linking and phase separation kinetics of monomer/LC mixtures govern the archi-tecture of the obtained polymer networks.

At least two absorption bands are available to monitor polymerization/cross-linking processes and to evaluate the conversion of the acrylic double bonds of TPGDA. One of the most characteristic absorption bands that is quite often used in FTIR analysis of acrylates is the one corresponding to the -CH=CH- vibration at 810cm^{-1}. However, since the LC E7 alone exhibits a strong absorption band near 810cm^{-1} originating from the vibration of the phenyl groups [9] another peak that appears at 1637cm^{-1} is used for the analysis of the monomer/E7 mixtures.

The calculation of the monomer conversion is made by considering the peak heights of the absorption band at 1637cm^{-1}. The conversion ratio C is calculated using the following equation:

$$C(\%) = 100 \times \frac{(A_{1637})_{(D=0)} - (A_{1637})_{(D)}}{(A_{1637})_{(D=0)}}$$

Were $(A_{1637})_{(D=0)}$ is the height of the peak at 1637cm^{-1} (i.e. irradiation dose D is zero), and $(A_{1637})_{(D)}$ is the corresponding result for the system exposed to a dose D.

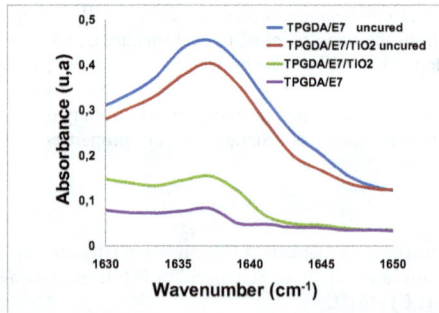

Fig.1: *Acrylic bond in 1637cm^{-1} of the different systems PDLC: doped and undoped for 5min.*

Dielectric Materials and Applications: ISyDMA'2016   Materials Research Forum LLC
Materials Research Proceedings **1** (2016) 131-134   doi: http://dx.doi.org/10.21741/2474-395X/1/33

The evolution of the acrylic double-bond conversion as a function of UV light exposure time under UV-TL08 represented by Fig1 was showed a decrease in absorption intensity bond at 1637cm^{-1} in presence of TiO$_2$ nanoparticles.

In figure 2 the monomer conversion is around 100%.

The effect of insertion of nanoparticles TiO$_2$ on the morphology of TPGDA/E7 films is investigated. The polarized optical microscope results from these films are plotted in Fig.3. This indicated a decrease in the nematic phase of LC with the addition TiO$_2$ NPs.

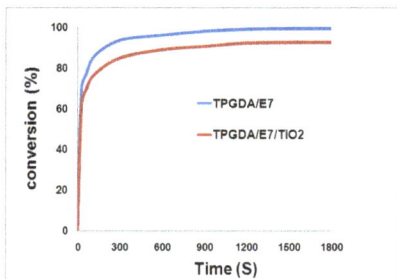

Fig.2: *Photopolymerization kenitics of doped and undoped PDLC films*

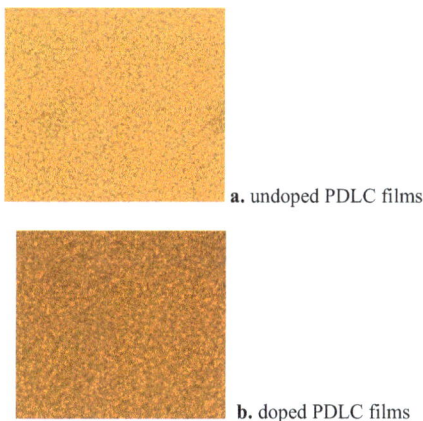

a. undoped PDLC films

b. doped PDLC films

Fig.3 (a-b): *Photos MOP 40X*

Conclusions

The morphology and phase separation of doped and undoped PDLC films are studied with the addition of a TiO$_2$ nanoparticles in a polymer/LC. Overall, the results predicted that the incorporation of TiO2 nanoparticules slowed the phases separation kinetics and thus affect the morphology of our systems.

References

[1] M. Mucha, (2003). Prog. Polym. Sci, 28, 837. http://dx.doi.org/10.1016/S0079-6700(02)00117-X

[2] Lampert, C. M. (2004). Mater. Today, 7, 28. http://dx.doi.org/10.1016/S1369-7021(04)00123-3

[3] Z. Tianyi, X. Jun, (2007). Adv. Display, 73, 54.

[4] S. Miyama, T. Nishida, N. Sakai, (2006). J. Display, 2, 121.
http://dx.doi.org/10.1109/JDT.2006.872306

[5] Xu, J. Dong, J. Nie . (2008). Rare Met. Mat. Eng, 37, 277.

[6] K. Yoshikaw, N. Toshima. (2002). Appl. Phys. Lett, 81, 2845.
http://dx.doi.org/10.1063/1.1511282

[7] T.Yu, S. Chang, J. Chia, K. Yan , (2008). Opt. Lett, 33, 1663.
http://dx.doi.org/10.1364/OL.33.001663

[8] Nolan, P., & Coates, D. (1995). U. S Patent, 5, 476.

[9] R. Bhargava, S. Wang, J. Koenig. (1999) Macromol. 32.

Dielectric Materials and Applications: ISyDMA'2016   Materials Research Forum LLC
Materials Research Proceedings 1 (2016) 135-138   doi: http://dx.doi.org/10.21741/2474-395X/1/34

Dielectric properties in large temperature and frequency ranges of ferroelectric ceramics

H. AIT LAASRI[1,2], A. TACHAFINE*[1], D. FASQUELLE[1], N. TENTILLIER[1],
J.-C. CARRU[1], M. ELAATMANI[2], L.C. COSTA[3], A. OUTZOURHIT[4]

[1]UDSMM Laboratory, ULCO University Calais, France

[2]ESMIA team, Cadi Ayyad University, Marrakech, Morocco

[3]I3N and Physics Department, University of Aveiro, Aveiro, Portugal

[4]LPSCM Laboratory, Cadi Ayyad University, Marrakech, Morocco

Corresponding author: E-mail: tachafin@univ-littoral.fr

Keywords: Materials, BaTiO$_3$, SrTiO$_3$, Permittivity, Capacitor, Ferroelectric, Relaxor

Abstract. Lead-free ferroelectric materials derived from BaTiO$_3$, SrTiO$_3$ and CaZrO$_3$ were synthetized by the solid state reaction. Effects of ionic substitutions in A cationic sites between Ba, Ca and Sr and in B cationic sites between Ti and Zr on structural and dielectric properties were investigated. Structure and morphology of all the ceramics were determined by X-ray diffraction and SEM. Dielectric measurements were investigated in the temperature range from 80 K to 800 K and in frequency from 100 Hz to 1.8 GHz. Dielectric permittivity shows a good stability over all the frequency range for BTO, BCT, BZT, SCT and CZT ceramics. Barium substitution by Ca and Zr decreases the dielectric permittivity and moreover reduces the loss tangent over 1MHz by a factor of 10 compared to BTO

Introduction

Due to the multiplication of standards and norms of telecommunication systems, an important need for frequency tunable components and multifunctional devices appears. Our work suggests that it could be possible to realize electronic systems for radiofrequencies and microwaves applications. The simplest structure allowing to profit from this study is the tunable capacitor with DC voltage that can be afterwards integrated in tunable or reconfigurable microwaves devices such as resonators and high frequency filters. In order to do so, we must use materials presenting low losses and high dielectric permittivity stable in frequency and temperature. The BaTiO$_3$ material has good dielectric properties, which make it the most used based material to elaborate high dielectric permittivity capacitors [1]. However, the variations of its dielectric permittivity with temperature are too excessive for practical applications. One can then carry out phases mixing to obtain a high dielectric permittivity, low losses and a Curie temperature above a range of useable temperatures, typically from -25°C to 75°C. In particular, substitutions in the BaTiO$_3$ perovskite cell can modify its dielectric characteristics in favour of the stability properties sought [2-4]. Thereafter, it is possible to obtain four electric states at room temperature: ferroelectric, paraelectric, relaxor and antiferroelectic.

Materials and methods

Ceramic samples with 3 mm of diameter and 3 mm of thickness as well as of 6 mm of diameter and 1 mm of thickness were prepared by the conventional solid-state reaction method. These ceramics are at room temperature: ferroelectric BaTiO$_3$ (BTO) and Ba$_{0.5}$Ca$_{0.5}$TiO$_3$ (BCT), paraelectric Ba$_{0.6}$Sr$_{0.4}$TiO$_3$ (BST) and SrTiO3 (STO), relaxor BaZr$_x$Ti$_{1-x}$O$_3$ (BZT), anti-ferroelectric Sr$_{1-x}$Ca$_x$TiO$_3$ (SCT) and CaZr$_{1-x}$Ti$_x$O$_3$ (CZT). X-ray diffraction (XRD) analysis was perfomed at room temperature on the powdered samples. Microstructures were examined by scanning electronic microscopy (SEM). Temperature and frequency dependences of dielectric permittivity and loss-tangent of these ceramics were investigated from 80 K to 800 K and from 10 Hz to 1.8 GHz.

Dielectric Materials and Applications: ISyDMA'2016   Materials Research Forum LLC
Materials Research Proceedings **1** (2016) 135-138   doi: http://dx.doi.org/10.21741/2474-395X/1/34

Results.

A Structural caractérisations

Figure 1a shows the XRD patterns of the BTO, BST, BZT and BCT powders at ambient temperature. All the peaks of the XRD diagrams were indexed to the pure perovskite phase according to ICDD files. The peaks are intense and very narrow showing a good crystallinity of the samples. No peaks were detected which could be assigned to secondary phases or unreacted oxides. Figure 1b shows the SEM micrograph of $SrTiO_3$ ceramic. The micrograph shows a homogeneous microstructrure and well-developed grain morphology. Compactness of all the samples were over 90%.

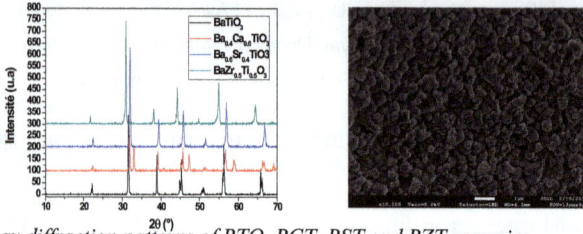

Fig. 1a: *X-ray diffraction patterns of BTO, BCT, BST and BZT ceramics*

Fig. 1b: *SEM micrograph of $SrTiO_3$ ceramic*

B Electrical characterizations

1. Room temperature and large frequency ranging

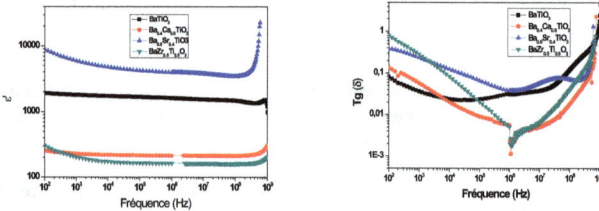

Fig. 2: *Frequency dependences of the dielectric permittivity and loss-tangent of BTO, BCT, BZT and BST ceramics*

Figure 2 shows the frequency dependence of the dielectric permittivity and loss-tangent of BTO, BCT, BZT and BST samples from 100 Hz to 1.8 GHz at room temperature. Dielectric permittivity shows a good stability over all the frequency range for BTO, BCT and BZT ceramics. Ba substitution by 40% of Sr increases the dielectric permittivity up to 5000 at 100 kHz. Ba substitution by Ca and Zr decreases the dielectric permittivity but drastically reduces the loss tangent over 1MHz by a factor 10 compared to BTO.

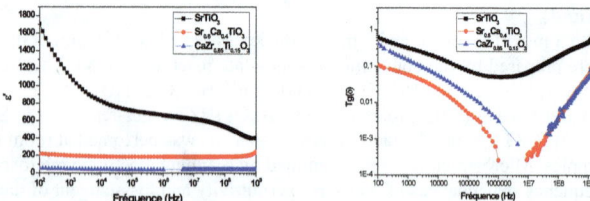

Fig. 3: *Frequency dependence of the dielectric permittivity and loss-tangent of STO, SCT and CZT ceramics*

Figure 3 shows the frequency dependences of the dielectric permittivity and loss-tangent of STO, SCT and CZT samples from 100 Hz to 1.8 GHz at room temperature. For SCT and CZT ceramics, the dielectric permittivity shows a good stability over all the frequency range whereas STO presents dispersion in low frequencies. SCT ceramic exhibits a constant dielectric permittivity (ε'=200) at room temperature which is much lower than the STO ceramic but higher than the CZT one. The SCT ceramic loss-tangent is lower than 0.1 in all the frequency range while the loss-tangent for ST is the highest one for all our SCT ceramics ($0.1 \leq x \leq 1$).

2. Low frequency and variable temperature

Figure 4 shows the temperature dependence of the dielectric permittivity of the BTO ceramic from 80 K to 500 K at 1 kHz, 10 kHz, 100 kHz and 1 MHz. At low temperature, the BTO ceramic exhibits two structural phase transitions: tetragonal-orthorhombic at T_1=300K and orthorhombic-rhomboedric at T_2=209K [5]. BTO ceramic shows a normal ferroelectric behavior as temperature of the maximum of dielectric permittivity does not depend on frequency. The maximum of dielectric permittivity is about 6500 at 1 MHz and at 393 K corresponding to the Curie temperature. Figure 5 shows the temperature dependence of the dielectric permittivity of BZT (a) and SCT (b) ceramics at 1 kHz, 10 kHz, 100 kHz and 1 MHz. BZT ceramic exhibits relaxor behavior as temperature of the maximum of dielectric permittivity T_m increases when the frequency is increased. Moreover a higher dispersion of the dielectric permittivity is observed for temperatures below T_m [6]. Dielectric permittivity of SCT ceramic is constant from 20°C to 450°C for frequencies higher than 100 kHz : no ferroelectric transition is observed.

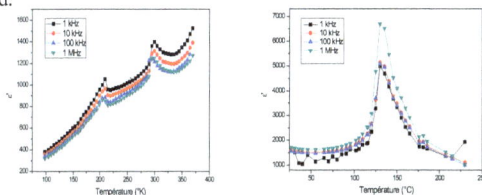

Fig. 4. Temperature dependence of the dielectric permittivity of BaTiO$_3$ ceramic from 80 K to 500 K at different frequencies

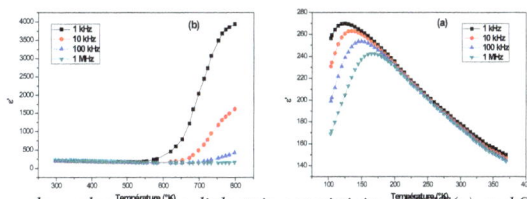

Fig. 5: Temperature dependence of the dielectric permittivity of BZT(a) and SCT(b) ceramics at different frequencies

Conclusion

BaTiO$_3$ and SrTiO$_3$ doped ceramics were prepared using the conventional solid state reaction. XRD analysis revealed perovskite-type structure for all ceramics. Dielectric permittivity shows a good stability from 10 Hz to about 1 GHz for BCT, BZT, SCT and CZT ceramics at 20°C. Ba substitution by Ca and Zr reduces drastically the loss tangent above 1 MHz by a factor 10 compared to BTO. The dielectric permittivity of SCT ceramic is constant in a wide temperature and frequency ranges, what is important for electronic applications.

ACKNOWLEDGMENT

This work was supported by the Toubkal PHC program n° MA/14/308

Dielectric Materials and Applications: ISyDMA'2016   Materials Research Forum LLC
Materials Research Proceedings **1** (2016) 135-138   doi: http://dx.doi.org/10.21741/2474-395X/1/34

References

[1] 1. A. J. Moulson and J. M. Herbert, Electroceramics, Chapman and Hall Press, NewYork (1996).

[2] J.-C. Carru and al., *"Down scaling at submicron scale of the gap width of interdigitated $Ba_{0.5}Sr_{0.5}TiO_3$ capacitors"*, IEEE Transactions on Ultrasonics, Ferroelectrics and Frequency Control, vol. 62, n°2, pp. 247-254 (february 2015).

[3] A. Tachafine, J.-C. Carru and al., *"Relaxor behavior in $(Ba_{1-1.5x}Bi_x)(Zr_YTi_{1-Y})O_3$ ceramics"*, Ceramics International, vol. 37, pp. 2069-2074 (2011). http://dx.doi.org/10.1016/j.ceramint.2011.04.124

[4] A. Aoujgal, H. Ahamdane, M. P. F. Graça, L. P. Costa, A. Tachafine, J.-C. Carru, and A. Outzourhit, *"Structural and relaxor behavior of $Ba[Zr_xTi_{1-x-y}](Zn_{1/3}Nb_{2/3})_yO_3$ ceramics obtained by solid state reaction"*, Solid State Communications, vol. 150, pp. 1245-1248 (2010). http://dx.doi.org/10.1016/j.ssc.2010.03.035

[5] S. Mahajan, O.P. Thakur, C. Prakash and K. Sreenivas, *"Effect of Zr on dielectric, ferroelectric and impedance properties of $BaTiO_3$ ceramic"*, Bull. Mater. Sci., vol. 34, n°7, pp. 1483-1489 (december 2011). http://dx.doi.org/10.1007/s12034-011-0347-2

[6] A Tachafine L. C. Costa, J.-C. Carru and al.,*"Classical and Relaxor Ferroelectric Behavior of Titanate of Barium and Zirconium Ceramic"*, Spectroscopy Letters, vol. 47, pp. 404-410 (2014). http://dx.doi.org/10.1080/00387010.2013.872664

Dielectric Materials and Applications: ISyDMA'2016
Materials Research Proceedings **1** (2016) 139-142

Materials Research Forum LLC
doi: http://dx.doi.org/10.21741/2474-395X/1/35

Optical properties of zinc sulfide (ZnS) thin films prepared by chemical spray pyrolysis

Nour El houda TOUIDJEN*[1], Farida MANSOUR[1], Bouteina BENDAHMANE[1], Mohammed Salah AIDA[2]

[1]LEMEAMED, Electronic Department, UniversityFrères Mentouri Constantine, Constantine, Algeria

[2]LCMI, Physic Department, University Frères Mentouri Constantine, Constantine, Algeria

*Corresponding author: E-mail: houdatouidjen@yahoo.fr

Keywords: Insulator Thin Films, Spray Pyrolysis, Zinc Sulfide, ZnS, Optical Characterization, UV-Visible Spectroscopy, XRD, Raman Scattering

Abstract. In this work we study the properties of zinc sulphide (ZnS) thin films, deposited by chemical spray pyrolysis technique at an optimized deposition parameters such as scanning nozzle speed, pressure and time of 5 mm/s along X and Y-axis, 1 bars and 10 min respectively. The glass substrates temperature was varied from 300°C to 450°C, using a starting solution of 0.1M dehydrated zinc acetate and 0.05M thiourea. Studies on optical, structural and morphological properties were performed on the samples using UV-VIS spectrophotometer, X-ray diffraction (XRD), and Raman spectroscopy. XRD revealed amorphous structure of ZnS films. Raman spectroscopy investigation was used for identification of the phonons in the spectra of ZnS thin films. The optical properties of these films indicated a large band gap of 3.23 eV to 3.68 eV, and relative optical transmittance in the visible spectra of more than 80%, making ZnS films suitable as electroluminescent, buffer layer in thin films solar cells and dielectric filters.

Introduction

Zinc sulphide (ZnS) is an II-VI semiconductor with a wide direct band gap and *n*-type conductivity. These properties make ZnS very attractive for optoelectronic device applications, such as window layer in heterojunction photovoltaic solar cells, dielectric filter and a light emitting diode in the blue to ultraviolet spectral region [1-2]. Many growth techniques have been reported to prepare ZnS thin films, such as electron beam evaporation, thermal evaporation, and sol-gel processing [3-4]. Chemical spray pyrolysis was the technique used in the deposition of ZnS thin films [5]. As it provides uniform and homogeneous films. In this work, we report the influence of substrate temperatures on the optical and structural properties of ZnS thin films for applications in photovoltaic and optoelectronic, by using UV-VIS spectrometry, X-ray diffraction (XRD), and Raman spectroscopy.

Experimental details

Thin films of ZnS were deposited by chemical spray pyrolysis method. The starting solution was 0.1M dehydrated zinc acetate ($C_4H_6O_4Zn$, $2H_2O$) as source of Zn and 0.05M thiourea (CH_4N_2S) as source of S in 100 ml deionized water. The fixed parameters of the deposition process are time, pressure, spray nozzle- substrate distance and the scanning nozzle speed of 10 minutes, 1bars, 18 cm and 5 mm/s along X and Y-axis respectively. The glass substrates temperature were varied in the range of T= 300°C to T=450 °C with a shift of 50°C.

Dielectric Materials and Applications: ISyDMA'2016   Materials Research Forum LLC
Materials Research Proceedings **1** (2016) 139-142   doi: http://dx.doi.org/10.21741/2474-395X/1/35

Results
A. Structural characterizations

Pattern of ZnS films formed by spray pyrolysis was obtained via X-ray diffractometer by using CuKα radiation (α = 1.54045 Å) with a diffraction angle (2θ) range from 20° to 80°. Figures 1 and 2 indicate that the obtained films are amorphous. However, the increase in the intensity (400 a.u to 800 a.u) from 300°C to 450°C is due to thickness films increase, which is estimated from 15 nm to 75nm respectively.

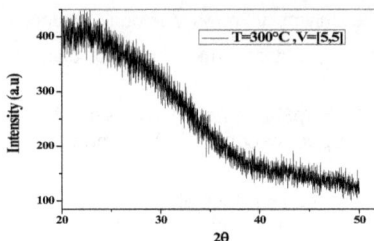

Fig.1. XRD pattern of ZnS thin film deposited at 300°C

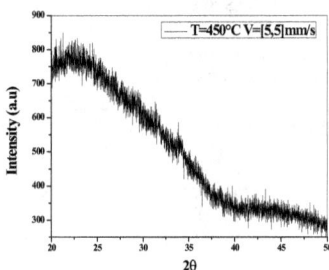

Fig.2. XRD pattern of ZnS thin film deposited at 450°C

Furthermore, the Raman scattering characterization was performed at room temperature to examine the crystal quality and vibration properties of ZnS films. Figure 3 shows the Raman spectra of ZnS which recorded in the range 400-1300 cm^{-1}. A typical Raman band of ZnO lattice at 438 cm^{-1} and 582 cm^{-1} assigned to the E2 and A1 LO modes [6, 7]. Additional peaks at about 1120 cm^{-1} and 1320 cm^{-1} assigned LO mode of ZnS. This result suggests that a single phase ZnS occurs but the crystals formed are too small.

Dielectric Materials and Applications: ISyDMA'2016   Materials Research Forum LLC
Materials Research Proceedings **1** (2016) 139-142   doi: http://dx.doi.org/10.21741/2474-395X/1/35

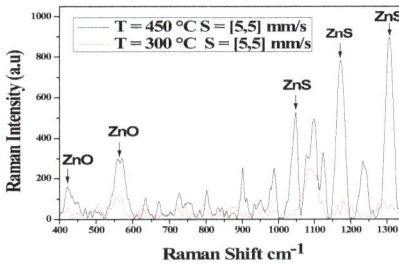

Fig .3.Raman spectra of ZnS thin film deposited at 300°C and 450°C

B. Optical properties

Optical properties for the ZnS thin films deposited on glass substrates at temperatures from 300°C to 450°C were measured in the range of (200-900 nm) using a spectrophotometer.

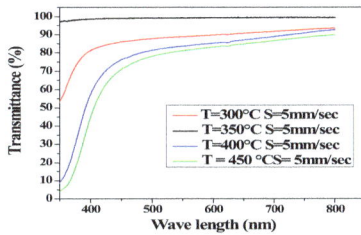

Fig.4. Optical transmittance spectra of ZnS thin films prepared at various substrate temperatures

The relative optical transmittance observed in Figure 4, represents a high transmittance in the visible region more than 80% confirming a good material for optoelectronic devices. We notice that the increase in temperature slightly decreases the transmittance from more than 90% to 80%. Figures 5 and 6 indicate, the wide direct band gap energy of 3.23eV to 3.68eV calculated from the intersection of the straight-line portion of the $(\alpha h\upsilon)^2$ and the $(h\upsilon)$ on the abscissa axis. The reduce of transition from 90 % to 75 % with the increase of temperature; this is due of film thickness increase. These results favorite the use of the ZnS thin films in several applications fields such as optoelectronics, like a light emitting diode in the blue to ultraviolet spectral region.

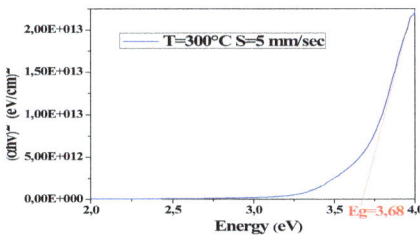

Fig.5. Plot of $(\alpha h\upsilon)$ ² versus photon energy at T_d=300°C

Dielectric Materials and Applications: ISyDMA'2016   Materials Research Forum LLC
Materials Research Proceedings 1 (2016) 139-142   doi: http://dx.doi.org/10.21741/2474-395X/1/35

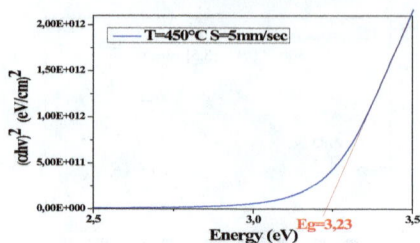

Fig.6.Plot of (αhυ) ² versus photon energy at T_d=450°C

Conclusion

Transparent and homogeneous ZnS thin films were successfully deposited using chemical spray pyrolysis method from the mixture of dehydrated zinc acetate and thiourea at substrate temperatures of 300, 350, 400, and 450°C. XRD showed that ZnS films have amorphous structure. The Raman scattering measurements were used for identification of the phonons in the spectra of ZnS thin films. The optical properties exhibited a high transmittance in the visible region with more than 80%. The optical energy gap values of the thin films were estimated between 3.23eV and 3.68 eV and they are inversely proportional with the deposition temperature. The wide energy band gap and high transmittance make the ZnS film well suitable as buffer layer in thin solar cells and dielectric filters.

References

[1] Geeta Rani,P.D.Sahare, ''Spectroscopy of Nickel-Doped Zinc Sulfide Nanoparticles''. Spectroscopy Letters, 46, (6), (2013).

[2] Nguyen Hutuan, Soonillee, Nguyen Nang Dinah" Investigation of pure green-colour emission from inorganic-organic hybrid LEDs based on colloidal CdSe/ZnS quantum dots''. Int. Journal of Nanotechnology, 10, (3), 304-312, (2013).

[3] Sagadevan Sureshb,''Synthesis, structural and dielectric properties of zinc sulfide nanoparticles''. International Journal of Physical Sciences, 8(21), 1121-1127 (2013).

[4] M.Espindola-Rodriguez, M.Placidi, O.Vigil-Galán, V.Izquierdo-Roca, X.Fontané, A.Fairbrother D.Sylla, E.Saucedo , A.Perez-Rodríguez, '' Compositional optimization of photovoltaic grade Cu_2ZnSnS_4 films grown by pneumatic spray pyrolysis ".Thin Solid Films, (535), 67–72, (2013).

[5] P.O.Offor, B.A.Okorie, B.A.Ezekoye, V.A.Ezekoye, j.I.Ezema , ''Chemical spray pyrolysis synthesis of zinc sulphide (ZnS) thin films via double source precursors".Journal of Ovonic Research, 11, (2), 73–77, (2015).

[6] A.Brayek, M.Ghoul, A.Souissi, I.BenAssaker, H.Lecoq, S.Nowak, S.Chaguetmi, S.Ammar, M.Oueslati, R.Chtourou '', Structural and optical properties of ZnS/ZnO core/shell nanowires grown on ITO glass''. MaterialsLetters, 29, 142–145, (2014). http://dx.doi.org/10.1016/j.matlet.2014.04.192

[7] A.Khan Alim, A.Vladimir Fonoberov, Shamsa.Manu, and A. Balandin Alexander,"Micro-Raman investigation of optical phonons in ZnO nanocrystals'', Journal of Applied Physics, **97**, 124313, (2005). http://dx.doi.org/10.1063/1.1944222

Dielectric Materials and Applications: ISyDMA'2016   Materials Research Forum LLC
Materials Research Proceedings **1** (2016) 143-146   doi: http://dx.doi.org/10.21741/2474-395X/1/36

Evaluation of critical volume in strongly inhomogeneous electric fields

I. ADJIM*, N. ADJIM

Faculté de Génie Electrique USTO-MB, Oran, Algeria

Corresponding author: E-mail: Adj_ilhem@yahoo.com

Keywords: Electric Field, Free Electrons, Critical Volume, Net Ionization Coefficient, Electrical Discharge

Abstract. On the basis of an experimental study of the dielectric strength of air with different densities and for inter-electrode distances of 1 and 2 cm, we evaluated by simulation, the variation of critical volume based on the of electric potential difference applied to a   system "point-plane" electrodes in which the field is highly non-homogeneous. The boot discharge delay time, the mode of evolution of the landfill and the probability of the dielectric strength are resulting from the value of this volume. We also took into account the change in net ionization coefficient on the likely path of the discharge between the two electrodes. The results confirm that in strongly inhomogeneous field, the critical volume in positive polarity is slightly different from the volume of the ionization zone and increases exponentially with the applied voltage.

Introduction

The study of electrical discharges in the air is of a growing interest in basic research. It is also of great interest in industrial problems concerning isolation, network protection and energy distribution (high and very high voltage transformer lines, generators, appliances ...). In the future the increasing demand in electrical power needs the construction of very high (1 MV) voltage networks. Hence, for economical considerations, it is important to deeply understand the mechanism of initialization and developments of the landfill through experimental models.

The concept of critical volume was introduced in our previous articles [1 - 4]. We defined the borders of a critical volume so that a free electron (electron germ) present in this space has a certain probability to multiply in order to give an avalanche of critical size leading to electric discharges.

For a positive polarity of the active electrode the critical volume is limited by two surfaces S1 and S2 (fig1). A free electron located on surface S1 can develop an avalanche of critical size when reaching the anode. S2 represents the iso-field surface where the gradient of the electric potential is equal to the field of ionization of the gas.

Numerical simulations were carried out using the same values of discharge obtained from a physical model. Laplace equation was solved by using the same boundary conditions obtained from the experimental model. The distribution of the electric field in the space between the electrodes in the absence of any load of space at the time of application of the voltage was determined. The iso-field S2 surface of the critical volume was deduced from this distribution. In the other hand, the surface S1 was determined using the law of Meek-Raether.

Dielectric Materials and Applications: ISyDMA'2016   Materials Research Forum LLC
Materials Research Proceedings 1 (2016) 143-146   doi: http://dx.doi.org/10.21741/2474-395X/1/36

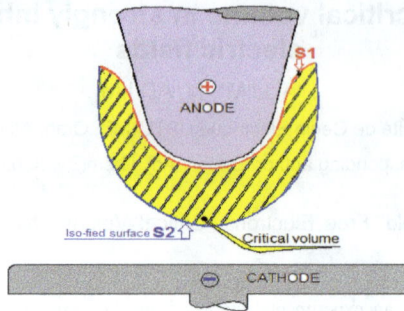

Fig 1: Critical Volume in positive polarity

Results of Simulations

The experimental model is composed of an electrode charged of a soft steel rod with a diameter of 1 cm and a length of 2 cm. The tip is machined in the shape of a cone on 1cm long and completed by a spherical carrot of 1 mm radius of curvature. The plane electrode is carried out in the same material in the form of a cylinder with a diameter of 4 cm and a thickness of 4 mm. The sharp edges have been chamfered and polished to mitigate the effect of point [5]. The forms of revolutions are symmetrical with respect to the main axis of the electrodes.

The model has been numerically simulated as "axesymetries" in two dimensions (2D) and the actual dimensions of the physical model have been introduced in the numerical model. The meshing was made sufficiently fine in order to obtain an acceptable accuracy with a moderate computational time. The meshing was made finer on the areas most likely to ionizations and who are affected by the trajectory of the evolution of the discharge.

For example, we have shown in figure 2 the limits of the surface S2 of the critical volume for voltages of 17 kV, 20 kV, 24 kV and 28kV. Iso-field surfaces obtained by simulation are limited by cones inscribed with the electrode at the lower end and their tops are formed by a half ellipsoid of equation $(X^2 + Z^2) / a + Y^2 / b = 1$. The values of "a" and "b" are determined for each breakdown voltage.

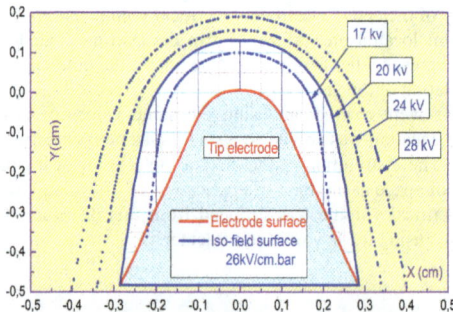

Fig 2: Evolution of the iso-field surface for different voltages in strongly inhomogeneous field

Calculations and simulations were carried out along the main axis of symmetry of the electrodes where the electric field is strongest in order to determine the boundaries of the surface S1

Figure 3 shows the numerical solution of Meek-Raether integral [6 and 7], such as:

Dielectric Materials and Applications: ISyDMA'2016   Materials Research Forum LLC
Materials Research Proceedings **1** (2016) 143-146   doi: http://dx.doi.org/10.21741/2474-395X/1/36

$$\int_0^{x_c} (\alpha - \eta) \ dx = Log \ N_c = \ Log 10^8 = 18,5$$

Where $(\alpha\text{-}\eta)$ being the net ionization coefficient.

The value of Xc from which a free electron reaches the anode with a critical mass of 10^8 electrons are easily deduced from figure 3. The critical distance is very small; we can consider the value the critical volume approximately equal to the volume of the ionization zone. This result is due to the electric field which is strongly inhomogeneous.

The results of calculations of the critical volume in highly inhomogeneous field are summarized in the graph of figure 4. It should be noted that the variation of this volume as a function of the applied voltage is an exponential curve [8].

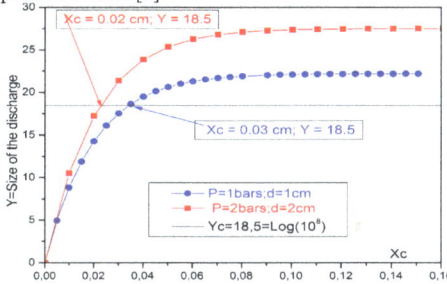

Fig 3 : Size of the discharge at a distance Xc from the tip electrode

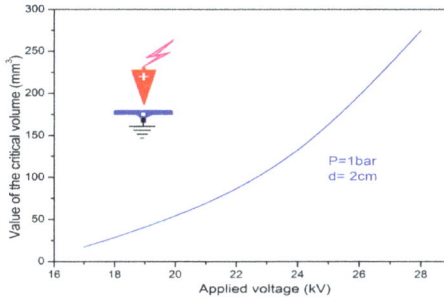

Fig 4: Variation of the critical volume depending on the applied

Conclusion

The results show that the critical volume increases exponentially with the applied voltage. It is found that this volume is slightly different from the volume of the gas ionization zone for a system with a positive polarity of the active electrode and of very small radius of curvature. The boot delay time and the probability of the dielectric strength depend entirely on this volume.

References

[1] N. ADJIM, M. KAMLI, B, SENOUCI, Identification des électrons germes dans l'air en champ électriques non homogènes (Identification of electron germs in the air in inhomogeneous electric field). 5ème Conférence sur la haute tension USTMB. Oran 2003.

[2] N. ADJIM, M. SEBBANI et M. RAHLI, Influences de la polarité de l'électrode active sur la tension d'amorçage et sur le taux de production des électrons initiaux dans l'air (Influences the

Dielectric Materials and Applications: ISyDMA'2016   Materials Research Forum LLC
Materials Research Proceedings **1** (2016) 143-146   doi: http://dx.doi.org/10.21741/2474-395X/1/36

polarity of the active electrode on the ignition voltage and on the production rate of the initial electrons in air). CNHT 2007, Algerian Journal of Technology, Number Special (2007)

[3] N. ADJIM, Identification et evolution de la population d'electrons germes en champ electrique fortement non homogene (Identification and evolution of the population of germ electrons in a strongly inhomogeneous electric field.). Thèse de doctorat. USTO-MB (2008)

[4] N. ADJIM, A. ALLALI and B. SENOUCI, Experimental study of the rate of initial the electrons sources in the air between 1 and 3 absolute bars in no homogeneous electric field. International Journal of Applied Engineering Research, ISSN 0973-4561 Volume 2, Number 3 September (2007)

[5] M. AGUET, M. IANOZ, Haute tension – traite d'électricité d'électronique et d'électrotechnique (High voltage - Treaty of electricity, electronics and electrotechnical engineering). Publié sous la direction de J. Neirynck, Dunod.

[6] J.M. MEEK, A theory of spark discharge. Physical review, volume 57 (1940). http://dx.doi.org/10.1103/PhysRev.57.722

[7] J.M. MEEK, J.D. Craggs, Electrical Breakdown of Gases, Clarendon, Oxford (1953).

[8] G.BERGER « Retard à la formation de la décharge couronne positive dans l'air ». Thèse de Doctorat ès-sciences, Université Paris-Sud, 1980.

Dielectric Materials and Applications: ISyDMA'2016
Materials Research Proceedings 1 (2016) 147-150

Materials Research Forum LLC
doi: http://dx.doi.org/10.21741/2474-395X/1/37

Consequences of dielectric mismatch on the engineering band gap of Pbs/*CdS* core/shell quantum dots

A. ZOUITINE*[1], A.IBRAL[2], E. FEDDI[1], E. ASSAID[2]

[1]Groupe Optoelectronique des boites quantiques des semiconducteurs, ENSET de Rabat, Université Mohamed V, Rabat, Marocco

[2]Equipe d'Optique et Electronique du Solide, Département de Physique, Faculté des Sciences, Université Chouaïb, Doukkali, B. P. 20 El Jadida principale, El Jadida, Marocco

*Corresponding author: E-mail: aszouitine@gmail.com

Keywords: Dielectric Mismatch, Effective Mass Mismatch, Electron and Hole Energies, Core/Shell Nanostructure, Polarization Charges

Abstract. The combined effects of double dielectric mismatch, electron and hole effective mass mismatch and quantum confinement on the ground state energy and on the wave function of an electron and a hole confined in a spherical core/shell nanostructure, embedded in a dielectric matrix, or suspended in an organic solution or water are studied in the framework of effective mass approximation. The core/shell nanostructure is made by a spherical core fabricated with a small band gap semiconductor with dielectric constant epsilon1, coated with another spherical semiconductor having a large band gap with dielectric constant epsilon2. The structure is embedded in a dielectric matrix or suspended in an organic solution. Due to band offsets between core, shell and host matrix, electron and hole are confined in the smallest band gap material. The charge carriers are in interaction with a self-polarization charges appearing at the boundaries of each semi-conductor. The developed theoretical approach is applied to determine electron and hole energies and the variation of band gap as fuction of core and shell sizes in the cases of *PbS/CdS* single quantum dot.

Introduction

The recent progress achieved in crystal growth technology has made possible the fabrication of zero dimensional 0D systems such as clusters, quantum dots QD and quantum crystallites. They may be obtained by precipitation in either isolating [1] or semiconducting [2] matrices, or synthesized in organic liquids [3,4]. In these systems, ultimate quantum confinement effects restrict the motions of the optically excited electrons and holes in the three spatial directions.

Nowadays, the experimental techniques of growth have made it possible to fabricate high-quality semiconductor quantum dots within a large range of sizes for different fields of applications [5-9]. For the last ten years, it has been possible to process a new class of spherical quantum dots called quantum dot-quantum well or inhomogeneous quantum dots (IQDs) composed of two semiconductor materials. One of them, that with the smaller bulk band gap, is embedded between a core and outer shell of the material with the larger band gap. The experimental investigations in "coated" nano-spheres: ZnS(core,shell)/CdSe(well) [10], CdS/PbS [11] have shown that these structures can exhibit some remarkable and interesting phenomena associated with the redistribution of the electron and hole wave function.

In the present paper we focus on type I. this kind of structure is composed by a spherical narrower band gap semiconductor nanocrystal over coated with a wide band gap semiconductor. We determine the single particles ground state energies and wave functions. To examine the influence of double dielectric constant mismatch, we solve analytically Poisson equation in the case of a point charge placed anywhere in a spherical core/shell nanostructure [12]. Other authors have also studied this effect in particular Fonoberov and Li Tsung [13,14].

Dielectric Materials and Applications: ISyDMA'2016   Materials Research Forum LLC
Materials Research Proceedings 1 (2016) 147-150   doi: http://dx.doi.org/10.21741/2474-395X/1/37

Background Theory
In the framework of the effective mass approximation and as- summing isotropic, parabolic and non-degenerated bands, the Hamiltonian of a single-particle reads as:

$$H_{0i} = (\frac{\hbar}{j}\nabla_i)\frac{1}{2m_i^*(r_i)}(\frac{\hbar}{j}\nabla_i) + V_{wi}(r_i) \quad (i = e,h)$$

(1)

The first term stands for the hermitian kinetic energy operator for a position dependent effective mass particle proposed for the first time by BenDaniel and Duke [15].

In the present study, we focus on single particles ground states energies corresponding to the following quantum numbers $n = 1$, $l = 0$ and $m = 0$. Hence the nanostructure effective gap is related to the shell semiconductor gap by the following equation:

$$E_g^{Core/Shell} = E_g^{Shell} + E_e^{1,0,0} + E_h^{1,0,0}$$

(2)

We consider three different cases according to the value of the core radius R_C. In the first case, the single particle energy $E_i < V_{0i}$ and the radial part of the single particle wave function writes:

$$R^{1s}_i(r_i) = \begin{cases} A_{1i}\dfrac{\sin(k_{1i}r_i)}{r_i}, & 0 < r_i < R_C \\ A_{2i}\dfrac{sh(k_{2i}(r_i - R_S))}{r_i}, & R_C < r_i < R_S \end{cases} \quad (i = e,h)$$

(3)

$$k_{1i} = \sqrt{2m_{1i}^* E_i / \hbar^2} \quad \text{and} \quad k_{2i} = \sqrt{2m_{2i}^*(V_{0i} - E_i)/\hbar^2}$$

In the second case, $E_i = V_{0i}$. The radial part of the single particle reads:

$$R^{1s}_i(r_i) = \begin{cases} A_{3i}, & 0 < r_i < R_C \\ A_{4i}\dfrac{sh(k_{4i}(r_i - R_S))}{r_i}, & R_C < r_i < R_S \end{cases} \quad (i = e,h)$$

(4)

where $k_{4i} = \sqrt{2m_{2i}V_{0i}/\hbar^2}$.
In the last case, $E_i > V_{0i}$.

$$R^{1s}_i(r_i) = \begin{cases} A_{5i}\dfrac{\sin(k_{1i}r_i)}{r_i}, & 0 < r_i < R_C \\ A_{6i}\dfrac{\sin(k_{2i}(r_i - R_S))}{r_i}, & R_C < r_i < R_S \end{cases} \quad (i = e,h)$$

(5)

$$k_{5i} = \sqrt{2m_{1i}^* E_i / \hbar^2} \quad \text{and} \quad k_{6i} = \sqrt{2m_{2i}^*(E_i - V_{0i})/\hbar^2}$$

Results and discussion
In the following, we have used the effective units $a_D = \hbar^2 \varepsilon_m / m_e^* e^2$ as our unit of length and the effective Rydberg $R_D = e^2 / 2\varepsilon_m a_D$ as our unit of energy. We use PbS/CdS. In figure.(fig.1) we drawn the variations of electron energie without .(fig.1) and with.(fig.2) self-polarization effect. to calculate the energy taking into account the effect of polarization we use the expression found in the reference [12].

Table1: CdS and PbS physical parameters used innumerical calculations

	CdS[16]	PbS[16]
m_e^*/m_0	0.2	0.085
m_h^*/m_0	0.7	0.085
Eg(ev)	2.5	0.4
$\varepsilon/\varepsilon_0$	5.5	17
V0e(ev)	1.2	-

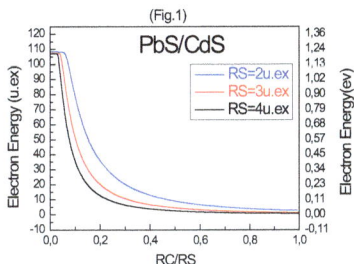

Fig 1: Electron energy drawn versus core to shell radii ratio Rc/RS for following values of shell radius RS: 2ex.units, 3ex.units 4ex.units.

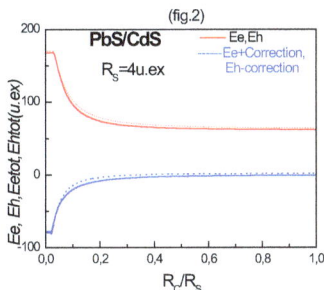

Fig 2: Electron energy, hole energy with and without self-polarization effect drawn versus core to shell radii ratio Rc/RS for RS=4ex.units.

To summarize, in the framework of the envelope function approximation, we determined analytically the ground state energies eq.(2) and wave functions of an electron and a hole confined in a PbS/CdS core/shell nanostructure eq.(3,5), as functions of the inner to outer radii ratio.

We showed that electron or hole energy decrease monotonously when core radius increases from 0 to shell radius.

References

[1]. Ekimov, A. I., Kudryavtsev, I. A., Ivanov, M. G. and Efros, A., L., J.Luminescence 46,83(1990). http://dx.doi.org/10.1016/0022-2313(90)90010-9

[2]. Yoffe, A. D., Adv. Phys.42,173(1993). http://dx.doi.org/10.1080/00018739300101484

[3]. Rossetti, R.,Ellison,J.L., Gibson,J.M.and Brus,L. E.,J.Chem.Phys.80, 4464 (1984).[4]. Brus, L., Appl. Phys.A53, 465 (1991).

[5] Zhenda Lu, Chuanbo Gao, Qiao Zhang, Miaofang Chi, Jane Y. Howe, Yadong Yin, Nano Lett.11(8), 3404 (2011). http://dx.doi.org/10.1021/nl201820r

[4] L. E. Brus, J. Chem. Phys. 80(9), 4403 (1984). http://dx.doi.org/10.1063/1.447218

[6] W. Russ Algar, Mario G. Ancona, Anthony P. Malanoski, Kimihiro Susumu, Igor L. Medintz, ACS Nano, 6 (12), 11044 (2012).

[7] Roberto Trevisan, Pau Rodenas, Victoria Gonzalez-Pedro, Cornelia Sima, Rafael Sánchez Sánchez, Eva M. Barea, Ivan Mora-Sero, Francisco Fabregat Santiago, Sixto Gimenez, J. Phys. Chem. Lett., 4 (1), 141 (2013). http://dx.doi.org/10.1021/jz301890m

[8] Jui-Ming Yang, Haw Yang and Liwei Lin, ACS Nano, 5 (6), 5067 (2011). http://dx.doi.org/10.1021/nn201142f

[9] Igor L. Medintz, Thomas Pons, James B. Delehanty, Kimihiro Susumu, Florence M. Brunel, Philip E. Dawson and Hedi Mattoussi, Bioconjugate Chem, 19 (9), 1785 (2008).

[10] A.R. Kortan, R. Hull, R.L. Opila, M.G. Bawendi, M.L.Steigerwald, P.J. Carroll, L. Brus, J. Am. Chem. Soc. 112 (1990)1327. http://dx.doi.org/10.1021/ja00160a005

[11] H.S. Zhou, I. Honma, H. Komiyama, J. Phys. Chem. 97 (1993) 895. http://dx.doi.org/10.1021/j100106a015

[12]. Asmaa Ibral, Asmaa Zouitine, El Mahdi Assaid, Hicham El Achouby, El Mustapha Feddi, Francis Dujardin. Physica B 458 (2015) 73–84. http://dx.doi.org/10.1016/j.physb.2014.11.009

[13]. V. A. Fonoberov and E. P. Pokatilov,A. A. Balandin Phys. Rev. B66 085310 (2002). http://dx.doi.org/10.1103/PhysRevB.66.085310

[14]. L. Li Tsung and J. Kuhn Kelin, Phys. Rev. B B47, 12760 (1993).

[15] D. J. BenDaniel, C. B. Duke, Physical Review **152**(2), 683 (1966). http://dx.doi.org/10.1103/PhysRev.152.683

[16]. Joseph W.Hauss,H.S.Zhou, I .Honma, and H.Komiyama, Phys. Rev .B 13 ,15 (1993).

Dielectric Materials and Applications: ISyDMA'2016
Materials Research Proceedings 1 (2016) 151-153

Materials Research Forum LLC
doi: http://dx.doi.org/10.21741/2474-395X/1/38

Ge on porous silicon/Si substrate analysed by Raman spectroscopy and atomic force microscopy

R. MAHAMDI[1], S. GOUDER*[2], S. ESCOUBAS[3], L. FAVRE[3], M. AOUASSA[3], A. RONDA[3], I. BERBEZIER[3]

[1]LEA, Electronics Department, Batna 2, University, Batna, (05000) Algeria

[2]LARBI TÉBESSI University, Tébessa, (12000) Algeria

[3]IM2NP Aix-Marseille, UMR CNRS n°7334, Faculte des Sciences St-Jerome-Case142, 13397 Marseille Cedex 20 France

*Corresponding author: E-mail: soraya.gouder@gmail.com

Keywords: Porous Silicon, Germanium, Raman Spectroscopy, AFM, Nanostructure

Abstract. In this study, single crystal Ge layers have been deposited by molecular beam epitaxy on PSi substrate, with different thicknesses (40 nm and 80 nm) at the growth temperature of 400°C. Raman and Atomic force microscopy (AFM) have been applied for investigation of photoluminescence, structural and morphological properties of the Ge on PSi layers. The results show a stronger Raman intensity of PSi due to change of its optical constant. Similarly the Si/Ge/PSi sample shows a peak at 399 cm^{-1} but with lower intensity compared with that of PSi probably due to the Si emission partially covered by the Ge inside the pores. Besides that a sharp Raman peak at 298 cm^{-1} is observed which reflects Raman active transverse optical mode of the introduced Ge which indicates the growth of Ge with good crystallinity. AFM characterization shows the rough silicon surface which can be regarded as a condensation point for small skeleton clusters to form, with different size of pores. These changes are highly responsible for its photoluminescence in the red wavelength range. This study explores the applicability of prepared Ge/PSi layers for its various applications in advanced optoelectronics field and silicon-on-insulator applications.

Introduction

Nowadays, PSi is considered a strong candidate material for applications in the optoelectronic industries. In addition, PSi is used as a material for the sensing layer in a chemical sensor, biological sensors and sacrificial layer in micromachining, and light emitting diode (LED) in optoelectronic devices [1, 2].

In our previous work [3], we reported the use of PSi as a stressor layer for the growth of planar and fully relaxed monocrystalline Ge membranes. Based on those results, we suggested that the different samples can be used as relaxed pseudo-substrate and can be integrated in conventional microelectronic technology. However, to suggest the use of these samples in optoelectronic technology, further work is needed. To help address these issues, this work reports AFM and the Raman analysis of light emitting PSi and Ge on PSi samples.

Experimental

PSi layers were formed by the electrochemical etching of boron-doped (p-type) silicon wafers (orientation: <001>, resistivity: ~0.01 Ω.cm) in an electrolyte containing hydrofluorid acid (HF 35 %), ethanol and water at constant current density. The Ge layers were grown on PS substrate by Molecular Beam Epitaxy. Layers with different thicknesses 40 and 80 nm were deposited at 400 °C. During this process, layers were deposited with a rate of Ge growth about 5 A°/min. Measurements of surface topography of prepared samples were carried out with an atomic force microscope (AFM) PSIA XE-100 in the tapping mode. Raman measurement using a Bruker spectrophotometer (SENTERIA) was done. A beam of 532 nm line from argon laser at 10 mW output power was used for excitation.

Dielectric Materials and Applications: ISyDMA'2016   Materials Research Forum LLC
Materials Research Proceedings 1 (2016) 151-153   doi: http://dx.doi.org/10.21741/2474-395X/1/38

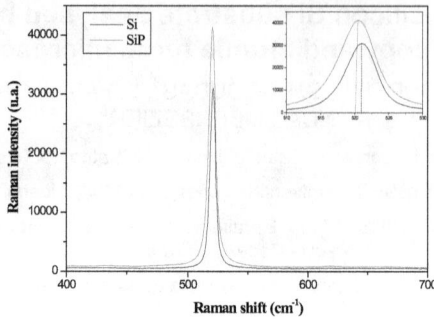

Fig. 1: *Raman spectra of Si and PSi*

Results and discussion

Representative Raman spectra for PSi layer and silicon are presented in figure 1. The Raman spectrum for crystalline silicon consists of one sharp peak situated at 521 cm^{-1}. However, the PSi spectrum shows a strong peak at 520 cm^{-1}, this little shifting of Raman peak is attributed to the reduction in the phonon energy as a result of disturbances in the silicon lattice due to porous structure [4] and to a low porosity. On the other hand, the stronger Raman intensity of PSi is due to change of its optical constant [5]. In addition, the absence of other peak in Raman spectrum confirms that the prepared sample retains the crystallinity of bulk silicon wafer.

Figure 2 shows Raman spectrum of Ge (40nm)/PS/Si. Similarly the Si/Ge/PSi sample shows a peak at 514 cm^{-1} but with lower intensity compared with that of PSi probably due to the Si emission which can be partially covered by the Ge inside the porous. Besides that a sharp Raman peak at 298 cm^{-1} is observed which reflects Raman active transverse optical mode (TO) of the introduced Ge which indicate the growth of Ge layers with good crystallinity [6, 7]. Presence of peaks at 393 cm^{-1} shows an evidence of Si-Ge alloy mode between 300 cm^{-1} and 520 cm^{-1} indicating that intermixing at Ge/Si interfaces are important. Also, The Raman spectra showed that good crystalline structure of the Ge can be produced inside silicon pores.

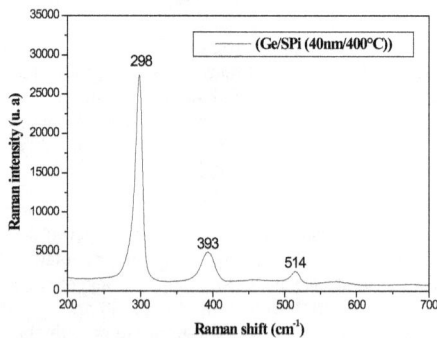

Fig. 2: *Raman spectrum of Ge (40nm) on PSi at 400°C*

Figure 3 shows the AFM images of prepared samples. Samples show the surface roughness and pyramid like hillocks surface.

AFM characterization shows the rough silicon surface which can be regarded as a condensation point for small skeleton clusters to form. This cluster like porous silicon surface plays an important role for the strong visible luminescence.

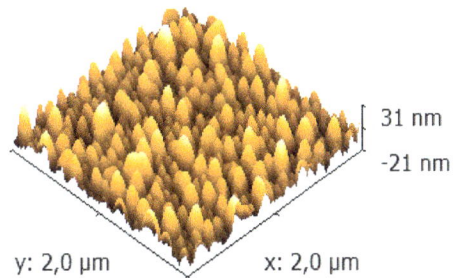

Fig. 3: 3D AFM image of Ge on PSi at 400°C

Conclusion

Various samples of Ge on PSi deposited by molecular beam epitaxy were characterized and studied by Raman spectroscopy and AFM. From Raman study it was revealed that the silicon optical phonon line shifted somewhat to lower frequency from 520.5 cm^{-1} out the less possibility of quantum confinement effect. This shifting of peak attributed the reduction in the phonon energy as a result of disturbances in the silicon lattice due to porous structure. This study explores the applicability of prepared Ge/PSi layers for its various applications in advanced optoelectronics field field such as relaxed pseudo-substrate [3].

Acknowledgments

The authors are grateful to the laboratory STM for providing the PSi samples.

References

[1] A.T. Fiory and N.M. Ravindra, "Light emission from silicon: Some perspectives and applications", Journal of Electronic Materials, vol. 32, issue 10, pp. 1043-1051, 2003. http://dx.doi.org/10.1007/s11664-003-0087-1

[2] V. Mulloni, L. Pavesi, "Porous microcavities as optical chemical sensors", Appl. Phys. Lett, vol. 76, pp. 2523-2525, 2000. http://dx.doi.org/10.1063/1.126396

[3] S. Gouder, R. Mahamdi, M. Aouassa, S. Escoubas, L. Favre, A. Ronda, and I. Berbezier. "Investigation of microstructure and morphology for the Ge on Porous Silicon/Si substrate hetero-structure obtained by Molecular Beam Epitaxy". Thin solid Films, vol. 550, pp 233-238, 2014. http://dx.doi.org/10.1016/j.tsf.2013.10.183

[4] R. S. Dubey, D. K. Gautam, "Synthesis and Characterization of Nanocrystalline Porous Silicon Layer for Solar Cells Applications," J of Opto and Biom Mat, Vol 1, Issue 1, p. 8-14, March 2009.

[5] Yang, Min, D Huang, P Hao, F Zhang, X Hou, X Wang. "Study of the Raman peak shift and the linewidth of light-emitting porous silicon" J. Appl Phys, 75(1): 651-653, 1994. http://dx.doi.org/10.1063/1.355808

[6] Liu, Feng-Qi, Zhan-Guo Wang, Guo-Hua Li, Guang-Hou Wang. "Photoluminescence from Ge clusters embedded in porous silicon". J. Appl. Phys 83(6), 3435-3437. 1998. http://dx.doi.org/10.1063/1.367139

[7] Maeda, Yoshihito, N Tsukamoto, Y Yazawa, Y Kanemitsu, Y Masumoto. "Visible photoluminescence of Ge microcrystals embedded in SiO[sub 2] glassy matrices", Appl. Phys Lett, 59(24), 3168-3170, 1991. http://dx.doi.org/10.1063/1.105773

Dielectric Materials and Applications: ISyDMA'2016   Materials Research Forum LLC
Materials Research Proceedings 1 (2016) 154-158   doi: http://dx.doi.org/10.21741/2474-395X/1/39

Original solution for a photovoltaic installation at a remote site

Meriem CHADEL[1,2,*], Moustafa Mohammed BOUZAKI[1,2], Boumediene BENYOUCEF[1], Asma CHADEL[1], Michel AILLERIE[2,3],

[1] University of Tlemcen, URMER, Tlemcen, Algeria

[2] Université de Lorraine, LMOPS-EA 4423, 57070 Metz, France

[3]CentraleSupelec, LMOPS, 57070 Metz, France

*Correspondingauthor: e-mail: ch_meriem_ph@yahoo.fr

Keywords: Generator PV, Solar Radiation, Power, Temperature, Series and Parallels Resistances

Abstract. We study a solution for photovoltaic plants located in an isolated site in the aim to minimize losses in PV panels due to the effect of temperature changes and intensity irradiation variations. The influence of the temperature of the solar cells is an important focus in research. We can say that the temperature has a detrimental effect on the characteristics of the solar cell. When photons of low energy are lost, there will be a loss of light output. Pmax is proportional to intensity irradiation. The series resistances reduce the efficiency of the solar cell which is not preferable. The parallel resistors increase the efficiency of the solar cell which is preferable.

Introduction

The photovoltaic energy is one of the fundamental renewable energy due its easy availability [1]. The solar cell is the electrical generator for an installation PV standalone and the connected in the network [2].

The solar cells are generally associated in series and in parallel and then encapsulated to obtain a photovoltaic module [3]. A PV array consists of modules interconnected to unit producing high continuous power compatible with the usual electrical equipment. Photovoltaic modules are usually connected in series-parallel to increase the voltage and current at the output of the generator.

The I-V characteristic of the GPV depends on the level of illumination and the cell temperature and of the aging of the assembly [2]. In addition, its operating point of the GPV directly depends on the load it supplies. To extract every moment the maximum power available across the GPV.

Modeling of solar panel

The characteristic I-V of a photovoltaic generator is based on the same principle of a solar cell that was modeled by the equivalent circuit in Figure 1 [4, 5]. This circuit introduces a current source [6], a diode in parallel and series resistances Rs, parallel (shunt) Rsh to account for dissipative phenomena at the cellular level.

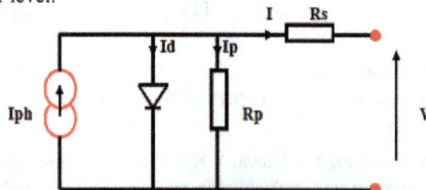

Fig. 1: Circuit equivalent of a photovoltaic cell.

Dielectric Materials and Applications: ISyDMA'2016   Materials Research Forum LLC
Materials Research Proceedings 1 (2016) 154-158   doi: http://dx.doi.org/10.21741/2474-395X/1/39

The series resistance is due to the contribution of the base resistances and the front of the junction and contacts the front and rear. Parallel resistance realizes effects such as the current losses through the edges of the cell, it is reduced by the penetration of metal impurities in the junction (especially if this penetration is deep) [7]. This circuit can be used both for a unit cell), for a module or a panel made up of several modules.

$$I = I_{ph} - I_0 \left(e^{\frac{q}{nKT}(V+R_s I)} - 1 \right) - \frac{V+R_s I}{R_p} \tag{1}$$

Where: I_{ph}, I_0 respectively the photo- current, the reverse saturation current of the diode and thermal stress with: n the ideality factor of the diode , q the electron charge , k is Boltzmann's constant, T the temperature of the cell which varies in function of the illumination and the ambient temperature. $I_{Rp}=I_{sh}$

$$I_0 = \frac{I_{cc}-(V_{co}/R_p)}{e^{\frac{q}{nKT}(V+R_s I)}} \tag{2}$$

$$I_d = I_0 (e^{\frac{q}{nKT}(V+R_s I)} - 1 \tag{3}$$

$$I_{Rp} = \frac{V+R_s}{R_p} \tag{4}$$

$$I = I_{ph} - I_d - I_{Rp} \tag{5}$$

Results of the simulation and validation of the model

1.   Effects of irradiation in the characteristic IV

First, we will present the results of the simulation of a solar panel with the following values of resistances series and parallelRs = 0.01Ω and Rp = 1000 Ω.
    When the illumination increases, the intensity of the photovoltaic current rises, the I-V curves shift to increasing values for the module to produce a larger electrical power. The current-voltage characteristic of the module shows a maximum power point voltage which is between 0.6 and 1.2 Volts at 25 ° C.

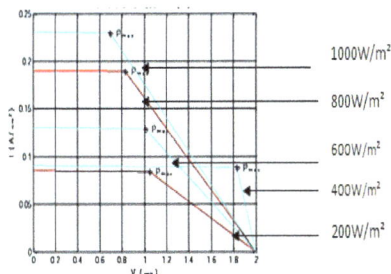

Fig2: IV characteristic of a different irradiation a Module PV.

When the radiation level is high generation of electron hole pairs will be important to increase the intensity of the current produced by the panel that is clear in the figure 2.

Dielectric Materials and Applications: ISyDMA'2016   Materials Research Forum LLC
Materials Research Proceedings 1 (2016) 154-158   doi: http://dx.doi.org/10.21741/2474-395X/1/39

For each curve, the value of I is constant at the beginning in more or less different intervals', however, it will decrease in order to achieve the value of the V_{CO}.

The current (I) is inversely proportional to voltage V.

With the increase in the value irradiation, the coordinates (Im, Vm) of the workings of the points shift to higher values, the same for Pmax (maximum power).

2. Characterization I-V for Different Temperatures

The temperature of the junction is proportional to the surface temperature of the solar cells.

The influence of temperature on the characteristic I = f (V) (Fig. 3) is low when the junction is a same as ambient temperature. The variation becomes significant if the temperature of the junction increases.

The influence of temperature on the characteristic quantities of the cell becomes significant when the temperature of the junction different of the ambient temperature [3].

The solar flux and encapsulation of cells causes an increase in the temperature of the junction. When cells operate at high junction temperatures, they lose their characteristics and aging rapidly.

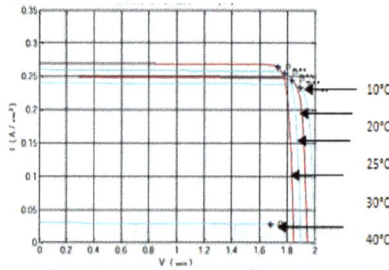

Fig 3: IV characteristics for different temperatures

It is clear that when changing the temperature varied from the junction to higher values, there will be a slight increase in short circuit current. The curves intersect with an insignificant decrease in the values of the V_{CO}.

For each curve, the value of I is constant at the beginning, however, it will decrease in order to achieve the value of the V_{CO}.

3. Characteristic IV for Different Resistances Series

Resistance series characterise the losses by Joule effect of the resistance of the semiconductor and losses through the collection grids and bad Ohmic contacts of the cell. The influence of Rs on the characteristics of a cell is very important to the high polarization levels (near Vco) (Fig. 4).

Current technology tries to minimize the R_S value by control of the illuminated region, the geometry of the cell and the concentration of impurities. The cells currently marketed Si generally have series resistances:Rs = 0.01Ω–0.2Ω.

To: VCO = 1.75v And ICC = 0.15A.

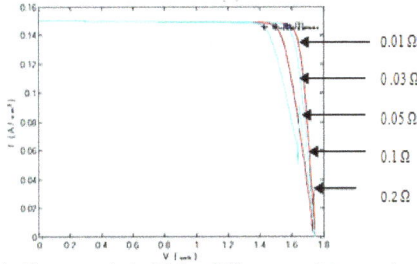

Fig 4: Characteristic IV for different resistances in series.

It is clear that when changing the resistance Rs to higher values, the short-circuit current and open circuit voltage remain constant, there will be a slight operating point decrease (reduction in maximum power)

The current I is inversely proportional to V(Fig. 4).

4.   Characteristic I-V for Different Resistances Shunt

The parallel resistance (or shunt) characterised recombination losses due to the thicknesses of the regions P and N and the load area and space.

Rp is linked directly to the manufacturing process, the existence of defect structures and surface states. Rp also affects the characteristic $I = F(v)$ its influence is very important for low bias levels (near Icc). The cells currently marketed If generally have a parallel resistance $Rp = 10^2\Omega - 10^4\Omega$.

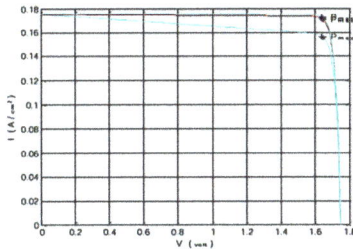

Fig 5: I-V Characteristics of a module for variation R_p.

With the increase of Rp value the coordinates of the Maximum power point shift to higher values, the same for Pmax.

It is clear that when changing the resistance Rp to higher values, the short -circuit current and open circuit voltage remain constant, it will increase MPP (increase in maximum power).

The current I is inversely proportional to V (figure 5).

Conclusion

In this paper, the impact of temperature, irradiation and resistance series-parallel on the performance of a PV generator were studied.

The maximum power of a solar panel is related to the amount of light it receives; that is to say, the number of photons that can set an excitation to form electron hole pairs. When photons of low energy are lost, there will be a loss of light output. Pmax is proportional to intensity irradiation.

The influence of the temperature of the solar cells is one of the important researches. Increasing the temperature of the junction is due to exposure of the latter to the sun which affects performance. We can say that the temperature has an adverse effect on the characteristics of the solar cell. The series resistances reduce the power of the solar cell which is not preferable. The parallel resistances increase the efficiency of the solar cell which is preferable.

References

[1] M. El Ouariachi et al, Analysis, optimization and modeling of electrical energies produced by the panels and systems, Revue des energies renouvelables, Vol 14 N°4, 2011, page 707-716.

[2] Series resistance effects on solar cell measurements Advanced Energy Conversion, Vol 3, pp, 455-479.

[3] Subhash Chander et al, Impact of temperature on performance of series and parallel connected mono-crystalline silicon solar cells, Energy reports 1 (2015) 175-180. http://dx.doi.org/10.1016/j.egyr.2015.09.001

[4] Han, L., Koide, N., Chiba, Y., &Mitate, T. (2004). Modeling of an equivalent circuit for dye-sensitized solar cells. *Applied Physics Letters*, *84*(13), 2433-2435. http://dx.doi.org/10.1063/1.1690495

[5] Koide, N., Islam, A., Chiba, Y., & Han, L. (2006). Improvementofefficiencyofdye-sensitized solar cells based on analysis of equivalent circuit. Journal of Photochemistry and Photobiology

[6] *A: chemistry*, *182*(3), 296-305.

[7] Meriem Chadelet al., Study of a photovoltaic system connected to the network with different technologies the panel PV centred, IJAER, Volume 10, Number 18 (2015),pp 38931-38936.

[8] M. Sabry, Influence of temperature on methods for determining silicon solar cell series resistance, solar research department PV laboratory, journal of solar energy engineering, Vol 129, 2007.

Dielectric Materials and Applications: ISyDMA'2016
Materials Research Proceedings 1 (2016) 159-162

Materials Research Forum LLC
doi: http://dx.doi.org/10.21741/2474-395X/1/40

FITR and physical properties studies of xBi$_2$O$_3$-(100-x)P$_2$O$_5$ glasses

Zahra RAMZI*[1], Wafaa NACHIT[1], Khalil BENKHOUJA[1], Samira TOUHTOUH[2]

[1]E2M_LCCA, Chemistry Department, Faculty of Sciences, UCD University, P.O. Box 20, 24000, El Jadida, Morocco

[2]LabSIPE, Laboratory of Engineering, National School of Applied Sciences, UCD University, El Jadida, Morocco

*Corresponding author: E-mail: Zahramzi80@gmail.com

Keywords: Phosphate Glasses, Infrared Spectroscopy, Physical Properties, Glass Matrix

Abstract. Glasses of xBi$_2$O$_3$ - (100-x)P$_2$O$_5$ system with ($15 \leq$ x ≤ 30) mol% have been prepared by using melt-quench technique. Density (d) is measured and molar volume (Vm), average molecular weight (M), oxygen packing density (O), for all the glass samples have been calculated. An XRD confirms the amorphous nature of the glasses. IR technique is used to investigate the structure of glasses in order to obtain information about the competitive role of Bi$_2$O$_3$ in the formation of glass network.

I. Introduction

Many researchers have focused on phosphate glasses in the last years due to their diversified applications in technology, medicine, as biomaterials and in clinical dosimetry [1, 2]. Phosphate glass exhibit very important physical properties such as low melting temperature, high thermal expansion coefficient, low glass transition temperature, low softening temperature and high ultraviolet(UV) transmission[3, 4]. Combining bismuth oxide (Bi$_2$O$_3$) with phosphorus pentoxide allows one to tune the optical properties in a wide range depending on the glass composition. During the last two decades, many binary ,ternary, and Quaternary bismuth based glass have been done by various authors like, Bi$_2$O$_3$-GeO$_2$[5], ZnO-Bi$_2$O$_3$-SiO$_2$[6],Na$_2$O–B$_2$O$_3$–Bi$_2$O$_3$–MoO$_3$[7].For example Shashidhar Bale et al. [8] studied the role of Bi$_2$O$_3$ in the formation of the glass network by Raman and IR techniques. Chahine et al. [9] reported IR and Raman spectra of sodium-bismuth-copper phosphate glasses, reflecting the structural role of bismuth.

In this study, Infrared measurements were made on xBi$_2$O$_3$–(100-x) P$_2$O$_5$ (where x is in mol %, ranging from 15 to 30 in steps of 5) glass systems to determine the changes that appears in the local structure of the phosphate glass matrix with the gradually addition of Bi$_2$O$_3$. The amorphous nature of this glass was confirmed by X-ray diffraction technique and their densities were measured by the Archimedes methods. Other physical properties, such as molar volume and Oxygen packing density are also evaluated. The relationships between composition and properties were demonstrated.

II. Experimental Procedure

A. preparation of glasses

The glass samples were prepared using high purity commercial materials α Bi$_2$O$_3$, NH$_4$H$_2$PO$_4$ of analytical grade (Aldrich 99.9%). The batches of suitable proportions of starting products were mixed in an agate mortar and then heated in air at 1100°C (2hours). The melted batches are quenched to room temperature. The amophous nature of the products are identified by X-ray diffraction. All glass samples were transparent and Brown in color as shown in Fig. 1.

Dielectric Materials and Applications: ISyDMA'2016   Materials Research Forum LLC
Materials Research Proceedings 1 (2016) 159-162   doi: http://dx.doi.org/10.21741/2474-395X/1/40

Fig. 1. Photograph image of 70%P_2O_5- 30%Bi_2O_3 glasses

The infrared spectra of the glasses were recorded at room temperature, using KBr disc technique.

B.   Density measurement

The density (d) of the prepared glass samples was determined by Archimedes method using diethyl orthophtalate as the immersion fluid at room temperature. The density is calculated using the formula,

$$d = d_{orth} \frac{W_{air}}{W_{orth}}$$
(1)

Where, W_{air} is the weight of the glass sample in air, W_{ortho} is the weight of the glass sample in orthophtalate and d_{orth} is the density of orthophtalate .

C.   Molar volume

The molar volume (V_m) is calculated using the relation:

$$V_m = \sum \frac{X_i M_i}{d}$$
(2)

where, x_i is the molar fraction and M_i is the molecular weight of the ith component.

D.   Oxygen packing density (0)

The oxygen packing density of the glass samples were calculated using the following relation,

$$O = n \frac{d}{M}$$
(3)

III.   Result and discussion

A.   Physical properties

The change in atomic geometrical configuration, co-ordination number, cross-link density and the dimensions of the interstitial space in the glass network decides the density. Hence, the density is a tool in revealing the degree of change in the structure with the glass composition. Figure 2 shows the molar volume and the density increase proportional to the bismuth content. The molar mass of the bismuth (iii) oxide (465.96 g/mol) is heavier than the molar mass of phosphate oxide (142g/mol) and hence, the glass matrix becomes more dense when Bi^{3+} ions are added into the glass network [10]. Usually, the density of the glass changes in the inverse direction of the molar volume, but in this study, the density and molar volume increase with the bismuth contents. In the present glass system oxygen packing density decreases as the concentration of Bi_2O_3 increases. This indicates that the structure becomes loosely packed with increase in the concentration of Bi_2O_3
B.FT-IR spectra analysis

The vibration mode of the bimute phosphate glass network shows the presence of six band regions. the absorption bands and their assignments are summarized in Table 1.

Dielectric Materials and Applications: ISyDMA'2016   Materials Research Forum LLC
Materials Research Proceedings 1 (2016) 159-162   doi: http://dx.doi.org/10.21741/2474-395X/1/40

TABLE I. Vibration types of main absorption bands in samples

Absorption band/cm^{-1}	Vibration type
450	Bi-O bonds in BiO$_6$ octahedra Units
620	vibration of Bi—O band in [BiO$_3$]
780	symmetric stretching vibrations of P-O-P
1085 and 970	v of P–O$^-$ groups (chain terminator)
1180	PO$_2$ symmetric stretching vibration
1270	the symmetric stretching of the P=O terminal oxygens

The infrared transmittance studies at different compositions indicate that the structure of the glass consists of four main units including [BiO$_3$], [BiO$_6$], [PO$_3$] and [PO$_4$].

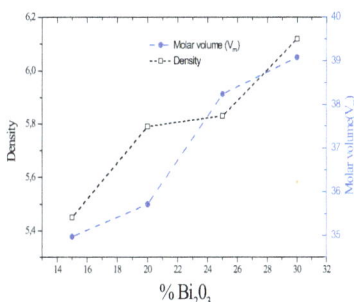

Fig. 2. Variation of density and molar volume of Bismuth phosphate glasses with the Bi$_2$O$_3$content

The intensity of the vibrations observed for the glasses is reduced and broadened because of their disordered structure. The occurrence of symmetric vibration bands clearly suggests that stronger Bi-O bonds are present in the glass composition.

VI.   Conclusion

The variation of density and molar volume with the addition of Bi$_2$O$_3$ content indicates that the effect of Bi$_2$O$_3$ on the glass structure is dependant on its concentration. By analyzing infrared spectra of Bi$_2$O$_3$- P$_2$O$_5$ glasses it is found that Bi^{3+} cations are incorporated in the glass network as [BiO$_3$] pyramidal and [BiO$_6$] octahedral units

References

[1] Z.M. Da Costa, W.M. Pontuschka, J.M. Giehl, C.R. Da Costa, J. Non-Cryst. Solids 3523663(2006).

[2] S. Fan, C. Yu, D. He, K. Li, L. Hu, Radiat. Meas. 46 ,46(2011). http://dx.doi.org/10.1016/j.radmeas.2010.09.002

[3] Franks, K., I. Abrahams, G. Georgiou and J.C. Knowles, Biomaterials, 22, 497, (2001). http://dx.doi.org/10.1016/S0142-9612(00)00207-6

[4] Talib, Z.A., Y.N. Loh, H.A.A. Sidek, M.D.W. Yusoff, W.M.M. Yunus and A.H. Shaari, Ceramic International, 30, 1715.(2004). http://dx.doi.org/10.1016/j.ceramint.2003.12.146

Dielectric Materials and Applications: ISyDMA'2016   Materials Research Forum LLC
Materials Research Proceedings **1** (2016) 159-162   doi: http://dx.doi.org/10.21741/2474-395X/1/40

[5] Maylon M. Martins, Diego S. Silva, Luciana R. P. Kassab Sidney J. L. Ribeiro and Cid B. de Araújo, J. Braz. Chem. Soc., Vol. 26, No. 12, 2520-2524 (2015)

[6] Rajni Bala, Ashish Agarwal, Sujata Sanghi, and S. Khasa,J Integr Sci Technol, 3(1), 6-13(2015).

[7] Yasser B. Saddeek, K.A. Alya, A. Dahshanb, I.M.El. Kashefc, Journal of Alloys and Compounds, 494, 210–213(2010)

[8] Bale, S., Rahman, S., Awasthi, A.M. and Sathe, V. Journal of Alloys and Compounds, 460, 699-703 (2008) http://dx.doi.org/10.1016/j.jallcom.2007.06.090

[9] Chahine, A., Et-Tabirou, M. and Pascal, J.L. Materials Letters, 58, 2776-2780(2004). http://dx.doi.org/10.1016/j.matlet.2004.04.010

[10] Sidek, H.A.A.; Rosmawati, S.; Talib, Z.A.; Halimah, M.K.; Daud, W.M. Am. J. Appl. Sci., 6, 1489–1494. (2009). http://dx.doi.org/10.3844/ajassp.2009.1489.1494

Dielectric Materials and Applications: ISyDMA'2016
Materials Research Proceedings 1 (2016) 163-166

Materials Research Forum LLC
doi: http://dx.doi.org/10.21741/2474-395X/1/41

Study of impurities effect on the dielectric properties of ammonium phosphate fertilizer using the dielectric impedance spectroscopy

Nacira LEBBAR[*1], Redouane LAHKALE[1], Wafaa ELHATIMI[1], Elmouloudi SABBAR[*1], Anass HAFNAOUI[2], Khouloud MEHDIB[2], Rachid BOULIF[2]

[1]Laboratory of Material's Physical- Chemicals. Department of Chemistry, University of Chouaîb Doukkali, El Jadida, Morocco

[2]Fertilizer Laboratory, Direction of research and developpement, OCP group.SA, El Jadida, Morocco

*Corresponding author: e-mails: n.cerocp@yahoo.fr & esabbar@yahoo.fr

Keywords: Ammonium Phosphate Fertilizers, Dielectric Properties, Impedance Spectroscopy

Abstract. Ammonium phosphate fertilizers are obtained from neutralization reactions of phosphoric acid with ammonia. Several works were carried out to define and to control the impurities effect on the physicochemical properties of these fertilizers. However the description of these effects remains an area of disagreement among researchers. Among these studies, those were conducted to investigate the dielectric properties of certain phosphate materials, such as KH_2PO_4 and $NH_4H_2PO_4$. Many of these crystal phases are isostructural to the paraelectric phase (EP), by cooling they exhibit a phase transition (PE) to a ferroelectric phase (FE), this usually para-ferroelectric transition occurs at low temperature. In this work our interest is, particularly, to study the impurities effect on the physical properties related to phase transitions in the fertilizer matrix.

Introduction

The reactivity and control of granular materials such as fertilizers, as well as the particle size and /or shape accordance with standards are generally important industrial concerns that affect their conditions of use.

A review of the literature revealed that the impurities of phosphoric acid such as iron, aluminum, and magnesium were judged to be primarily responsible for the variation of the physico-chemical quality of fertilizers [1].

The aim of the current work is to control the morphology and reactivity of fertilizer granules, and study the effects of impurities on their physico-chemical quality.

Several characterization and laboratories techniques [2] were used such as X-ray diffraction (XRD), the infrared spectroscopy (IR), the thermogravimetric analysis (TGA), scanning electron microscopy (SEM), atomic absorption

Spectroscopy (AAS), inductively-coupled plasma spectrometry (ICP) spectroscopy and dielectric impedance.

So, the impedance measurements we carried out with a Modulab MTS (Solartron Analytical) between 0,01Hz and 1MHz at room temperature.

Results

1. Synthesis of ammonium phosphate

After chemical characterization of phosphoric acid used, two acid solutions were prepared, one for blank test, and the other for the test doped with a mix of iron and aluminum impurities with equal proportion.

Dielectric Materials and Applications: ISyDMA'2016   Materials Research Forum LLC
Materials Research Proceedings **1** (2016) 163-166   doi: http://dx.doi.org/10.21741/2474-395X/1/41

Ammonium phosphates are produced via the reaction between gaseous ammonia and prepared phosphoric acid solution in a pre-neutraliser reactor. The fertilizers are produced by controlling the mole ratio (MR) of ammonia to phosphoric acid [**3**]. When the MR is 1.0, MAP is formed according to the following equation (1):

$$NH3 + H3PO4 \rightarrow NH4H2PO4 \ (MAP) \tag{1}$$

The resulting slurry was dried in an oven maintained at 60°C and then, crushed and ground to finer than 500 microns, and then granulated by a suitable granulation device, the granulated product was then dried at a temperature of 60°C.

2. Granulation and solubility yield

Yield granulation of the fertilizer product is determined from the results of particle size analysis (particle size between 2.5 and 4 mm).

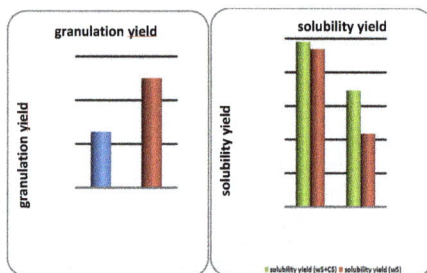

Fig. 1: Impurities effect on granulation and solubility yields of finished product

The effect of doping has a positive impact on the physical quality of the studied fertilizer, manifested by the improvement of granulation yield, however, these doping causes a decrease in the fertilizer solubility yield.

3. X-ray diffraction

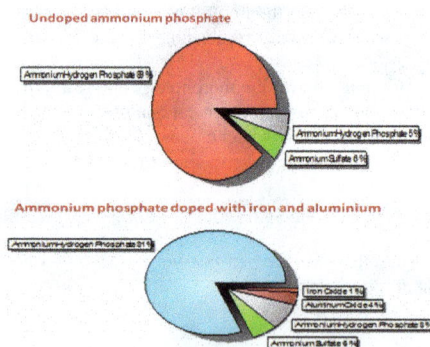

Fig.2: Quantification of identified components of ammonium phosphate undoped and doped with iron and aluminums impurities

Dielectric Materials and Applications: ISyDMA'2016
Materials Research Proceedings 1 (2016) 163-166

Materials Research Forum LLC
doi: http://dx.doi.org/10.21741/2474-395X/1/41

The doping effect is seen by the appearance of other phases that are identified by the X-ray diffraction.

Dielectric measurements

1. Dielectric properties

The complex impedance spectra for all studied compositions were performed using an MTS Modulab (Solartron Analytical) in LPCM Laboratory (Chemistry Department, El Jadida, Morocco). The measurements are performed at ambient temperature on thin wafers, under low pressure between 0.01 Hz to 1MHz [4].

2. Nyquist diagrams analysis

The Nyquist diagrams of a doped fertilizer and undoped fertilizer are characterized by the appearance of two semicircular who's the curvature rays decrease depending the doping (Fig.3).

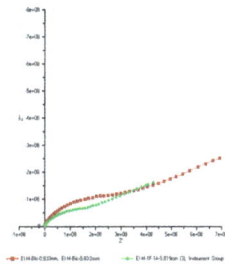

Fig. 3: Nyquist diagram of prepared fertilizers.

3. Conductivity analysis

Evolution of the real part of complex conductivity versus frequency for both fertilizers (a) shows a slow decrease of Z 'at low frequencies and fast beyond 100Hz, resulting in an increase of the conductivity as a function of the frequency (b).

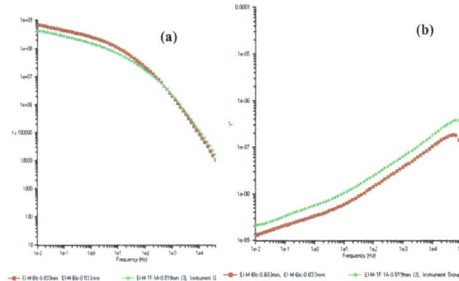

Fig. 4: (a) Z ' versus frequency , (b) Y' versus frequency

The effect of doping is clearly manifested by an increase in conductivity.

4. Modulus analysis

In the polycrystalline material, this formulation increases the intrinsic properties related to grain by reducing the effects of polarization at the electrodes as well as all other interfacial phenomena solid electrolytes.

Indeed, the doping of the fertilizer is manifested by a shift of the relaxation frequency towards lower frequencies, for the undoped fertilizer this results in a fast relaxation (Fig. 5).

Dielectric Materials and Applications: ISyDMA'2016   Materials Research Forum LLC
Materials Research Proceedings 1 (2016) 163-166   doi: http://dx.doi.org/10.21741/2474-395X/1/41

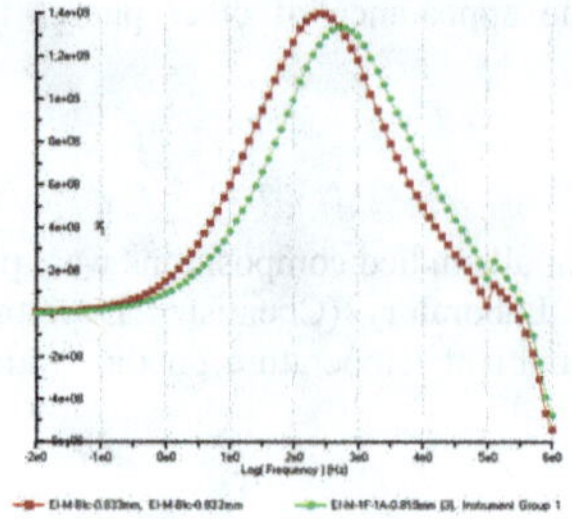

Fig. 5: Relaxation frequency variation of M " versus frequency.

5. Modeling of complex impedance spectra

The experimental data are adjusted by the equivalent circuit (Fig.6). Each contribution (grains, grain boundaries) is represented by a parallel combination of resistance (R) and the constant phase element (CPE).

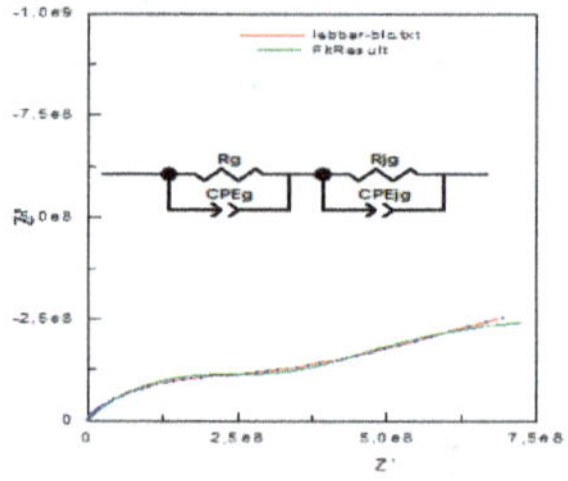

Fig. 6: equivalent circuit model

Conclusion

• The effect of doping has a positive impact on the physical quality of the studied fertilizer, manifested by the improvement of granulation yield and hardness of granules, however, this doping causes a decrease in the fertilizer solubility yield.

• The effect of doping is clearly manifested by an increase in conductivity.

References

[1] G.R. Campbell, Y.K. Leong, C. C. Berndt, J. L. Liow. Chemical Engineering Science. 61(2006) 5856 -5866. http://dx.doi.org/10.1016/j.ces.2006.05.010

[2] E. Sabbar, M.E. De Roy, F. Leroux. Microporous and mesoporeux Materials. 103 (2007) 134-141. http://dx.doi.org/10.1016/j.micromeso.2007.01.044

[3] F. P. Achorn, E. F. Dillard, A. W. Frazier and D. G. Salladay, Effect of Impurities in Wet-Process Phosphoric Acids on DAP Grades, ISMA Technical Conference, Austria, 1980

[4] A. El Melouky, R. El Moznine, R. Lahkale, R. Sadik, E. Sabbar, E. Chahid, E. Choukri, D. Mezzane, A. Belboukhari, Journal of Optoelectronics and Advanced Materials. 15 (2013) 1239-1247.

Dielectric Materials and Applications: ISyDMA'2016   Materials Research Forum LLC
Materials Research Proceedings 1 (2016) 167-170   doi: http://dx.doi.org/10.21741/2474-395X/1/42

Charging process and trapped charge characterization in insulating materials using an influence current method (ICM) in adapted SEM

O. MEKNI*, D. GOEURIOT[1], S. JOAO SAO[1], C. MEUNIER[1], B. ASKRI[2], K. RAOUADI[2], G. DAMAMME[3]

[1]Laboratory Georges Friedel CNRS UMR 5307, Ecole des Mines de Saint Etienne, Saint Etienne, France

[2]Laboratory of Materials, Organization and Properties LR99ES17, University of Tunis El Manar, Tunis, Tunisia

[3]CEA, Gramat, France

*Corresponding author: E-mail: omar.mekni@emse.fr

Keywords: Ceramics, Space Charge, Dielectric Breakdown, Trapping, SEM

Abstract. The choice of insulating material for applications requiring an electrical insulation matter in the field of microelectronics (e.g substrate) or high-voltage energy transport is related to the corresponding breakdown voltage value which limits their use in some applications. To improve the resistance to dielectric breakdown, it is imperative to understand and control the cause of this catastrophic phenomenon reducing the reliability of some instrumentation. It is well known that breakdown is correlated with the presence of space charge within the insulators. In this context, we are interested in this work to characterize the trapped charge and its stability in ZrO_2/Y_2O_3 sintered polycrystalline ceramic using a special arrangement in the SEM chamber. The firing temperature effect is studied and discussed.

Introduction

The microscopic damage processes in insulating materials under electrical stress are not yet understood but are indisputably related to energy localization on defects [1]. Electric breakdowns are present in all applications using insulating materials, ranging from microelectronics [2] to technologies associated with cabling and isolators in energy transport and other applications for example in aerospace [3]. These breakdowns are commonly associated with electron trapping [4]. In fact, it was shown that breakdowns correlate with fast electrons detrapping [5,6] which is provoked by an external field, a mechanical shock or under electron irradiation. It is noteworthy that detrapping produces an increase in temperature high enough to reach the materials melting point, producing a shock waves and a solid dielectric - plasma transition. During a non-radiative transition leading to electron trapping, the energy gained by the system is communicated to the phonons and when the material internal energy reaches a critical value, the statistical motion of phonons represents the energy source of detrapping. In short, detrapping depends on electron – phonon interactions. To extend these research directions, it is necessary to handle characterization techniques of charging phenomena. The charge injection and the dielectrics characterization by electron beam techniques are particularly suited to the damage process study. In the remainder of this work, we focused on measuring the influence current in order to study the trapped charge dynamics during and after electron irradiation. We present our experimental setup, some results and their discussion.

Experimental procedures

Samples preparation

The used samples are polycrystalline yttria stabilized zirconia with dimensions of 1 mm thickness and 16 mm diameter. ZrO_2 stabilized with 4 mol % of Y_2O_3 samples were treated by sintering from

Dielectric Materials and Applications: ISyDMA'2016   Materials Research Forum LLC
Materials Research Proceedings 1 (2016) 167-170   doi: http://dx.doi.org/10.21741/2474-395X/1/42

spray dried powder manufactured by BAIKOWSKI International. The starting powder was uniaxially pressed (400 MPa) into disks (20 mm in diameter) using a compression machine INSTRON. Before sintering, organic compounds were eliminated by heating (debinding) at 1 °C/min up to 600 °C with dwell duration of 1 hour. The Firing temperatures are indicated in the figures.

Experimental setup

Contrary to the SEMME method (Scanning Electron Microscope Mirror Effect) which quantifies the total trapped charge amount, the Influence Current Method (ICM) used in this study allow tracing back to the trapped charge dynamic evolution during (charging) and after (charge decay) electron irradiation. Our measurements are made using an adapted Scanning Electron Microscope (Carl Zeiss SUPRA 55 VP). The experiments were carried out in the configuration shown in Fig. 1. This special arrangement in the SEM chamber allows measuring separately and simultaneously the influence (I_{inf}) and conduction (I_c) currents via two picoammeter. Electrons are injected in spot mode with a defocused beam of 335 μm in diameter in order to average the recorded signal over several grains of the sample. The technique reliability was verified and the results are reproducible.

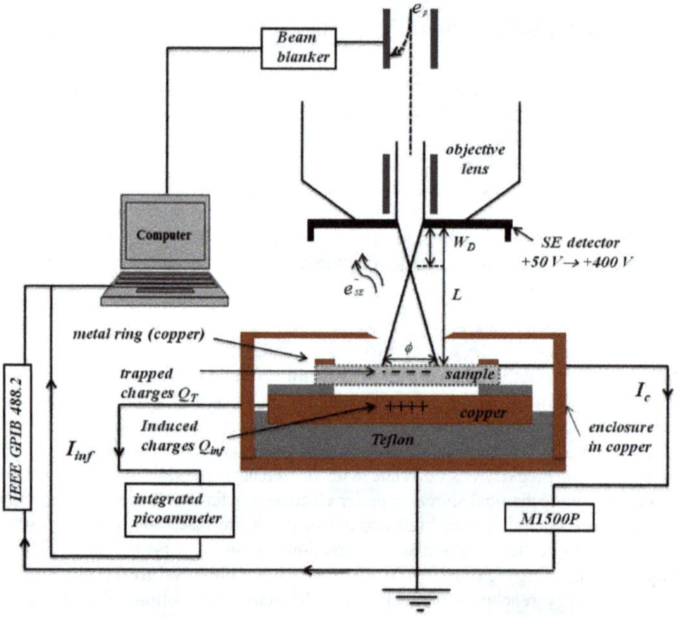

Fig. 1. Descriptive Scheme of the experimental design to measure the influence and conduction currents.

Results and discussions

By electron irradiation, a part of injected charges remains trapped within the insulator. This trapped charge Q_t produces an image charge Q_{im} with opposite sign in the copper probe and measured via an integrated picoammeter.

Dielectric Materials and Applications: ISyDMA'2016
Materials Research Proceedings 1 (2016) 167-170

Materials Research Forum LLC
doi: http://dx.doi.org/10.21741/2474-395X/1/42

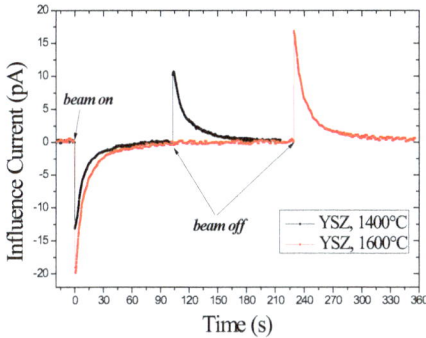

Fig. 2. Firing temperature effect on the influence current measurements for yttria stabilized zirconia (YSZ) polycrystalline ceramic.

The trapped charge Q_t is obtained by integrating the influence current and by taking into account the electrostatic influence factor K. The I_{inf} temporal evolutions for two different firing temperatures (1400 and 1600 °C) are reported in Fig. 2. The current intensities tend towards a zero when an equilibrium state between trapped charge, conduction current and secondary electron current is reached. Unlike other materials, such as alumina, a charge decay phase is observed after blanking the beam. The corresponding trapped charge evolution (Fig. 3) increases according to a first order exponential law and then reaches a plateau corresponding to the amount of trapped charge saturation Q_s. After blanking the beam (*beam off* in the figure), Q_t shows an exponential decay until stabilized again but this time at a lower value Q'_s that represents the stabilized trapped charge within the sample. The rest of the trapped charge ΔQ_s is evacuated from the irradiated volume. The Firing temperature effect was studied and it turned out that it greatly influences the trapping, the spreading of charges and subsequently the breakdown voltage. More the firing temperature increases more the amount of stabilized trapped charge increases. This is directly related to the grains size. In Fact, this tendency is due to the eventual decrease in the grains boundary density in which the oxygen vacancies are concentrated.

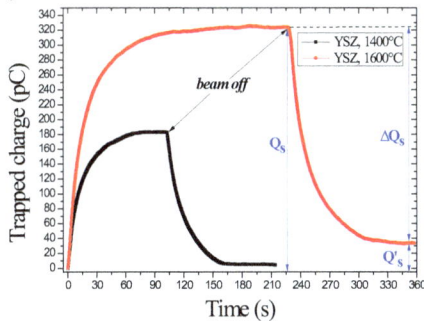

Fig. 3. Firing temperature effect on trapped charge evolution during and after electron irradiation for yttria stabilized zirconia.

Dielectric Materials and Applications: ISyDMA'2016
Materials Research Forum LLC
Materials Research Proceedings **1** (2016) 167-170
doi: http://dx.doi.org/10.21741/2474-395X/1/42

Conclusion

Internal polarization energy, associated with the lattice deformation around the trapped charge can be released during a relaxation process. This event can be produced when the trapped charge is suddenly detrapped, causing consequently the breakdown of the insulator. Therefore, an understanding of the physical mechanisms of the space charge formation responsible for this effect was paramount. The developed technique allows measuring the influence current and access to the trapped charge dynamic evolution and therefore enhancing the understanding of trapping phenomenon, spreading and stability of charges in insulating materials.

References

[1] N. Bourne, Materials in Mechanical Extremes: Fundamentals and Applications, Cambridge University Press, Cambridge, UK, 2013. http://dx.doi.org/10.1017/CBO9781139152266

[2] R. Stoklas, D Gregušová, K Hušeková, J Marek and P Kordoš, Trapped charge effects in AlGaN/GaN metal-oxide-semiconductor structures with Al_2O_3 and ZrO_2 gate insulator, Semicond. Sci. Technol., **29** (2014) 4. http://dx.doi.org/10.1088/0268-1242/29/4/045003

[3] T. Paulmier, B. Dirassen, M. Belhaj, and D. Rodgers, Charging properties of space used dielectric materials, IEEE Trans. Plasma Sci., **43** (2015) 9. http://dx.doi.org/10.1109/TPS.2015.2453012

[4] C. Bonnelle, G. Blaise, C. Le Gressus, D. Tréheux, Physique de la localisation des porteurs de charge, Applications aux phénomènes d'endommagement, Les isolants, Lavoisier, 2010.

[5] G. Blaise, C. Le Gressus, Flashover in insulators related to the destabilization of a localized space charge, C. R. Acad. Sci. Paris, **314** (1992) 1017-1024.

[6] G. Blaise, Charge detrapping induced dielectric relaxation. Application to breakdown in insulating films, Microelectron. Eng. **28** (1995) 55-62. http://dx.doi.org/10.1016/0167-9317(95)00015-Z

Dielectric Materials and Applications: ISyDMA'2016
Materials Research Proceedings 1 (2016) 171-174

Materials Research Forum LLC
doi: http://dx.doi.org/10.21741/2474-395X/1/43

Current characteristics in corona charged polyimides films

Zehira ZIARI*, Hala MALLEM, Nesrine AMIOUR and Salah SAHLI

Laboratoire de Microsystèmes et Instrumentation, Faculté des Sciences de la Technologie,
Université des Frères Mentouri Constantine, Constantine 25000, Algeria

*Corresponding author: E-mail: zziari_zahira@yahoo.fr

Keywords: Charging Current, Corona Discharge, Polyimides, Surface Potential Decay

Abstract: In this study, an experimental investigation on current characteristics during negative corona charging of polyimide (PI) has been carried out. The effect of the applied voltage on the corona current is studied to investigate the PI films electrical properties. The experimental results show clearly that the charging voltage affects the charging currents. These later presented a monotonically decreasing appearance over time and their amplitudes increase with increasing the value of the corona applied voltage. Current measurements during corona charging confirm the interpretation of surface potential measurements after corona charging where the increase of the charging voltage induces an increase in current level leading to the increase of surface potential decay rate.

Introduction

Aromatic polyimide (PI) is the most important class of high temperature thermostable polymer. The material offers good radiation and chemical resistance, which make it useful as an electrical insulator in hostile environments [1, 2].

Many experimental devices have been developed in recent years to determine the main mechanisms responsible for the transport of charges in insulators. The measurement of surface potential decay of polymer films after deposition of negative charges through corona discharge is frequently used to study the electrical properties and to understand electric charge transport phenomena in polymers. This simple and low cost technique remains powerful in many applications, including the study of thin insulator layers [3-6].

The work carried out on polyimide films which is widely used in aerospace applications [7, 8], highlights the importance of the surface potential decay measurement as an analytical tool in electrical properties and shows its complementarily with the measurement of the charging current.

In this paper, the effects of the applied voltage on the charging currents measured on polyimide thin films during negative corona charging and surface potential decay measured after corona charging in a point-grid-plate configuration has been investigated.

Experimental Details

The material employed in this work for the charging currents and the surface potential decay measurements was 50 µm thick polyimide films (PI; Kapton® HN of Goodfellow).

The experimental setup for corona charging discharge and surface potential decay measurement used in our experiments is divided into two parts: the corona charging part and surface potential measurement part. During charging, the surface potential is controlled by a metal grid between the corona tip and PI sample for initial surface potential monitoring.

After corona charging, the sample was immediately transferred in a controlled manner under a Monroe vibrating probe connected to an electrostatic voltmeter.

The charging currents characteristics were recorded using a Keithley 6517A electrometer via a GPIB interface.

All experiments were performed at ambient air. The relative humidity of the atmosphere was varying between 35 and 45% and the temperature value was close to 25 °C.

Dielectric Materials and Applications: ISyDMA'2016   Materials Research Forum LLC
Materials Research Proceedings 1 (2016) 171-174   doi: http://dx.doi.org/10.21741/2474-395X/1/43

Experimental Results and Discussion

The effect of the applied grid voltages on the charging currents measured on polyimide thin films during negative corona charging was studied. The difference between voltages of tip and grid V_p-V_g = 4 kV, the distance between grid and sample $d_{grid/film}$ = 5 mm and the distance between tip and grid $D_{tip/grid}$ = 1 cm.

In Fig. 1 is presented the variation of the charging current recorded on a polyimide film as a function of the charging time with negative grid voltage of about 800 V. We find that the electric current decreases with the charging time (the current decreases from -1.12.10^{-7} to -4.3.10^{-8} A after 150 s of the charging corona). This current evolution presented two distinct variation zones: at short time, a rapid current transient is observed followed by more gradual variation for long time, and then stabilization at an almost constant value.

Fig. 2 presents the variation of the charging current as a function of the charging time for different values of grid voltages varying from -400 to -2800 V. We note that the charging currents exhibit a decrease over charging time and their initial amplitudes increase with increasing the value of the applied grid voltage (the value of the initial current measured is about -5.9.10^{-8} A with a grid voltage of -400 V and increases around -1.7.10^{-6} A for a grid voltage of -2800 V). We observed in Fig. 2 that the evolution of the charging current depending on the charging time does not maintain the same appearance when varying the applied grid voltage. This difference of the charging current evolution is probably due to the difference of the amount of charges deposited on the polyimide film, when varying the grid voltage [4, 6].

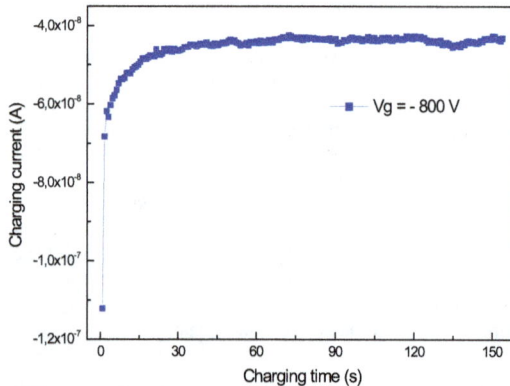

Fig. 1. Evolution of charging current as a function of charging time.

Dielectric Materials and Applications: ISyDMA'2016   Materials Research Forum LLC
Materials Research Proceedings **1** (2016) 171-174   doi: http://dx.doi.org/10.21741/2474-395X/1/43

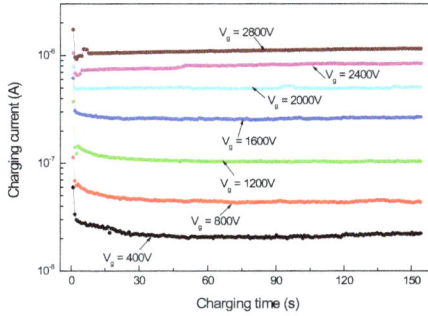

Fig. 2. Evolution of charging current as a function of charging time for different charging levels
of polyimide films.

In Fig. 3 is presented the decay rate variation of the surface potential as a function of the initial surface potential values varying from 407 to 2575 V. It appears that the surface potential decay rate increases non-linearly with the initial surface potential. For the low values of initial potential, the decay rate is relatively low. On the other hand, for the high values of initial potential, the decay rate increase rapidly. This behavior is directly related to the internal electric field, which is much larger when the initial surface potential is high [4].

The results of the surface potential decay measurements are in good agreement with those of the charging current measurements. Indeed, we have shown that increasing the grid voltage induces an increase in current level which increases the surface potential decay rate when the initial potential increases. In general, a charge that moves slowly induce a very low current, while with surface potential measurement a longer time is necessary to detect any effect of slow charge movement on surface potential.

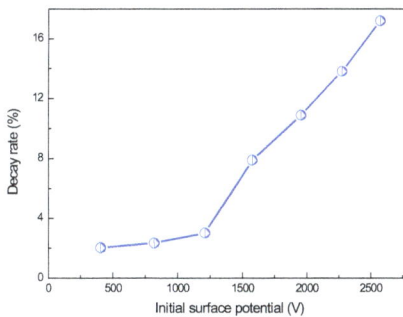

Fig. 3. Decay rate variation of polyimide films with initial surface potential.

Dielectric Materials and Applications: ISyDMA'2016   Materials Research Forum LLC
Materials Research Proceedings 1 (2016) 171-174   doi: http://dx.doi.org/10.21741/2474-395X/1/43

Conclusion

The experimental study of the current charging evolution by corona discharge in polyimide film showed that the charging currents present a decrease with charging time and their amplitudes increase with increasing the applied grid voltage. The study of surface potential decay shows that increasing the charging level significantly influences the behavior of the decay.

References

[1] T. Liang, Y. Makita and S. Kimura, "Effect of film thickness on the electrical properties of polyimide thin films", Polym., Vol. 42, pp. 4867-4872, 2001. http://dx.doi.org/10.1016/S0032-3861(00)00881-8

[2] T. Agag, T. Koga and T. Takeich, "Studies on thermal and mechanical properties of polyimide-clay nanocomposites", Polym., Vol. 42, pp. 3399-3408, 2001. http://dx.doi.org/10.1016/S0032-3861(00)00824-7

[3] J. Zha, G. Chen, Z. Dang and Y. Yin, "The influence of TiO2 nanoparticle incorporation on surface potential decay of corona-resistant polyimide nanocomposite films", J. Electrostatics, Vol. 69, pp. 255-260, 2011. http://dx.doi.org/10.1016/j.elstat.2011.04.001

[4] Z. Ziari, S. Sahli, A. Bellel, Y. Segui and P. Raynaud, "Simulation of Surface Potential Decay of Corona Charged Polyimide", IEEE Trans. Dielect. Electr. Insul., Vol. 18, N°. 5, pp. 1408-1415, 2011. http://dx.doi.org/10.1109/TDEI.2011.6032809

[5] D. Min, M. Cho, A.R. Khan and S. Li, "Charge transport properties of dielectric revealed by isothermal surface potential decay", IEEE Trans. Dielect. Electr. Insul., Vol. 19, N°. 4, pp. 1465-1473, 2012. http://dx.doi.org/10.1109/TDEI.2012.6260024

[6] Y. Zhuang, G. Chen, M. Rotaru, "Charge injection in gold ground electrode corona charged polyethylene film: surface potential decay and corona charging current measurement", 14th International Symposium on Electrets (ISE), 28-31 Aug. 2011, pp. 125-126. http://dx.doi.org/10.1109/ise.2011.6085014

[7] P. Molinié, P. Dessante, R. Hanna, T. Paulmier, B. Dirassen, M. Belhaj, D. Payan and N. Balcon, "Polyimide and FEP charging behavior under multienergetic electron-beam irradiation", IEEE Trans. Dielect. Electr. Insul., Vol. 19, N°. 4, pp. 1215-1220, 2012. http://dx.doi.org/10.1109/TDEI.2012.6259993

[8] D. Hastings and H. Garrett, Spacecraft-environment interactions, Cambridge, Cambridge University Press, pp. 132-177, 1996. http://dx.doi.org/10.1017/CBO9780511525032

Dielectric Materials and Applications: ISyDMA'2016
Materials Research Proceedings 1 (2016) 175-178

Materials Research Forum LLC
doi: http://dx.doi.org/10.21741/2474-395X/1/44

Evaluation of thermal characterization of PMMA/PPy composite materials

I. BOUKNAITIR*[1], N. ARIBOU[1], S. A. E. KASSIM[1], M. E. ACHOUR[1], L. C. COSTA[2]

[1]LASTID Laboratory, Physics Department, Faculty of Sciences, IbnTofail University, 14000 Kenitra, Morocco

[2]Physics Department, University of Aveiro, 3810-193 Aveiro, Portugal

*Corresponding author: E-mail: ilham.bouknaitir@gmail.com

Keywords: Specific Heat Capacity, DSC, Polypyrrole, Polymethymethacrylate, Composite, Conducting Polymer

Abstract. In this study, the specific heat Cp formed by PMMA with different concentrations of PPy are measured. The specific heat is measured by using differential scanning calorimetry method (DSC). We notice that in this nano-composite (PMMA/PPy) our sample containing 2 - 8 wt% of PPy filler material. For modelisation, we have used the mixture law for thermal properties. As the results show, we have obtained a good agreement between the thermal properties.

Introduction

The prediction of the properties of a composite, using those of their constituents is an important problem for theoretical and experimental physics. The mixing laws, which relate the macroscopic properties of composite materials to those of their individual constituents, have been a subject of study since the end of the 19[th]century.

Thermal properties of materials consisting of conducting particles embedded in an insulating matrix have been extensively studied in the past [1-8].

It is well known that specific heat capacity C_p, as a function of conducting particle concentration, undergoes an insulator/conducting transition. This occurs at a critical concentration where the particles contact each other and, as a consequence, a continuous electrical path of the doped particles is built throughout the polymer matrix. That is, when the filler content is low, the mean distance between conducting particles is large and the conductivity is restricted by the presence of the insulating matrix, but increasing the conductive phase content, the conducting particles get closer and, at a critical concentration, known as percolation threshold, the electrical properties are dominated by them [9].

In the mean field theory, the inclusion conducting phase and the insulating host phase play a symmetrical role. Each part of the system, inclusion and/or host particles, interacts with the other particles by the mean field created around each of them [10].Nevertheless, for a lot of composites this theory does not fit accurately the experimental results.

In this article, we present the results of experimental study on Specific heat capacity Cp of PPy particles in polymer composites. The experimental Cp values were compared with those obtained by Bueche law [11].

Theoretical model

The physical quantities as Cp of many composites have been discussed by mixture law, this lost suggested that the heat capacity Cp for different concentration, can be expressed as:

$$C_p = \phi C_{pf} + (1 - \phi) C_{ph} \tag{1}$$

Dielectric Materials and Applications: ISyDMA'2016   Materials Research Forum LLC
Materials Research Proceedings 1 (2016) 175-178   doi: http://dx.doi.org/10.21741/2474-395X/1/44

Where Cp is the specific heat capacity of PPy/PMMA, Cp_f is the specific heat capacity of PPy, Cp_h is the specific heat capacity of PMMA and ϕ is a concentration of PPy

Experimental

A   Materials

PPy powder was obtained by doping intrinsic PPy with tosylate anion TS-. The doping rate was controlled by the XPS technique and was found to be of the order of one sulfur (S) for four nitrogen (N), ie one tosylate ion TS- for four M pyrrole monomers [12]. The average grain size of PPy is in the range of 10–15μm. The two powders, PMMA and PPy, are mixed in several proportions and pressed at 5 000 Kg/cm^2 and 150 °C. After pressing, the samples are allowed to cool freely to room temperature to give solid disc-shaped samples ready for measurements. All discs had a diameter of 13 mm and 3-4 mm thickness. A set of samples was made with PPy weight concentrations wt. The DC conductivities of the PPy and the PMMA matrix are 54 and 3.10^{-15} $(\Omega.m)^{-1}$; and the densities are 1.20 and 1.14-1.20 g.cm^{-3} respectively, the percolation threshold ϕ_c for this series of samples was approximately 3.85 %. The glass transition temperature of PMMA polymer is $T_g \approx 115$ °C [13].

B   DSC Measurements

The DSC was carried out using a Shimadzu DSC-50 Differential Scanning Calorimeter programmed between 20 and 500°C at a heating rate of 10°C/min and under a nitrogen flow of 30 ml/min. The sample is placed in an aluminum cell or plate with a lid and the reference cell is empty. The measurements are taken under an atmosphere of nitrogen with a flow rate of 30ml/min, at temperatures between 20 and 800°C and a heating rate of 5°C/min. The sample was placed in a platinum cell.

Results and discussions

Different measurement methods exist for determining the specific heat (Cp) by differential scanning calorimetry (DSC). In DSC the heat flux is directly proportional to the specific heat, Cp can thus be directly calculated from the DSC signal. The following equation is used for calculating the specific heat:

$$C_p = |DSC|/m*\beta \tag{2}$$

Where C_p is the heat capacity, DSC is the heat flux, β is the Heating rate and m is the mass of the sample.

The specific heat values obtained are plotted against temperature in Figure 1, and compared to the mixture distribution in Figure 2. Where Cp_f and Cp_h have the respective value among the experimental data and the law of mixtures, eq(1) .

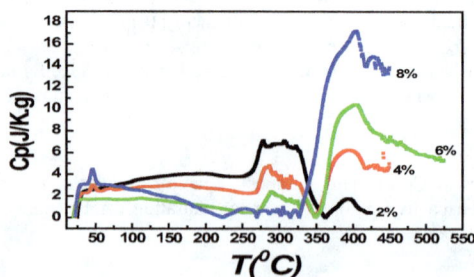

Fig. 1. Specific heat Cp for 4 samples of composite PPy/PMMA for various temperatures

Dielectric Materials and Applications: ISyDMA'2016   Materials Research Forum LLC
Materials Research Proceedings 1 (2016) 175-178   doi: http://dx.doi.org/10.21741/2474-395X/1/44

The measured specific heat capacity values of PMMA/PPy nanocomposites are given in Table 1 and Fig. 2.

As expected, a linear dependence of specific heat upon the weight filler fraction is observed for these nanocomposites, and the experimental values measured for the specific heat of PMMA/PPy nanocomposites follow quite well the calculated values from Eq.1, see Fig. 2 [14].

In figure 2 shows the variation of the specific heat capacity of the composite PMMA / PPy according to the concentration of PPy. From this figure there is a linear increase in the specific heat capacity of the composite according to the concentration of PPy. We notice that the theoretical values of specific heat capacity of PMMA / PPy composite are a good agreement with experimental values.

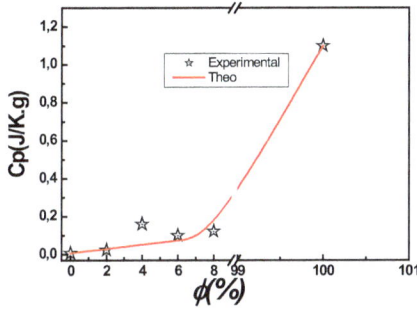

Fig. 2. Specific heat Cp of composite PPy/PMMA versus the concentration

Table I The Specific heat Cp of PPy/PMMA for several concentrations.

ϕ(%)	Cp (J/(kg.K) Experimental	Cp(J/(kg.K) Theorical
0	$7,11.10^{-3}$	$7,11.10^{-3}$
2	$2,4.10^{-2}$	$2,9.10^{-2}$
4	$1,6.10^{-1}$	$5,1.10^{-2}$
6	10^{-1}	$7,3.10^{-2}$
8	$1,22.10^{-1}$	$9,4.10^{-2}$
100	1,1	1,1

Conclusion

In this work, we report some thermal properties of the composites consisting of polypyrrole (PPy) particles in a polymethacrylate (PMMA) matrix. The study under specific heat capacity the nanocomposites was measured using deferential scanning calorimetry (DSC). We notice that specific heat capacity increased with increasing PPy particle content following the mixture law [15].

Reference

[1] J.C. Garland, D.B. Tanner, Electrical transport and optical properties in inho- mogeneous media, in: AIP Conference Proceedings 40, American Institute of Physics, New York, 1977.

[2] S. Torquato, Random Heterogeneous Materials: Microstructure and Macro- scopic Properties, Springer-Verlag, New York, 2002. http://dx.doi.org/10.1007/978-1-4757-6355-3

Dielectric Materials and Applications: ISyDMA'2016   Materials Research Forum LLC
Materials Research Proceedings 1 (2016) 175-178   doi: http://dx.doi.org/10.21741/2474-395X/1/44

[3] A.H.Sihvola,ElectromagneticMixingFormulasandApplications,Institutionof Electrical Engineers, London, 1999.

[4] M.E. Achour, Electromagnetic properties of carbon black filled epoxy polymer composites, in: C. Brosseau (Ed.), Prospects in Filled Polymers Engineering: Mesostructure, Elasticity Network, and Macroscopic Properties, Transworld Research Network, Kerala, 2008, pp. 129-174.

[5] J. Belattar, M.P.F. Graça, L.C. Costa, M.E. Achour, C. Brosseau, J. Appl. Phys. 107 (2010) 124111-124116. http://dx.doi.org/10.1063/1.3452366

[6] L.C. Costa, F. Henry, M.A. Valente, S.K. Mendiratta, A.S. Sombra, Eur. Polym. J. 38 (2002) 1495-1499. http://dx.doi.org/10.1016/S0014-3057(02)00044-7

[7] L. Flandin, T. Prasse, R. Schueler, W. Bauhofer, K. Schulte, J.Y. Cavaillé, Phys. Rev. B 59 (1999) 14349-14355. http://dx.doi.org/10.1103/PhysRevB.59.14349

[8] M.T. Connor, S. Roy, T.A. .Ezquerra, F.J.B. .Calleja, Phys. Rev. B 57 (1998) 2286-2294. http://dx.doi.org/10.1103/PhysRevB.57.2286

[9] M. El Hasnaoui, A. Triki, M.P.F. Graça, M.E. Achour, L.C. Costa, M. Arous, J. Non-Cryst. Solids 358 (2012) 2810-2815. http://dx.doi.org/10.1016/j.jnoncrysol.2012.07.008

[10] F. Henry, L.C. Costa, Phys. B: Condens. Matter. B 387 (2007) 250-258. http://dx.doi.org/10.1016/j.physb.2006.04.041

[11] I. Tavman , Y. Aydogdu , M. Kök , A. Turgut , A. Ezan. " Measurement of heat capacity and thermal conductivity of HDPE/expanded graphite nanocomposites by differential scanning calorimetry,"Archives of materials science and engineering, vol 50, pp. 56-60, July 2011.

[12] A. Belhadj Mohamed, J. L. Miane, H. Zangar. Polym. Int. 50, 773 (2001). http://dx.doi.org/10.1002/pi.686

[13] N. Aribou, A. Elmansouri, M. E. Achour, L. C. Costa, A. Belhadj Mohamed, A. Oueriagli, A. Outzourhit. Spectro. Lett. 45, 477 (2012). http://dx.doi.org/10.1080/00387010.2012.667035

[14] T. Chelidze,. Y.Gueguen. Pressure – induced percolation transitions in composites. J. Phys. D: Applied Phys. 1998. v.31. _ PP. 2877_2885.

[15] T. Chelidze,. Y.Gueguen. Electrical spectroscopy of porous rocks: a review – I. Theoretical models // Geophys. J. Int. 1999. Vol. 137. PP. 1 – 15.

Dielectric Materials and Applications: ISyDMA'2016
Materials Research Proceedings 1 (2016) 179-182

Materials Research Forum LLC
doi: http://dx.doi.org/10.21741/2474-395X/1/45

Numerical study of influence of protective materials on photovoltaic cell efficiency: comparison between glass and teflon

Abdellah LAAZIZI*[1], Hicham LABRIM[2], Hamid EZ-ZAHRAOUY[3], Khalid NOUNEH[4]

[1]Laboratory of Materials, Metallurgy and Process Engineering, ENSAM, University Moulay Ismail, Marjane II, BP: 15290 El Mansour, Meknès, Morocco

[2]National Energy Center of Sciences and Nuclear Techniques, CNESTEN – Rabat, Morocco

[3]Faculté des Sciences, BP 1014, Av Ibn Battouta, Rabat, Morocco

[4]Faculté des Sciences, B.P. 133, University Ibn Tofail, Kenitra, Morocco

*Corresponding author: E-mail: a.laazizi@ensam.umi.ac.ma

Keywords: Numerical Simulation, Photovoltaic, Efficiency, Explicit Scheme, Heat Transfer

Abstract. Predicting the temperature field in protective materials can contribute to enhance performance of photovoltaic panels. Materials as Glass or Teflon, on which protective layers have been made and are often employed to protect photovoltaic cells, have an effect on heat transfer through photovoltaic cells. Teflon is known as an excellent dielectric. In this context, heat transfer was simulated by using Finite Difference Method with explicit scheme. The temperature field was computed for the two different materials. First results showed that for both studied materials, the temperature field as well as the rate of heat transfer during daytime and of cooling during night are very different. With this knowledge, engineers can design new system to improve the efficiency of solar panels that operate in non-optimal conditions.

Introduction

Usually, photovoltaic panels are formed of many different layers (Fig. 1). PV panels are more efficient at lower temperatures; the current and voltage output of a PV panel is affected by changing weather conditions [1]. This is due to an increase in resistance of the circuit that results from an increase in temperature. Engineers design systems with active and passive cooling. Therefore, increasing the amount of generated electricity over the useable lifetime of the module PV can reduce the cost of electricity. Cooling the PV panels allows them to function at a higher efficiency and produce more power [2-5].

Hasana et al. [6] have developed a photovoltaic–phase change material (PV–PCM) system to reduce photovoltaic temperature dependent power loss. The system has been evaluated outdoors with two phase change materials; a salt hydrate and a eutectic mixture of fatty acids, capric acid–palmitic acid in two different climates of Dublin, Ireland and Vehari, Pakistan. Both the integrated PCMs maintained lower PV panel temperature than the reference PV panel. Salt hydrate maintained lower PV temperature than capric–palmitic acid at both the tested sites. The lower temperatures affected by the use of the PCMs prevented the associated PV power loss and increased PV conversion efficiencies.

For polycrystalline PV panels, if the temperature decreases by one Celsius degree, the voltage increases by 0.12 V then the temperature coefficient is 0.12 V/C. The general equation for estimating the voltage of a given material at a given temperature is (1):

$$V_{oc,amb} = \alpha_T \times (T_{STC} - T_{amb}) + V_{oc,rated} \tag{1}$$

Dielectric Materials and Applications: ISyDMA'2016   Materials Research Forum LLC
Materials Research Proceedings 1 (2016) 179-182   doi: http://dx.doi.org/10.21741/2474-395X/1/45

While it is important to know the temperature of a solar PV panel to predict its power output, it is also important to know the PV panel material because the efficiencies of different materials have varied levels of dependence on temperature.

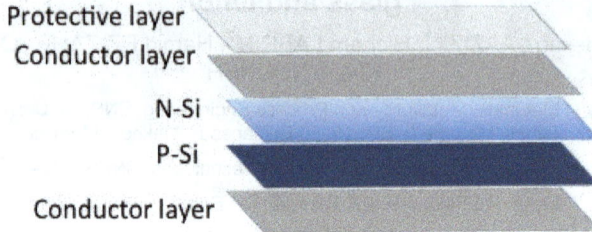

Fig. 1. Typical photovoltaic panel components.

In this project, we proposed to study the effect of temperature field on photovoltaic panel efficiency, especially on protective layer of panel component often is made of Glass or TEFLON. This comparison will help us to choose the material that will not heat up rapidly during exposure of PV panel. This solution can be considered as passive solution.

Heat transfer modelling
In this study, the daily solar radiation was taken as $\varphi = 5.3$ KWh/m^2 which represents the average value in Morroco. The governing equation was energy conservation equation (2). The upper part of the cell, which is in glass, shalt be in contact with air, which will create a heat exchange by convection (3). The finite-difference method was used to solve (2) and (3). The next step consists of the discretization of space and time variables. The explicit finite difference scheme built an iterative calculation in time and has to respect the following relation in order to be stable (5). Therefore, the choice of Δt is limited to it [7]. The heat conduction effects has been developed and simulated in MATALB environment.

$$\rho Cp \frac{\partial T}{\partial t} = k\left(\frac{\partial^2 T}{\partial x^2} + \frac{\partial^2 T}{\partial z^2}\right) \qquad (2)$$

$$z = 0 \longrightarrow -k\frac{\partial T(t,x,0)}{\partial z} = -h(T(t,x,0) - T_0) + S_0 \qquad (3)$$

$$T(t+1,j,k) = T(t,j,k) +$$

$$\frac{\Delta t\, k}{\rho C_p}\left[\begin{array}{c} \dfrac{T(t,j-1,k) - 2T(t,j,k) + T(t,j+1,k)}{\Delta x^2} + \\[2mm] \dfrac{T(t,j,k-1) - 2T(t,j,k) + T(t,j,k+1)}{\Delta z^2} \end{array}\right] \qquad (4)$$

$$\frac{1}{\rho C_p}\Delta t\left[\frac{1}{\Delta x^2} + \frac{1}{\Delta z^2}\right] \le \frac{1}{2} \qquad (5)$$

Dielectric Materials and Applications: ISyDMA'2016   Materials Research Forum LLC
Materials Research Proceedings **1** (2016) 179-182   doi: http://dx.doi.org/10.21741/2474-395X/1/45

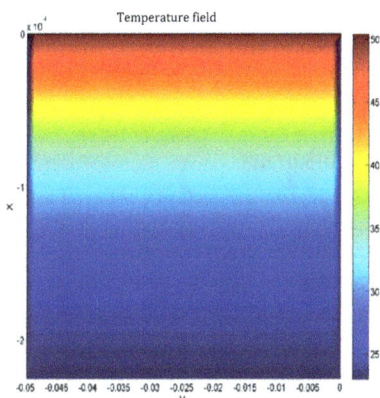

Fig. 2. Simulated temperature field in glass after 10 h

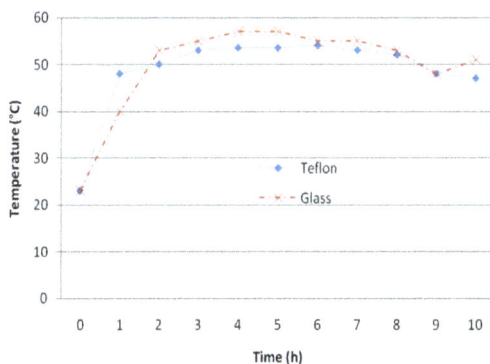

Fig. 3. Evolution of maximum temperature during a day for Teflon and Glass

Conclusion

In this work, a numerical model of the heat transfer was developed to calculate temperature field through two different protective materials used in photovoltaic panels Glass and Teflon. The numerical model solved the energy conservation equation by using a finite–difference method with the explicit scheme. Transfer by convection in panel' surface was taken into account. This model allowed us to follow the evolution of the temperature filed in PV panel during all daytime. These results enhanced our decision for the choice of protective materials and even for other PV component.

References

[1] J.S. Griffith, N.S. Rathod, and J. Paslaski, "Some tests of flat plate photovoltaic module cell temperatures in simulated field conditions", Proc. 15th IEEE Photovoltaic Specialists Conf., Kissimmee, FL, 1981, pp.822-830.

Dielectric Materials and Applications: ISyDMA'2016   Materials Research Forum LLC
Materials Research Proceedings 1 (2016) 179-182   doi: http://dx.doi.org/10.21741/2474-395X/1/45

[2] J.S. Cashmore et al., "Improved Conversion Efficiencies of Thin-Film Silicon Tandem (MICROMORPH) Photovoltaic Modules," J. Solar Energy Materials & Solar Cells 144, pp. 84–95, 2016. http://dx.doi.org/10.1016/j.solmat.2015.08.022

[3] A. Cortes et al., "Numerical evaluation of the effect of photovoltaic cell installation on urban thermal environment," J. Sustainable Cities and Society 19, pp. 250–258, 2015. http://dx.doi.org/10.1016/j.scs.2015.07.012

[4] A. Bai et al, "Technical and Economic Effects of Cooling of Monocrystalline Photovoltaic Modules under Hungarian Conditions," Renewable and Sustainable Energy Reviews 60, pp. 1086–1099, 2016. http://dx.doi.org/10.1016/j.rser.2016.02.003

[5] E. Urrejolaa et al., "Effect of Soiling and Sunlight Exposure on the Performance Ratio of Photovoltaic Technologies in Santiago, Chile," Vol. 114, pp. 338–347, 15 April 2016.

[6] A. Hasana et al., "Increased Photovoltaic Performance Through Temperature Regulation by Phase Change Materials: Materials Comparison in Different Climates," J. Solar Energy, Vol. 115, pp. 264–276, May 2015. http://dx.doi.org/10.1016/j.solener.2015.02.003

[7] A. Laazizi, B. Courant, F. Jacquemin and H. Andrzejewski, "Applied Multi-Pulsed Laser in Surface Treatment and Numerical–Experimental Analysis," J. Optics & Laser Technology 43, pp. 1257–1263, 2011. http://dx.doi.org/10.1016/j.optlastec.2011.03.019

Dielectric Materials and Applications: ISyDMA'2016
Materials Research Proceedings **1** (2016) 183-185

Materials Research Forum LLC
doi: http://dx.doi.org/10.21741/2474-395X/1/46

Analysis of the relationship between the distribution of a dielectric layer on a nano-tip apex and the distribution of emitted electrons

A.M. AL-QUDAH, Marwan S. MOUSA*

Department of Physics, Mu'tah University, Al-Karak, Jordan

*Corresponding author: E-mail: marwansmousa@yahoo.com & mmousa@mutah.edu.jo

Keywords: Electron Emission, Nano-Tip Apex, SEM Images, Dielectric Layer

Abstract. This paper analyses the relationship between the distribution of a dielectric layer on the apex of a metal field electron emitter and the distribution of electron emission. Emitters were prepared by coating a tungsten emitter with a layer of epoxylite resin (Clark Electromedical Instruments Epoxylite resin). A high-resolution scanning electron microscope was used to monitor the emitter profile and measure the coating thickness. Field electron microscope studies of the emission current distribution from these composite emitters have been carried out. The study found a correlation between the thickness distribution of the dielectric layer on the emitter apex and the distribution of electron emission. When the thickness distribution of the dielectric layer on the emitter apex is uniform and smooth, the distribution of electron emission takes the form of a bright single emission spot. When the thickness distribution of the dielectric layer is irregular, the electron emission image exhibits several emission spots.

Introduction

Field electron emission (FE) is the emission of electrons from the surface of a cathode under the influence of a high applied electrostatic field (typically about 3 V/nm) [1-2]. The field emitter is particularly attractive as an electron source, due to its suitable emission properties and simple operating principle [3]. Tungsten [4] is still one of the materials which are most frequently used for manufacturing field electron emitter tips. [5-7]. In manufacturing composite tungsten/CET-resin emitters (using Clark Electromedical Instruments Epoxylite resin), the role of the epoxylite resin has been to avoid degradation of the emitter tip resulting from ion sputtering processes during emission. The aims have been to obtain an electron emitter with long lifetime, and improve its emission characteristics [8]. During this work, six composite microemitters were prepared, with different apex radii ranging from 112 to 287 nm. Three emitters had a smooth substrate and a uniform dielectric layer, and produced bright single-spot field-electron-microscope (FEM) images, three emitters had irregular dielectric over layers, and produced multi-spot FEM images.

Materials and methods

The tungsten emitters used here were electrolytically etched from 0.1 mm high-purity (99.95%) tungsten wires (produced by Good Fellow Metals Company), using a 2M solution of sodium hydroxide (NaOH). The etched sample was then cleaned, to remove any residues of the NaOH solution on the tip surface, by immersing the sample in distilled water and subjecting it to an ultrasonic bath for a few minutes. Coating the emitters with epoxylite resin (Clark Electromedical Instruments (CEI-resin)) involves two steps. First, a tip is dipped into the resin very slowly, in the following way. A sample holder that keeps the samples vertical is mounted on a trolley that moves vertically and lowers/withdraws the sample into/from a flask of epoxylite resin, whilst keeping the 90° angle between the surface of epoxylite resin and the tip. Second, the coated emitters are carefully transferred to an oven and subjected to a curing cycle of thirty minutes at 100 °C to drive off the solvents, followed by another thirty minutes at 185 °C to complete the curing of the resin [9]. The composite emitter is then mounted in a standard field electron microscope (FEM) with an emitter-

Dielectric Materials and Applications: ISyDMA'2016   Materials Research Forum LLC
Materials Research Proceedings **1** (2016) 183-185   doi: http://dx.doi.org/10.21741/2474-395X/1/46

screen distance of 10 mm [10]. The emission images were photographed directly from a phosphor screen coated by tin-oxide layers. All experiments were performed under pressures of ~10^{-8} mbar. The radius of each emitter apex was measured from images taken in a 20 kV SEM, at magnifications up to 30000X.

Results

Typical multi-spot results obtained from these composite microemitters are shown below. These results include (i) SEM images [10], and (ii) FEM images taken by a digital camera. Fig. 1 shows a SEM image of an emitter with a non-uniform dielectric layer on its apex. This has apex radius $r_{SEM}=$ 141 nm and layer thickness $t_{SEM} = 21$ nm. The multi-spot FEM images are shown in Fig. 2.

Fig.1: SEM image for emitter 4 at magnification 30000X).

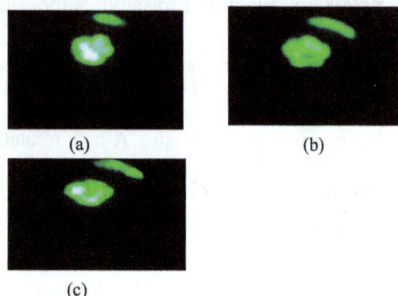

(a)

(b)

(c)

Fig. 2: FEM image structure for emitter 4 with: (a) 8 µA at 2500 V; (b) 7.4 µA at 2400 V; (c) 6.8 µA at 2300 V. Time separation between consecutive images is 15 minutes.

Discussion and conclusions

An electron source producing a single bright field emission image spot has been produced by employing a combination of a smooth emitter tip apex and a uniform epoxylite resin layer. The single-spot images appear to be associated with a uniform distribution of the resin layer on the emitter apex, and are thought to relate to the formation of a "conducting channel" through the resin. The multi-spot FEM images appear to be associated with an inhomogeneous resin-layer distribution and/or with substrate irregularities that could give rise to the formation of multiple emitting channels. The experimental conclusion is that the combination of a smooth emitter and a uniform resin layer generates a single bright emission spot. This makes these emitters interesting as potential electron sources.

References

[1] Mousa, M.S. (1996), Electron Emission from carbon fiber tips, Appl. Surf. Sci. 94/95, 129-135. http://dx.doi.org/10.1016/0169-4332(95)00521-8

[2] Forbes, R. G., Deane, J. H., Hamid, N., Sim, H. S. (2004), Extraction of emission area from Fowler-Nordheim plots. Journal of Vacuum Science and Technology B. 22, 1222-1226. http://dx.doi.org/10.1116/1.1691410

[3] Fischer, A., Mousa, M. S., and Forbes, R. G. (2013), Influence of barrier form on Fowler-Nordheim plot analysis. J. Vac. Sci. Technol. B 31, 032201. http://dx.doi.org/10.1116/1.4795822

[4] Young, R.D. (1959), Theoretical total -energy distribution of field emitted electrons, Phys. Rev. 113, 110-114. http://dx.doi.org/10.1103/PhysRev.113.110

[5] Latham, R.V., High Voltage Vacuum Insulation: The Physical Basis (London: Academic, 1981).

[6] Marrese, C.M. (2000), A review of field emission cathode technologies for electric propulsion systems and instruments, IEEE Aerospace Conference Proceedings, 4 85-98. http://dx.doi.org/10.1109/aero.2000.878369

[7] Marulanda, J.M. (Ed.), (Carbon Nanotubes InTech. 2010), pp. 311–340. ISBN: 978-953- 307-054-4.

[8] Mousa, M. S., Fischer A. and Mussa, K. O. (2012), Metallic and composite micropoint cathodes: Aging effect and electronic and spatial characteristics. Jordan J. Phys. 1, 21-26.

[9] Al-Qudah, A. M., Mousa, M. S., Fischer, A. (2015), Effect of insulating layer on the field electron emission performance of nano-apex metallic emitters. IOP Conf. Series: Materials Science and Engineering, 92, 012021. http://dx.doi.org/10.1088/1757-899x/92/1/012021

[10] Latham, R.V. and Salim, M.A. (1987) A microfocus cathode ray tube using an externally stabilised carbon- fiber field-emitting source, J. Phys. E. (Sci. Instr.), 20, 1083.

Dielectric Materials and Applications: ISyDMA'2016
Materials Research Proceedings **1** (2016) 186-189

Materials Research Forum LLC
doi: http://dx.doi.org/10.21741/2474-395X/1/47

Switch-on phenomena and field electron emission from MWCNTs encapsulated in glass

Marwan S. MOUSA*, Emad S. BANI ALI

Department of Physics, Mu'tah University, Al-Karak, Jordan

Corresponding author: E-mails:marwansmousa@yahoo.com & mmousa@mutah.edu.jo

Keywords: Cold Field Electron Emission, Fowler-Nordheim Plots, Field Electron Microscope, Carbon Nanotubes, Glass Microemitters

Abstract. Glass microemitters with internal carbon nanotubes show a switch-on emission current in the range of (1 to 20 μA) and stable saturation current. Nanocly ^{TM} NC 7000 Thin Multiwall Carbon Nanotubes with a high aspect ratio (>150) were used in this study. Measurements were made under ultra-high vacuum conditions at a base pressure of 10^{-9} mbar. Fowler-Nordheim plots of the current-voltage characteristics are shown, and current switch-on phenomena are noted.

Introduction

Carbon nanotubes (CNTs) are graphene sheets rolled into seamless hollow cylinders with diameters ranging from 1 nm to 50 nm [1]. They have the following properties: (1) high aspect ratio, (2) small radius of curvature at their tips, (3) high chemical stability and (4) high mechanical strength [1]. The Nanocyl ^{TM} NC7000 thin multiwall carbon nanotubes (MWCNTs) used in this work have 9.5 nm diameter and high aspect ratio (>150), and are made made by the process of chemical vapor deposition (CCVD). In field electron microscope (FEM) experiments, a high electric field is applied to the surface of a designed prepared glass micro point emitters [2], into which MWCNTs have been embedded. The distance from the emitter (cathode) to the FEM screen, which is coated by tin-oxide and phosphorus layers, was 10 mm.

Field electron emission (FE) is the emission of electrons induced from the surface of a material by a high electrostatic field (about 3 V/nm) [3, 4]. The field emitter is especially attractive as an electron source, due to its suitable emission properties and simple operating principle [5]. That CNTs should be good field emitters was apparent from the first article [6] (in 1995) that reported extremely low turn-on fields and high emission current densities.

In FE theory, cold field electron emission (CFE) is a regime where (i) the electrons in the emitting region are effectively in local thermodynamic equilibrium [7], and (ii) most electrons escape by deep tunneling from states close to the emitter's Fermi level [8]. The first scientist to put forward a presumed theory for CFE was Schottky, in 1923. Fowler and Nordheim (1928) developed the first appropriate theory for explaining CFE-related phenomena. This theory was later improved by Murphy and Good (1956) and many others [9]. CFE from emitters of moderate to large apex radius is described by approximate equations known as Fowler-Nordheim-type (FN-type) equations [2]. In this paper we present some techniques used to prepare glass microemitters with internal carbon nanotubes, and some results that are broadly consistent with earlier work on glass and composite microemitters [10, 11].

Experimental Arrangements

MWCNTs micro tips embedded in glass were used for this study. The composite emitters were prepared using a technique for pulling heated glass capillary tubes into fine points [10]. In this technique two bearings are located accurately on plates supported by three stainless steel rods fixed rigidly to the frame of the control unit. This frame is strong enough to serve as stable base for the instrument. The glass tube (OD = 1 mm , ID = 0.1 mm), where OD is outer glass tube diameter and ID is internal glass tube diameter, fits inside these bearings between the upper and lower chucks, with the furnace loop located around it. The lower chuck spindle sides vertically to pull down the

Dielectric Materials and Applications: ISyDMA'2016   Materials Research Forum LLC
Materials Research Proceedings 1 (2016) 186-189   doi: http://dx.doi.org/10.21741/2474-395X/1/47

glass tube under the gravity. As the temperature of the furnace is raised to the softening point of glass (1400 K for borosilicate glass) the tube neck lengthens [10]. The device is shown in Fig.1. The MWCNTs were entered into the opposite end of each glass tube so they would protrude at the tip with a wire plunger [11].

With the field electron microscope (FEM), the pre-bake-out pressure was ~10^{-7} mbar. This fell to ~10^{-9} mbar after the system was baked at about 200 °C for 12 h, in order to achieve ultra high vacuum (UHV) [12].

When the voltage applied to the emitter is increased slowly, a point is reached (at the switch-on voltage V_{SW}) where the emission current suddenly "switches-on" from a low value (about 1 nA) to a much higher saturated emission current I_{SAT}. Current-voltage (I-V) data are typically recorded using both direct (I vs V) plots and Fowler-Nordheim (FN) plots, which have the form $\ln\{J/V^2\}$ vs $1/V$, or $\log_{10}\{J/V^2\}$ vs $1/V$.

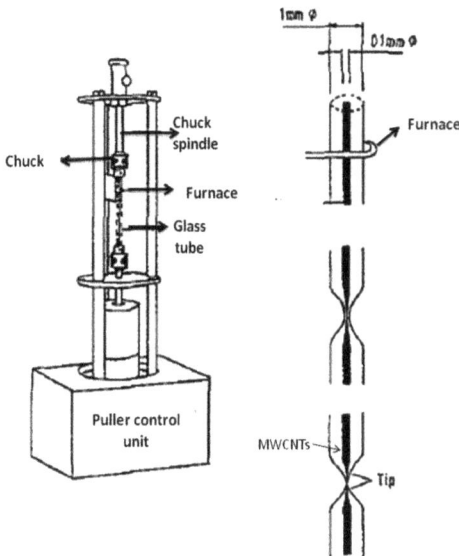

Fig. 1. A schematic diagram of the glass puller.

Results and Discussion

In addition to I-V plots and the related FN plots, electronic images have been recorded by a digital camera, in order to study the spatial distribution and stability of the emission current. Optical micrographs with magnification (50X) (e.g., Fig. 2(a)) allow us the possibility of studying the tip shape profile.

The I-V plot for the Fig. 2(a) tip is given in Fig. 2(b). The

(a)

(b)

(c)

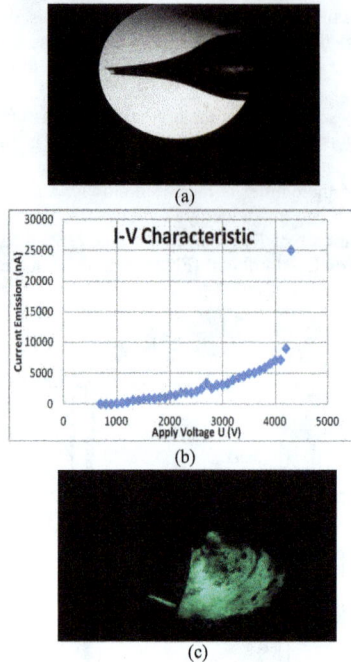

Fig. 2. (a) Optical micrograph of tip (CNT-A4), at 50X times magnification, allowing studies of the emitter (tip) profile. (b) The I-V plot of this tip shows a switch-on phenomenon, at the voltage and current values $V_{SW} = 4300$ V, $I_{SW} = 25$ μA. (c) FEM image after switch-on.

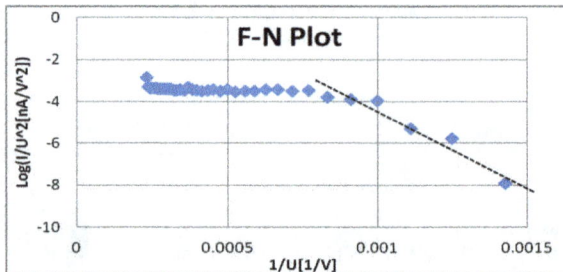

Fig. 3. The emission characteristics of this tip show the saturation effect for high emission current. The dashed line fitted to the low-current data gives a linear FN plot with slope −5970 decade V.

The related FN plot, in Fig. 3, exhibits saturation for high emission currents; at low fields the plot is linear, with slope about -5967 decade V. The emission image os this tip is shown in Fig. 2(c). Often, mages showed large bright spots, spread over most of the screen.

Materials Research Forum LLC
doi: http://dx.doi.org/10.21741/2474-395X/1/47

Conclusions

Emitter structures consisting of open end glass tips with internally filled MWCNTs have been fabricated. The emission characterization of these microemitters was studied using a FEM. Such emitters show several promising characteristics, e.g., high brightness and high maximum emission current.

References

[1] W.B. Choi, D.S. Chung, J.H. Lee, J.E. Jung, N.S. Lee, G.S. Park, and J.M. Kim, "Fully sealed, high-brightness carbon-nanotube field-emission display," Appl. Phys. Lett. vol. 75, pp. 3129-3131, 1999.

[2] R.G. Forbes, "Extraction of emission parameters for large-area field emitters, using a technically complete Fowler-Nordheim-type equation," Nanotechnology. vol. 23, 095706, 2012.

[3] M.S. Mousa, "Electron emission from carbon fiber tips," Appl. Surf. Sci., vol. 94/95, pp. 129-135, 1996.

[4] R.G. Forbes, J.H.B. Deane, N. Hamid, and H.S. Sim, "Extraction of emission area from Fowler-Nordheim plots," J. Vac. Sci. Technol B., vol. 22, pp. 1222-1226, 2004.

[5] A. Fischer, M.S. Mousa, and R.G. Forbes, "Influence of barrier form on the extraction of information from Fowler-Nordheim plots," J. Vac. Sci. Technol. B, vol. 31, 032201, 2013.

[6] L.A. Chernozatonskii, Yu.V. Gulyaev, Z.Ja. Kosakovskaja, N.I. Sinitsyn, G.V. Torgashov, Yu.F. Zakharchenko, E.A. Fedorov, and V.P. Val'chuk, "Electron field emission from nanofilament carbon films," Chem. Phys. Lett., vol. 233, pp. 63-68, 1995.

[7] M.S. Mousa, "A new perspective on the hot-electron emission from metal-insulator microstructures," Surf. Sci., vol. 231, pp. 149-159, 1990.

[8] M.S. Mousa and M. Al Share, "Study of the MgO-coated W emitters by field emission microscopy," Ultramicroscopy,vol. 79,pp. 195-202, 1999.

[9] R. G. Forbes, J. H. B. Deane, A. Fischer and M. S. Mousa, "Fowler-Nordheim plot analysis: a progress report," Jo. J. Phys., vol. 8, pp. 125-147, 2015.

[10] M. S. Mousa and D.B. Hibbert, "Analysis of some properties of metal-glass microemitters subjected to strong electric field," Appl. Surf. Sci. vol. 67, pp. 59-65, 1993.

[11] M.S. Mousa, "Effect of an internally conductive coating on the electron emission from glass tips," Surf. Sci. vol. 246, pp. 79-86, 1991.

[12] R.V. Latham and M.S. Mousa, "Hot electron emission from composite metal-insulator micropoint cathodes," J. Phys. D: Appl. Phys., vol. 19, pp. 699-713, 1986.

Dielectric Materials and Applications: ISyDMA'2016   Materials Research Forum LLC
Materials Research Proceedings 1 (2016) 190-192   doi: http://dx.doi.org/10.21741/2474-395X/1/48

Antenna gain and directivity improvement by using metamaterial unit-cell inspired in superstrate with zero refractive index

Mourad ELHABCHI*[1], Mohamed Nabil SRIFI[2], Rajae TOUAHNI[1]

[1]LASTID Laboratory, Department of physics, Faculty of Sciences, Ibn Tofail University, Kenitra, Morocco

[2]Electronics and Telecommunication Systems Research Group, National School of Applied Science (ENSA), Ibn Tofail University, Kenitra, Morocco

*Corresponding author: E-mail: mouradelhabchi@hotmail.fr

Keywords: Zero Index Metamaterial (ZIM), UWB Elliptical Antenna, Antenna Gain, Antenna Directivity, Effective Permittivity, Effective Permeability

Abstract. In this paper, a Zero Index Metamaterial (ZIM) is proposed for antenna gain and directivity enhancement. This ZIM is developed, characterized and employed for elliptical UWB operating on the frequency range 3.1Ghz-20GHz. The unique property of the metamaterial gathers the wave radiated from the antenna and collimates it towards the normal direction when used as a superstrate, the effective permittivity and permeability of the (ZIM) are designed to synchronously approach zero. This leads the ZIM to having an effective wave impedance matching with air and near-zero index simultaneously

Introduction

Metamaterials have many extraordinary properties, such as negative index, backward wave, inverse Doppler effect and backward Cerenkov radiation [1]. Metamaterials have been used to improve the performance of various microwave devices, especially antennas [2]. The computation method of retrieving the constitutive effective parameter, including permittivity, permeability and refraction index of metamaterial was discussed [3],[4]. Among the different kinds of metamaterials, near-zero index metamaterials (NZIM) have the ability to control the direction of antenna radiation [5]. A low profile high directivity antenna is designed [5], where Artificial Magnetic Conductor (AMC) substrate is used to reduce the height of antenna, and near-zero index metamaterial lens (NZIML) is used to increase the directivity of antenna. In 2009, Ma et al[6], was theoretically stated that an anisotropic ZIML with proper design had good impedance matching with air, so that anisotropic ZIMLS can efficiently enhance the antenna gain. Also Cheng et al. realized that ZIMLS composed of split ring resonator (SRR) array helps to achieve enhanced directivity for a line source [7]. However there is constraint in the operating bandwidth. Many research groups have realized low/zero-index metamaterials (LIM/ZIM) through experiments [8],[9].

According to Snells law[10], when the ray is incident from inside the LIM/ZIM into free space, the angle of refraction will be close to zero, so the refracted rays will be normal to the interface. This property provides a unique method of controlling the direction of radiation. A common method is to put a frequency selective surface (FSS) or an electromagnetic band-gap (EBG) metamaterial on top of a metallic ground plane containing a small antenna so as to form a Fabry-Perot resonant cavity, then the gain of the antenna is enhanced [1].

In this paper, an elliptical shaped monopole antenna with defected ground structure for UWB applications, which suffers from three major disadvantages: narrow bandwidth, low gain and lower power handling capability loaded by one-layer unit cell zero refrative- index metamaterial (ZIM) based high-gain and directivity superstrate will be developed.

Dielectric Materials and Applications: ISyDMA'2016   Materials Research Forum LLC
Materials Research Proceedings 1 (2016) 190-192   doi: http://dx.doi.org/10.21741/2474-395X/1/48

Antenna design

Fig. 1 shows the geometry of proposed metamaterial elliptical UWB antenna. The antenna is constructed on FR4 substrate with permittivity of 4.3 and dielectric loss tangent of 0.025 the dimension of substrate are 26×30×1mm³.This antenna is connected to a 50 SMA connector for signal transmission.The unit-cellmetamaterial consist of5×5 element.

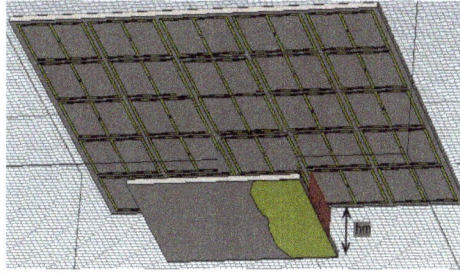

Figure 1 :the geometry of metamaterial antenna

Simulated results

Fig.2. and fig.3 shows the simulated gain of the reference UWB antenna and those of our proposed metamaterial antenna.Simulated results show that the gain of reference UWB antenna at 4Ghz is 2.36dB, have been enhanced on the proposed metamaterial antenna at the same frequency (4GHz) and are improved to 8.43dB. So, the antenna gain is enhanced by 6.07dB by using metamaterial.

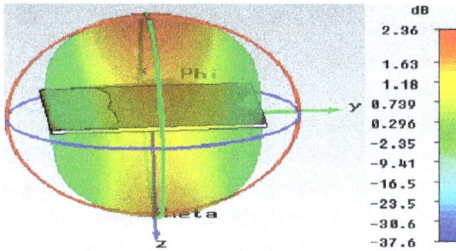

Figure2 : antenna without metamterial

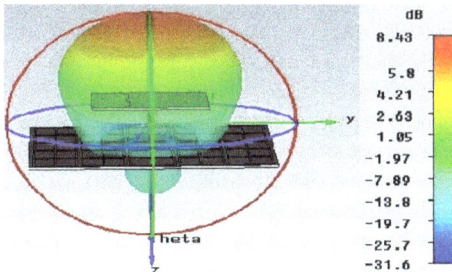

Figure 3 : antenna with metmaterial

Dielectric Materials and Applications: ISyDMA'2016   Materials Research Forum LLC
Materials Research Proceedings 1 (2016) 190-192   doi: http://dx.doi.org/10.21741/2474-395X/1/48

Summary results

TABLE I: Comparison antenna performances at 4GHz.

Parameter	antenna without metamterial	metamaterial antenna
Gain	2.36dB	8.43dB
Directivity	3.44dB	8.86dB
Efficiency	68.60%	95.14%

Conclusion

The simulation results of the proposed Single layer elliptical uwb antenna loaded by metamaterial superstrate giving exceptional results.This MTM shaped was designed to compose a near zero refractive index metamaterial , concentrates the radiation energy in direction close to the normal of the metamaterial structure. Consequently, the gain and directivity of antenna will be enhanced, and the the 3-dB beamwidth will be sharper.

References

[1] J. Lu, T. M. Gregorczyc, Y. Zhang, J. Pacheco, B. Wu, J. A. Kong, and M. Chen "Cerenkov radiation in materials with negative permittivity and permeability." Opt. Express, vol. 11, pp. 723–734, 2003. http://dx.doi.org/10.1364/OE.11.000723

[2] Y. Dong and T. Itoh," Metamaterial-based antennas", Proceedings of theIEEE, Vol. 100, No. 7, pp. 2271-2285, 2012.

[3] X. Chen, T. M. Grzegorczyk, Bae-Ian Wu, J. Pacheco, Jr. and J A. Kong, "Robust method to retrieve the constitutive effective parameters of metamaterials," Phys. Rev. E, vol. 70, 016608, 2004.

[4] D. R. Smith, D. C. Vier, T. Koschny, and C. M. Soukoulis, Electromag- netic parameter retrieval from inhomogeneous metamaterials, Physical Review E, vol. 71, no. 3, Article ID 036617, pp. 111, 2005.

[5] Enoch, S., G. Tayeb, P.Sabouroux, N. Guerin, and P. Vincent, "A metamaterial for directive emission", Physics Review Letters, vol. 89, no. 21, pp.1-4, 2002. http://dx.doi.org/10.1103/PhysRevLett.89.213902

[6] Ma, Y. G., P. Wang, X. Chen, and C. K. Ong, "Near-field plane-wave-like beam emitting antenna fabricated by anisotropic Metamaterial ", Applied Physics Letters, Vol. 94, No. 4, 044107(3), 2009.

[7] Ma, Y. G., P. Wang, X. Chen, and C. K. Ong," Near-field plane-wave-like beam emitting antenna fabricated by anistropic metamaterial", Applied Physics Letters, vol.94, no.4, 2009. http://dx.doi.org/10.1063/1.3077128

[8] G. Lovat, P. Burghignoli, F. Capolino, D. R. Jackson, and D. R. Wiltton, "Analysis of directive radiation from a line source in a metamaterial slab with low permittivity", IEEE Trans. Antennas. Propag., vol. 54, no. 3, pp. 10171030, Mar. 2006. http://dx.doi.org/10.1109/TAP.2006.869925

[9] G. Lovat, P. Burghignoli, F. Capolino, and D. R. Jackson, "Combinations of low/high perimittivity and/or permeability substrates for highly directive plannar metamaterial antennas", IET Microw Antennas Propag., vol. 1, pp. 177183, 2007. http://dx.doi.org/10.1049/iet-map:20050353

[10] RICARDO MARQUE´S, FERRAN MARTI´N, MARIO SOROLLA" Metamaterials with Negative ParametersTheory, Design, andMicrowave Applications"wiley interscience 2007.

Dielectric Materials and Applications: ISyDMA'2016   Materials Research Forum LLC
Materials Research Proceedings 1 (2016) 193-198   doi: http://dx.doi.org/10.21741/2474-395X/1/49

First principles studies of electronic and optical properties of Ternary semiconductors AgAlX$_2$ (X = S, Se, Te)

M.OULEDALI*[1], S. LOUHIBI-FASLA[2], B. AMRANI[3]

[1]Institute of Science and Technology, Department of Materials Science , University Centre of Tamanrasset, Algeria

[2]Laboratory of Micro and Nano-Physics (LaMiN), ENP-Oran, Algeria

[3]Laboratory of Physics of Thin Films and Materials for Electronics (LPC2ME), University of Oran Es Senia, Algeria

*Corresponding author: E-mail: med_ouled@yahoo.fr

Keywords: Chalcopyrites, FP-LAPW+lo, mBj, Structural Properties, Electronical Properties, Optical Properties

Abstract. Using first-principle, the electronic and optical properties of chalcopyrite AgAlX2 (X=S, Se, Te) are investigated. These materials have recently shown great interest and applications such as photovoltaic conversion. The calculations have been performed using the density functional theory (DFT) as implemented within the full potential linearized augmented plane wave plus local orbital (FP-LAPW+lo) method. For exchange and correlation energy treatment, we employed the generalized gradient approximation (GGA) proposed by Perdew et al. To calculate the accurate band structure, recently modified Becke Johnson (mBJ) potential was suggested as an alternative. Optical properties reveal that these compounds are suitable candidates for optoelectronic devices in the visible and ultraviolet (UV) regions. The results obtained were compared with experimental data from the literature.

Introduction

Semiconducting chalcopyrites with the formula $A^I B^{III} C_2^{VI}$ (where A^I=Ag; B^{III}=Al and C^{VI}=S, Se, Te) have been investigated because of their potential application in opto-electronic and nonlinear optical devices. These compounds are promising candidates for solar cells [1], photovoltaic detectors, light-emitting diodes [2], modulators, filters such as optical light eliminator filters [3] and optical frequency conversion applications in solid-state-based tunable laser systems [4]. Very few works have appeared in the literature in which the group I element is Ag [5, 6]. Horinaka and his group [3] have studied the optical properties of AgAlX$_2$ (X = S, Se, Te) and have shown that the optical isotropy properties of these materials satisfy the necessary conditions for optical light elimination filters.

S. Mishra and B. Ganguli [7] have studied the structural and electronic properties of AgAlX$_2$ (X= S and Se) chalcopyrite semiconductors using the TB-LMTO method. Jayalakshmi and Mageswari [8] have studied the dielectric constant and refractive index of AgAlX$_2$ (X=S, Se and Te) by using the FP-LMTO method. Verma et al. [9,10,11,12] have studied the several properties of the AgAlX$_2$ by semi-empirical models.

In this paper, we present a complementary effort in order to provide a basis for understanding future device concepts and applications.

Computational details

Full potential linearized augmented plane wave (FP-LAPW+lo) [13, 14] method within the generalized gradient approximation (GGA) [15], Engel-vosko generalized gradient approximation (EV-GGA) [16] and modified Becke-Johnson (mBJ) [17] exchange potential, as implanted in WIEN2K [18] code, are utilized to investigate the AgAlX$_2$ (S, Se and Te). Band gaps and electronic dispersion curves are usually underestimated by GGA while mBJ exchange potential presents very accurate electronic band structures for various types of semiconductors.

In the FP-LAPW method, the unit cell is divised into atomic sphere and interstitial regions. Inside the muffin-tin sphere, the wave function is expanded in spherical harmonics and the maximum l quantum number for the wave function expansion inside atomic spheres was confined to $l_{max}=10$. The plane wave cutoff of $K_{max}=7.0/R_{MT}$ is chosen for the expansion of the wave functions in the interstitial region while the charge density is Fourier expanded up to $G_{MAX} = 12 \ (Ryd)^{1/2}$. The calculations are performed using a mesh of 500 k-points in the Brillouin zone (BZ).

Results and discussion

Structural properties
These parameters can be optimized by calculating forces on the nuclei and using the damped Newton scheme [19] to find equilibrium atomic positions. The force on each atom after relaxation decreased to less than 1.0 mRyd au^{-1} for both crystals considered in this work. The total energies were then calculated as a function of volume and the obtained data fitted to the Murnaghan equation of state [20]. The structural parameters and bulk properties are listed in Table 1.

TABLE I: The calculated data is compared to the experimental and other theoretical data.

Compounds	Method	a(A°)	c(A°)	U	B (GPa)	B'
AgAlS$_2$	Present work	5.802	10.506	0.294	67.1396	4.3866
	Experiment [21]	5.72	10.13	0.290	-	
	Theo Ref [22]	5.73	10.3	0.3	-	
	Theo Ref (LDA) [7]	5.48	10.90	0.265		
AgAlSe$_2$	Present work	6.054	11.104	0.285	55.8458	4.3705
	Experiment [21]	5.95	10.75	0.270	59.48[2 3]	
	Theo Ref (LDA) [7]	5.78	11.52	0.263		
AgAlTe$_2$	Present work	6.432	12.284	0.268	44.9523	4.4230
	Experiment [21]	6.29	11.83	0.260	49.94[2 3]	
	Theo Ref (LDA) [7]	6.22	12.28	0.261		

***a* and *c*:** *The lattice parameters,*
U: *The anion displacement,*
B: *Bulk modulus*

Electronic properties
The electronicproperties of AgAlX$_2$ compounds arediscussed bycalculating band structures. The calculated band structuresof the compoundsusingmBJscheme are presented in Fig. 1.

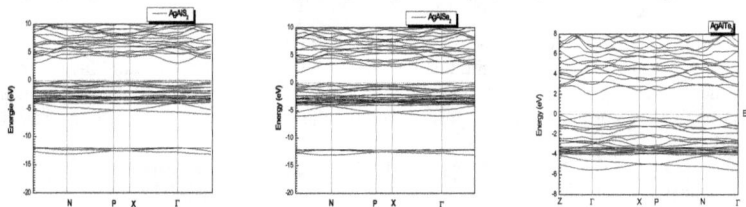

Fig. 1. Band structure for AgAlS$_2$, AgAlSe$_2$ and AgAlTe$_2$.

It is seen from the plots that the both the top and the bottom of the valence and conduction bands are positioned at the same symmetry points. The band gaps determined through GGA, EV–GGA and mBJ methods together with other theoretical and experimental results are given in Table 2.

Dielectric Materials and Applications: ISyDMA'2016   Materials Research Forum LLC
Materials Research Proceedings 1 (2016) 193-198   doi: http://dx.doi.org/10.21741/2474-395X/1/49

TABLE 2: Band gap values of chalcopyrite $AgAgX_2(X = S, Se, Te)$.

Compounds	Methods	XC	Eg (eV)
AgAlS2			
Present work		GGA-PBE	1.83
	FPLAW	GGA-EV	2.70
	(WIEN2k)	GGA-	3.13
Experiment [24]		MBJ	3.13
Theo Ref [7]			1.98
AgAlSe2			
Present work		GGA-PBE	1.16
	FPLAW	GGA-EV	1.84
	(WIEN2k)	GGA-	1.91
Experiment [24]		MBJ	2.55
Theo Ref [7]			1.59
AgAlTe2			
Present work		GGA-PBE	1.06
	FPLAW	GGA-EV	1.54
	(WIEN2k)	GGA-	1.94
Experiment [25]		MBJ	2.35
Theo Ref [7]			1.36

One can easily note that the computed band gap values obtained using mBJ match closely to experimental [24, 25] and other theoretical [7] data and are better than those calculated with GGA, EV–GGA.

Optical properties

In the present work, all the spectra are plotted with the incident radiation polarized both perpendicular and parallel to the tetragonal c-axis of the crystal. show the mBJ calculated real part of the dielectric function:

$$\varepsilon(\omega) = \varepsilon_1(\omega) + i\varepsilon_2(\omega) \tag{1}$$

which is related to the interaction of photons with electrons [26]. The imaginary part ε_2 (ω) of the dielectric function could be obtained from the momentum matrix elements between the occupied and unoccupied wave functions and is given by [27]:

$$\varepsilon_2(\omega) = \frac{4\pi^2 e^2}{V m^2 \omega^2} \sum \int \langle i|M|j \rangle f_i (1 - f_i) \times \delta\big(E_f - E_i - \hbar\omega\big) d^3k. \tag{2}$$

where ω is the electromagnetic radiation impinging the crystal, V is the unit cell volume, e and m are the charge and mass of the electron, M is the momentum operator in bracket notation, i and j are the initial and final states respectively, f_i the Fermi distribution function for the *ith* state and condition for the conservation of total energy is represented by $\delta(E_f - E_i - \hbar\omega)$ where, Ei and Ej are the energies of the electron in the *ith* and *jth* state.

The real part $\varepsilon_1(\omega)$ can be evaluated from $\varepsilon_2(\omega)$ using the Kramers–Kronig relations and is given by [28]:

$$\varepsilon_1(\omega) = 1 + \left(\frac{2}{\pi}\right) P \int_0^\infty \frac{\varepsilon_2(\omega)\omega' d\omega'}{\omega'^2 - \omega^2} \tag{3}$$

where*P* represents the principal value of the integral. All of the other optical properties, including the absorption coefficient $\alpha(\omega)$, the refractive index $n(\omega)$, the extinction coefficient $k(\omega)$, and the energy-loss spectrum $L(\omega)$, can be directly calculated from $\varepsilon_1(\omega)$ and $\varepsilon_2(\omega)$ [27,29].

Fig. 4 displays the real and imaginary parts of the electronic dielectric function $\varepsilon(\omega)$ spectrum for the photon energy ranging up to 40 eV, respectively. The real dielectric function $\varepsilon_1(\omega)$ gives the static

Dielectric Materials and Applications: ISyDMA'2016   Materials Research Forum LLC
Materials Research Proceedings 1 (2016) 193-198   doi: http://dx.doi.org/10.21741/2474-395X/1/49

dielectric constant in the zero frequency limit, which the main peaks of the real part for $AgAlS_2$, $AgAlSe_2$ and $AgAlTe_2$ are at 4.34, 3.37 and 2.43eV respectively. In these compounds there are three main peaks for $AgAlS_2$ and $AgAlTe_2$. The peaks are positioned around 4.34, 4.83 and 7.00 eV for $AgAlS_2$plus 2.43, 3.53 and 4.25 eV for $AgAlTe_2$. It is noted that both ε_1 and ε_2 display isotropic behaviors except the intermediate energies for $AgAlS_2$, $AgAlSe_2$ and $AgAlTe_2$ compounds. The zero frequency limits $\varepsilon_1(0)$ is the electronic part of the static dielectric constant. The values of $\varepsilon_1(0)$ for $AgAlS_2$, $AgAlSe_2$ and $AgAlTe_2$ are about 5.00, 5.81and 7.65 eV, respectively. Unfortunately, there are no available experimental data for comparison.

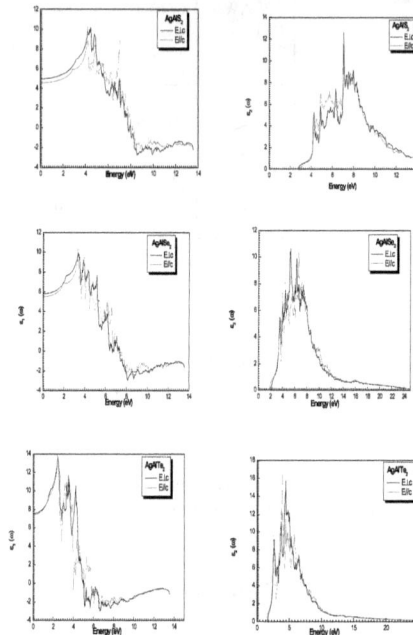

Fig. 2.Real part ($\varepsilon_1(\omega)$) and imaginary part ($\varepsilon_2(\omega)$) of the dielectric function AgAlX(X=S, Se, Te) compounds

Conclusion

In conclusion, we have performed systematic first-principles ab initio calculations for the structural, electronic, optical properties of chalcopyrite semiconductor $AgAlX_2$ using the density functional theory. The calculated structural parameters such as lattice constants, anion internal parameter (u) and Bulk modulus of our chalcopyrite are in good agreement with other available theoretical and experimentalwork. Analysis of the chalcopyrite band structure suggests a direct band gap semiconductor that is underestimated the experimental value;this result comes in agreement with all theoretical works using GGA calculations and improved by MBJ scheme. A substantial decrease of the band gap is found, which can improve photovoltaic conversion efficiency while the compress strain has no evident effect. This isalsoconfirmed by the increased visible light adsorption coefficient along c-axis direction under tensile strain. Hence it would be beneficial to verify experimentally the predictedresults which will hope to stimulate the succeeding studies.

References

[1] L.L. Kazmerski, Nuovo Cimento D 2 (1983) 2013. http://dx.doi.org/10.1007/BF02457903

[2] J.L. Shay, L.M. Schiavone, E. Buehier, J.H. Wernick, J. Appl. Phys. 43 (1972) 2805. http://dx.doi.org/10.1063/1.1661599

[3] H. Horinaka, S. Mononobe, N. Yamamoto, Jpn. J. Appl. Phys. 32 (1993) 109 (Suppl. 32–33). http://dx.doi.org/10.7567/JJAPS.32S3.109

[4] F.K. Hopkins, Laser Focus World 31 (1995) 83.

[5] S. Laksari, A. Chahed, N. Abbouni, O. Benhelal, B. Abbar, Comput. Mater. Sci. 38 (2006)223. http://dx.doi.org/10.1016/j.commatsci.2005.12.043

[6] L. Artus, Y. Bertrand, C. Ance, J. Phys. C: Solid State Phys. 19 (1986) 5937. http://dx.doi.org/10.1088/0022-3719/19/29/015

[7] S. Mishra, B. Ganguli / Solid State Communications 151 (2011) 523–528. http://dx.doi.org/10.1016/j.ssc.2011.01.024

[8] V. Jayalakshmi, S. Mageswari, B. Palanivel, AIP Conference Proceedings 147 2012, 1087. http://dx.doi.org/10.1063/1.4710385

[9] A.S. Verma, Philos. Mag. 89 (2009) 183. http://dx.doi.org/10.1080/14786430802593814

[10] A.S. Verma, R.K. Singh, S.K. Rathi, J. Alloys Compd. 486 (2009) 795. http://dx.doi.org/10.1016/j.jallcom.2009.07.067

[11] A.S. Verma, Solid State Commun. 149 (2009) 1236. http://dx.doi.org/10.1016/j.ssc.2009.04.011

[12] A.S. Verma, S.R. Bhardwaj, Phys. State Solidi B 243 (2006) 2858. http://dx.doi.org/10.1002/pssb.200642140

[13] G.K.H. Madsen, P. Blaha, K. Schwarz, E. Sjostedt, L. Nordstrom, Phys. Rev. B 64 (2001) 195134. http://dx.doi.org/10.1103/PhysRevB.64.195134

[14] K. Schwarz, P. Blaha, G.K.H. Madsen, Comput. Phys. Commun. 147 (2002) 71. http://dx.doi.org/10.1016/S0010-4655(02)00206-0

[15] Z. Wu, R.E. Cohen, Phys. Rev. B 73 (2006) 235116. http://dx.doi.org/10.1103/PhysRevB.73.235116

[16] E. Engel, S.H. Vosko, Phys. Rev. B 50 (1994) 10498. http://dx.doi.org/10.1103/PhysRevB.50.10498

[17] F. Tran, P. Blaha, Phys. Rev. Lett. 102 (2009) 226401. http://dx.doi.org/10.1103/PhysRevLett.102.226401

[18] P. Blaha, K. Schwarz, G.K.H. Madsen, D. Kvasnicka, J. Luitz, WIEN2k, An Augmented Plane Wave Plus Local Orbitals Program for Calculating Crystal Properties, Vienna University of Technology, Vienna, Austria, 2001; K. Schwarz, P. Blaha, G.K.H. Madsen, Comp. Phys. Commun. 147 (2002) 71. http://dx.doi.org/10.1016/S0010-4655(02)00206-0

[19] B. Kohler, S. Wilke, M. Scheffler, R. Kouba, C. Ambrosch-Draxl, Comput. Phys. Commun. 94 (1996) 31. http://dx.doi.org/10.1016/0010-4655(95)00139-5

[20] F.D. Murnaghan, Proc. Natl. Acad. Sci. USA 30 (1944) 244. http://dx.doi.org/10.1073/pnas.30.9.244

[21] H. Hahn, G. Frank, W. Klingler, A. Meyer, G. Stroger, Z. Anorg. Chem. 271 (1953) 153. http://dx.doi.org/10.1002/zaac.19532710307

[22] J. E. Jaffe and A. Zunger, Phys. Rev. B29, 1882 (1984). http://dx.doi.org/10.1103/PhysRevB.29.1882

[23] V Kumar, G M Prasad and D Chandra Phys. Stat. Solidi (b) 170 77 (1992). http://dx.doi.org/10.1002/pssb.2221700108

[24] A.S. Verma, Phys. Status Solidi B 246 (2009) 192. http://dx.doi.org/10.1002/pssb.200844242

[25] Hai. Xiao, Jamil. Tahir-Kheli, William.A. Goddard, Phys. Chem. Lett. 2 (2011) 212–217. http://dx.doi.org/10.1021/jz101565j

[26] J. Sun, H.T. Wang, N.B. Ming, J. He, Y. Tian, Appl. Phys. Lett. 84 (2004) 4544. http://dx.doi.org/10.1063/1.1758781

[27] S.A. Korba, H. Meradji, S. Ghemid, B. Bouhafs, Comput. Mater. Sci. 44 (2009) 1265. http://dx.doi.org/10.1016/j.commatsci.2008.08.012

[28] P.Y. Yu, M. Cardona, Fundamentals of Semiconductors, SpringerVerlag, Berlin, 1996. http://dx.doi.org/10.1007/978-3-662-03313-5

[29] M.Q. Cai, Z. Yin, M.S. Zhang, Appl. Phys. Lett. 83 (2003) 2805. http://dx.doi.org/10.1063/1.1616631

Dielectric Materials and Applications: ISyDMA'2016   Materials Research Forum LLC
Materials Research Proceedings **1** (2016) 199-202   doi: http://dx.doi.org/10.21741/2474-395X/1/50

Study of dielectric properties of polyvinyl chloride (PVC) by the thermally stimulated current technique (TSC)

Mohamed OUAZENE

University of science and technology Houari Boumediene : Algiers, DZ-16000, and National Preparatory School to ingéniorat studies: Algiers, DZ-16012, Algiers, Algeria

Corresponding author: E-mail: ouaz_moh@yahoo.fr

Keywords: Polymers, Polyvinyl Chloride (PVC), Thermally Stimulated Depolarization Currents, Vogel Theory

Abstract. Polyvinyl chloride (PVC) is one of the polymers used in a plasticized state in the cable insulation to medium and high voltage. The studied material is polyvinyl chloride (PVC 4000 M) from the Algerian national oil company. Industrial PVC 4000M is in the form of white powder. The aim of this paper is to study the effect of temperature by the method of thermally stimulated depolarization currents. Furthermore, it is inferred from this work that the phenomenon of physical aging can have important consequences on the properties of the polymer. It leads to a more compact rearrangement of the material and a reconstruction or reinforcement of structural connections.

Introduction

THE insulating properties of the polymers are widely used in electrotechnics or in the insulation of high and medium voltage cables and many other uses [1]. In recent decades, the insulation of underground cables in power systems is made with polymeric materials because they have significant advantages technically and economically [2]. However, although the insulation with polymeric materials has important advantages, there are some disadvantages such as the effect of

heat that may have a detrimental effect on the properties of polymers [3]. Polyvinyl chloride (PVC) is one of the polymers used in a plasticized state in the cable insulation to medium and high voltage. The studied material is polyvinyl chloride (PVC 4000 M) from the Algerian national oil company. Industrial PVC 4000M is in the form of white powder. The test sample is a pastille of 1 mm thick and 1 cm in diameter.

Experiment of tsc and maths

A   Experiment

The sample is placed between two planar electrodes of circular shape within a cylindrical enclosure in which a vacuum has been carried out by a pump system. The two electrodes are held by two rigid rods, which are used at the same time to bring the electric current. A temperature sensor made in platinum (TN 2AS) positioned at the immediate vicinity of material allows to track the temperature variations over time. The probe is placed in a hole made in the ground electrode. The rise in temperature takes place by using a heating resistor placed near the sample.

The electrical diagram of the thermally stimulated method depolarization current is given by Fig 1.

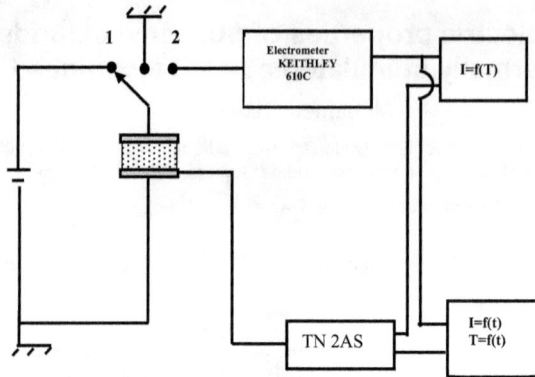

Fig 1. Circuit of measurement and recording

B   Relaxation in the polymers in the solid satete

In polymers, we observe a relaxation of large amplitude, called the primary relaxation, and secondary relaxations occurring at temperatures below Tg [4]. The first said α relaxation associated with the glass transition originates from a large-scale rotation mechanism of the segments of the main chain [5]. These rotational movements, favored by the rubbery state wherein the polymer is (T > Tg), however, are held back by the friction movements due to interactions between mobile segments and the environment to which they belong. The local movements of the lateral dipoles of the main chain are also involved in the relaxation mechanism.

One of the molecular theories leading to relaxation time leads to Vogel equation:

$$\tau = \tau_0 exp\, 1/\alpha_f (T - T_\infty) \tag{1}$$

where: α_f is the fraction of free volume, T is the temperature in K and T_∞ the temperature below which no rearrangement is possible. Representation of Log τ vs. 1 / T is linear.

C   Résultats and discussions

In a first step records the overall spectrum of the thermally stimulated depolarization current (TSC) in a PVC sample polarized at temperature $T_P = 81°C$. This is illustrated by Fig. 2. One notices a thermocourant peak at $T = 81°C$ preceded by a shoulder in the vicinity of 60°C.

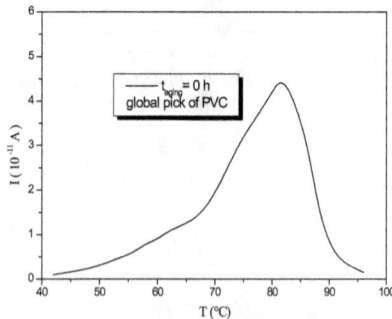

Fig. 2. Overall spectrum of TSDC of PVC at $T_P = 81\ °C$

Dielectric Materials and Applications: ISyDMA'2016
Materials Research Proceedings 1 (2016) 199-202

Materials Research Forum LLC
doi: http://dx.doi.org/10.21741/2474-395X/1/50

The sample of PVC has undergone within the measurement cell stays of zero hour, 0.3 hour, 0.6 hour, one hour and two hours at temperature $Tg = 80°C$. It is then biased at this same temperature for two minutes with an electric field of 3.10^5 V / m. The TSC technique is then applied and allows us to record the overall peaks corresponding to each time of aging as shown in Fig. 3.

One notices in this figure that the shape and position of the peaks do not seem to be affected by physical aging phenomenon. In the contrary, we note that the peak intensity decreases with the aging time. This result is similar to that found by Yianakopoulos G. et al. [2] by studying the aging P.V. C. For the isolated peak between 50 ° C and 60 ° C, there is a temperature corresponding to each T_∞ aging time and which allows tracing log τ vs. 1 / T as shown in Fig. 4.

Finally, note that the shape of the relaxation time spectrum does not change but there is movement of the latter towards the longest times. This leads us to say that the relaxation rate becomes more and more slow.

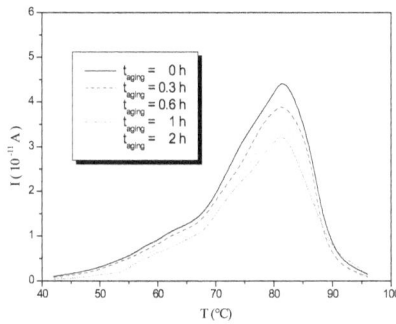

Fig.3. Overall peaks of aged PVC

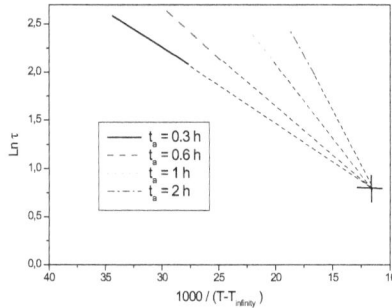

Fig.4. Vogel diagram of PVC aged at $T = 60°C$

Conclusion

The technique of thermally stimulated depolarization currents has proven that it is well suited to highlight the physical aging due to the temperature of the studied polymer (PVC).

References

[1] A. Boudet, les polymères: leurs structures et leurs propriétés, extract of the book: Voyage au cœur de la matière plastique, CNRS-France, Editions, 2003, pp. 31-33.

[2] N. Doulache and M Bendaoud: "Study of the Influence of the Speed of Cooling on the α-Relaxation in Polymers by the Thermostimulated Current Method," *International Journal of Polymer*

Dielectric Materials and Applications: ISyDMA'2016   Materials Research Forum LLC
Materials Research Proceedings **1** (2016) 199-202   doi: http://dx.doi.org/10.21741/2474-395X/1/50

Analysis and Characterization, Vol. 11(4), pp. 317-322, July 2006.
http://dx.doi.org/10.1080/10236660600754671

[3] S. Doulut, C. Bacharan, P. Demont, A. Bernès,, and C. Lacabanne, "Physical aging and tacticity effects on the α relaxation mode of amorphous polymers by thermally stimulated techniques," *Journal of Non-Crystalline Solids - Elsevier*, Vol. 235–237, pp. 645–651, August 1998.

[4] C.Lacabanne and D. Chatain, "Depolarization thermocurrents in amorphous polymers," Journal of Polymer Science: Polymer Physics Edition, Vol. 11 (12), pp. 2315–2328, 11March 2003.

[5] Y. Yorozu, M. Hirano, K. Oka, and Y. Tagawa, "Electron spectroscopy studies on magneto-optical media and plastic substrate interface," IEEE Transl. J. Magn. Japan, vol. 2, pp. 740-741, August 1987 [Digests 9th Annual Conf. Magnetics Japan, p. 301, 1982].
http://dx.doi.org/10.1109/TJMJ.1987.4549593

Dielectric Materials and Applications: ISyDMA'2016
Materials Research Proceedings 1 (2016) 203-206

Materials Research Forum LLC
doi: http://dx.doi.org/10.21741/2474-395X/1/51

Parallel plate waveguide with isotropic chiral medium route to metamaterial

Samia AIB[1], Fatiha BENABDELAZIZ[1], Chems EDDINE ZEBIRI[2],
Zinelabiddine MEZACHE[3]

[1]University Mentouri Constantine, Department of Electronics. Algeria

[2]University Ferhat Abbas, Setif, Department of Electronics, Algeria

[3]University des Frères Mentouri Constantine Department of Electronics, Algeria

*Corresponding author: E-mail:sammalak3@yahoo.com

Keywords: Isotropic Chiral, Metamaterial, Backward Wave

Abstract. Theoretical study of electromagnetic wave propagation in parallel plate waveguide with isotropic chiral medium is presented. This research work is based on the algebraic formulation of Maxwell equations according to the A formalism of constitutive relations. The dispersion modal equations is obtained and solved, the results of these equations confirmed the specificity of the bifurcation modes. The effect of chirality parameter (ξ), is considered when $\sqrt{\mu_r \varepsilon_r} \prec \xi$, (where μ_r, ε_r and ξ are respectively relative permeability and permittivity and chirality parameter). Results demonstrate the phenomena of backward waves in isotropic chiral medium.

Introduction
Chiral media are a subgroup of complex materials, which attracted the attention of several research groups working in the electromagnetic wave and optical areas. It can be called biisotropic media, (when the physical parameters of the medium are scalars) because of the coupling between electric field and magnetic field. This later appears in these constitutive relations as shown in [1]:

$$\begin{cases} B = \mu H + \left(\chi + j\xi\right)\sqrt{\mu_0 \varepsilon_0}\, E \\ D = \varepsilon E + \left(\chi - j\xi\right)\sqrt{\mu_0 \varepsilon_0}\, H \end{cases} \tag{1}$$

Where E, H, D and B represent electric field, magnetic field, electric flux density, and magnetic flux density respectively. ε and μ are respectively electrical permittivity and magnetic permeability and ε_0 and μ_0 are respectively permittivity and permeability of free space. χ and ξ are respectively non reciprocity (Tellegen) parameter and chirality (pasteur) parameter. This study is focused on wave's propagation in isotropic chiral medium so $\chi = 0$, where the effect of chirality is considered when $\sqrt{\mu_r \varepsilon_r} \prec \xi$, to create backward wave of one of the bifurcated modes (RCP: right circularly polarized or LCP: left circularly polarized).

Formulations
In this study the Maxwell equations are solved in parallel plate waveguide filled with isotropic chiral medium by the lightening of the constitutive equations "(1)", and with perfectly conducting planes placed at $x = \pm a$, as shown in [2]. Chirowaveguide propagation direction is along z axis and field quantities are all independent of y axis, physical parameter values are scalars. As a result we obtain the following set of coupled equations:

$$\frac{\partial^2}{\partial x^2}\begin{Bmatrix} E \\ H \end{Bmatrix} + \left(\left(\omega^2\mu\varepsilon + \beta_0^2\xi^2\right) - \beta^2\right)\begin{Bmatrix} E \\ H \end{Bmatrix} \mp j2\omega^2\sqrt{\mu_0\varepsilon_0}\,\xi\begin{Bmatrix} \mu H \\ \varepsilon E \end{Bmatrix} = 0.$$

(2)

After applying boundary condition on tangential components of electric fields which are zero at the surface of metallic walls, as shown in [3], [4], [5]:

$$\begin{cases} E_z = 0 \to x = +-a/2 \\ E_y = 0 \to x = +-a/2 \end{cases}$$

(3)

we obtain the following dispersion equations:

$$\Delta_{1,2} = \left(\frac{\sqrt{k_+^2 - \beta^2}}{k_+} + \frac{\sqrt{k_-^2 - \beta^2}}{k_-}\right)\sin\left(\frac{\sqrt{k_+^2 - \beta^2}\,a}{2} + \frac{\sqrt{k_-^2 - \beta^2}\,a}{2}\right)$$

$$\pm \left(\frac{\sqrt{k_+^2 - \beta^2}}{k_+} - \frac{\sqrt{k_-^2 - \beta^2}}{k_-}\right)\sin\left(\frac{\sqrt{k_+^2 - \beta^2}\,a}{2} - \frac{\sqrt{k_-^2 - \beta^2}\,a}{2}\right) = 0$$

(4)

Where:

β :is the propagation constant, $\beta_0 = \omega\sqrt{\varepsilon_0\mu_0}$

(5)

$K_\pm = \left(\omega\sqrt{\mu\varepsilon} \mp \beta_0\xi\right) = \beta_0\, n_\pm$

(6)

$n_\pm = \sqrt{\mu_r\varepsilon_r} \mp \xi$

(7)

$n_\pm$: are refractive indexes in chiral medium as mentioned in [6].
$K_\pm$: are numbers of right and left wave.
Cutoff frequencies f_c are obtained when $\beta = 0$:

$f_c = \dfrac{n}{2\sqrt{\varepsilon\mu}a}$

(8)

Results and discutions

In order to investigate the propagation characteristics of the guide made of isotropic chiral medium, three different type of chiral medium are chosen by considering various values of physical parameters.

We consider the first case: where $\varepsilon = \varepsilon_0$, $\mu = \mu_0$ and $\xi = -0.337$ that correspond to the case of Pelet and Engheta [3].in order to confirm our calculation.

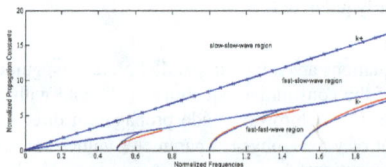

Fig. 1: Dispersion diagram for propagating modes (RCP and LCP wave) in a parallel plate chirowaveguide with $\varepsilon = \varepsilon_0$, $\mu = \mu_0$, $\xi = -0.337$

Dielectric Materials and Applications: ISyDMA'2016   Materials Research Forum LLC
Materials Research Proceedings 1 (2016) 203-206   doi: http://dx.doi.org/10.21741/2474-395X/1/51

In normalized propagation constant βa, a:is thickness of chiral medium, and normalized dimensionless frequency is

$$\Omega = \omega a \sqrt{\varepsilon_0 \mu_0} \Rightarrow \frac{\Omega}{2\pi} = \mathrm{fa}\sqrt{\varepsilon_0 \mu_0} \qquad (9)$$

For $\varepsilon = 0.51\varepsilon_0$, $\mu = \mu_0$ and $\xi = -\sqrt{2}/2$ the result is reported in fig (2).

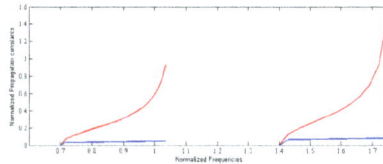

Fig. 2: Dispersion diagram for propagating modes (RCP and LCP wave) in a parallel plate chirowaveguide with $\varepsilon = 0.51\varepsilon_0$, $\mu = \mu_0$, $\xi = -\sqrt{2}/2$

The result for the case $\varepsilon = 0.41\varepsilon_0$, $\mu = \mu_0$ and $\xi = -\sqrt{2}/2$ is:

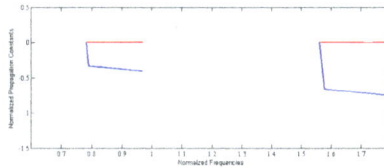

Fig. 3: Dispersion diagram for propagating modes (RCP and LCP wave) in a parallel plate chirowaveguide with $\varepsilon = 0.41\varepsilon_0$, $\mu = \mu_0$, $\xi = -\sqrt{2}/2$

In fig (1) there is a bifurcated modes which started from the same cutoff frequency. Propagation constant (i.e., RCP and LCP wave) are positive values corresponding to conventional propagation in chiral isotropic medium (i.e., right hand propagation). In fig (2), either RCP or LCP wave is near to zero and the other is positive. In fig (3) there is a negative propagation constant value of either RCP or LCP wave (which is near to zero in fig (2)) and the other wave is zero (which is positive in fig(2)) which correspond to left hand propagation (i.e., negative refractive index or metamaterial). What remains related to the change of $\sqrt{\mu_r \varepsilon_r} - |\xi|$ value, which is due to the change of ε value, in the second case. where $\sqrt{\mu_r \varepsilon_r} - |\xi| = 0.007$ and $\varepsilon = 0.51\varepsilon_0$, and in the third case where $\sqrt{\mu_r \varepsilon_r} - |\xi| = -0.066$ and $\varepsilon = 0.41\varepsilon_0$ which represent the threshold between the right hand and left hand medium. So if $\sqrt{\mu_r \varepsilon_r} - |\xi| \leq -0.066$, the wave propagation is in left hand medium, if $\sqrt{\mu_r \varepsilon_r} - |\xi| \succ -0.066$, the wave propagation is in right hand medium, and if $\sqrt{\mu_r \varepsilon_r} - |\xi| = 0$ there is no propagation.

Conclusion
The study of wave propagation in chiral isotropic medium is obtained, where negative refraction index can be achieved, of one of the bifurcated modes either (RCP or LCP wave), when this condition on the physical parameter is satisfied $\sqrt{\mu_r \varepsilon_r} \prec \xi$ (i.e., the chirality parameter value is increasing enough in front the $\sqrt{\mu_r \varepsilon_r}$ value).

Dielectric Materials and Applications: ISyDMA'2016   Materials Research Forum LLC
Materials Research Proceedings **1** (2016) 203-206   doi: http://dx.doi.org/10.21741/2474-395X/1/51

Knowing that metamaterial play a crucial role in lot of domains such as Biosensors, where metamaterial can be used to provide more sensitive guiding modes. Also in superlens (or perfect lens) which uses metamaterials to go beyond the diffraction limit where both propagating and evanescent waves contribute to the resolution of the image. And in cloaking device where Scientists are using metamaterials to bend light around an object, that causes it to be partially or wholly invisible to parts of electromagnetic (EM) spectrum. And it is simpler to realize metamaterial (or negative refraction index) by isotropic chiral medium than to realize it by both negative permeability and negative permittivity.

References

[1] Ougier, S., Chenerie, I., Sihvola, A., & Priou, A. (1994). Propagation in bi-isotropic media: effect of different formalisms on the propagation analysis. *Progress in Electromagnetics Research*, *9*, 19-30.

[2] Aib, S., Benabdelaziz, F., & Zebiri, C. E. Analysis of Parallel Plate Waveguide with Anisotropic Chiral Medium. Proceeding of the World Congress on New Technologies (New Tech 2015). *Barcelona, Spain – July 15 - 17, 2015 Paper No. 391*

[3] Pelet, P., & Engheta, N. (1990). The theory of chirowaveguides. *Antennas and Propagation, IEEE Transactions on*, *38*(1), 90-98. http://dx.doi.org/10.1109/8.43593

[4] Ghaffar A., Alkanhal MAS. (2014). Electromagnetic waves in parallel plate uniaxial anisotropic chiral waveguides. Optical Materials Express, 4(10), 1756-1761. http://dx.doi.org/10.1364/OME.4.001756

[5] Topa, A. L., Paiva, C. R., & Barbosa, A. M. (2010). Electromagnetic wave propagation in chiral H-guides. *Progress In Electromagnetics Research*, *103*, 285-303. http://dx.doi.org/10.2528/PIER10032106

[6] Tretyakov, S., Sihvola, A., & Jylhä, L. (2005). Backward-wave regime and negative refraction in chiral composites. *Photonics and Nanostructures-Fundamentals and Applications*, *3*(2), 107-115. 0509287-1. http://dx.doi.org/10.1016/j.photonics.2005.09.008

Dielectric Materials and Applications: ISyDMA'2016   Materials Research Forum LLC
Materials Research Proceedings 1 (2016) 207-213   doi: http://dx.doi.org/10.21741/2474-395X/1/52

RF/ microwaves biological effects and dielectric properties of human tissues

Mohamed Hamza QADDI*, Mohamed Nabil SRIFI

Electronics and Telecommunications Systems Research Group, National School of Applied Sciences, Ibn Tofail University, Kenitra, Morocco

*Corresponding author: E-mail: qaddihamza@gmail.com

Keywords: Biological Effects, Microwave Radiation, Mobile Phone, Human Tissues, Dielectric Materials

Abstract. Nowadays, the microwave and radiofrequencies are widely used (mobile phone, base station, wi-fi...) closed and near to the Human body (Children [1], pregnant [2], patients [3]...). The biological effects and health implications of exposure to radio frequency radiation have been widely investigated. Many factors affect the exposure of human body to radiofrequency radiations, this make the exposure assessment in a risk study very difficult to interpret. Ideally, the exact exposure dose should be measured, but there is no specified measure. In experimental situations, specific absorption rate (SAR), which is a measure of the maximum energy absorbed by a unit of mass of exposed tissue of a person, is used. Most of the research related to the RF and microwaves health effects neglects the most important parameter, which is the no uniformity of the biological tissues properties and theirs compositions. In this study, we will present the different models of the human body with its various Dielectric Properties of tissues, and see how it can affect the simulated and measured SAR values in human body.

Introduction

The biological effects and health implications of exposure to microwave and other radio frequency radiation have been under investigation for more than a decade by various investigators in many countries. Many factors can affect the exposure of human body to radiofrequency radiations, this make the exposure assessment in a risk study very difficult to interpret. Ideally, the exact exposure dose should be measured, but there is no specified measure. In experimental situations, specific absorption rate (SAR) is used. A SAR value is a measure of the maximum energy absorbed by a unit of mass of exposed tissue of a person, over a given time, or more simply the power absorbed per unit mass. SAR values are usually expressed in units of watts per kilogram (W/Kg) in either 1g or 10g of tissue. SAR provides a straightforward means for measuring the RF exposure of any radiation source.

To calculate SAR we should consider many parameters, we will group them in three. The first one is the frequency of the device, the second the position of the device; distance from the body... and the third one and the most important in this study is the dielectric properties of body tissues.

Experimental methods to evaluate SAR are quite costly and time consuming. It is against the law and moral to expose a human being to electromagnetic radiations for experimental purpose. Moreover, numerical analysis through simulations provides more flexibility to analyze the same.

In this paper, the history of research and development on the interaction between RF, microwave and human body is reviewed. We will gather different results of simulations and focus on dielectric properties of human body tissues and how it can increase or decrease SAR.

Radio frequency exposure and biologics

A biological effect occurs when a change can be measured in a biological system after the introduction of some type of stimuli. However, the observation of a biological effect, in and of itself, does not necessarily suggest the existence of a biological hazard [4]. A biological effect only becomes a safety hazard when it "causes detectable impairment of the health of the individual or of his or her offspring [5].

Dielectric Materials and Applications: ISyDMA'2016   Materials Research Forum LLC
Materials Research Proceedings 1 (2016) 207-213   doi: http://dx.doi.org/10.21741/2474-395X/1/52

The effects of radio frequency exposure on biologics, similar to other types of electromagnetic (EM) radiation, can be analyzed in two groups: thermal and non-thermal [6]–[8].

Biological effects that result from heating of tissue by RF energy are often referred to as "thermal" effects. It has been known for many years that exposure to high levels of RF radiation can be harmful due to the ability of RF energy to heat biological tissue rapidly [9].

Product heating depends on a few factors such as the frequency of the EM source, the dielectric constant, the contents of the water and the overall thickness. For instance, products greater in conductivity are more likely to absorb greater amounts of radiation which in turn generates more heat.

In addition, some studies suggest that there are reactions to RF exposure on substances that cannot be explained by thermal heating, since these non-thermal effects occur without a significant increase in the product temperature after the exposure [10].

When considering the biological effects of microwave radiation, the wavelength or frequency of the energy and its relationship to the physical dimensions of objects exposed to radiation become important factors [11].

The degree of temperature rise from exposure to microwaves is dependent on numerous physical and biologic factors:

- Intensity or power density;
- Duration of exposure; limited, however, by environmental and physiological factors.
- Frequency or wavelength of the radiation.
- Size and dimensions of the exposed object.
- Thermal regulatory capacity of the exposed subject.
- Thickness of tissue.
- Composition of tissue.

Definition of SAR

SAR (specific absorption rate)

Many factors can affect the exposure of human body to radiofrequency radiations, this make the exposure assessment in a risk study very difficult to interpret. Ideally, the exact exposure dose should be measured,

Transferring EM energy to tissues is a complex function of many variables. The external incident field intensity may be expressed in a variety of units. Exposure data may be expressed in terms of power density (mW/cm^2), external electric field strength (V/m), or magnetic field strength (A/m). None of these data provides investigators with sufficient insight into how fields interact with biological tissue. Consequently, the question arises as to the most suitable parameter for quantifying the interaction of EM fields and biological systems. In 1980, the National Council on Radiation Protection and Measurements (NCRP) designed Officially SAR as the suitable parameter. SAR is the basic parameter that is taken into consideration for the evaluation of the exposure hazards in the radio frequency and microwave range. "Specific" refers to the normalization to mass, and "absorption rate" refers to the rate of energy absorbed by the object.

SAR value is a measure of the maximum energy absorbed by a unit of mass of exposed tissue of a person, over a given time, or more simply the power absorbed per unit mass. SAR values are usually expressed in units of watts per kilogram (W/Kg) in either 1g or 10g of tissue.

SAR can be calculated by Equation (1) as follows:

$$SAR = \frac{\sigma E^2}{\rho} \tag{1}$$

Dielectric Materials and Applications: ISyDMA'2016   Materials Research Forum LLC
Materials Research Proceedings 1 (2016) 207-213   doi: http://dx.doi.org/10.21741/2474-395X/1/52

Where, σ is electrical conductivity (S/m), E is electric field intensity (V/m), and ρ is the density of the tissue (Kg/m^3). The first step in the analysis of electromagnetic radiation with human body is the determination of the induced internal electromagnetic field and its distribution.

Tissue Geometry and Size

The highest local SAR is usually at or near the surface of an externally exposed object. For curved surfaces and "resonant objects," high SARs ("hot spots") exist at various locations [12]. A complex biological system, such as a human body, consists of multiple layers of tissue. Each layer has different dielectric properties and forms an EM boundary. When exposed to an RF field, the field propagates within the multilayered object. A portion of the energy is reflected from each boundary, and a portion is transmitted into the next layer. The amount of transmission and reflection at each boundary depends on the difference in dielectric properties of the tissues (characteristic impedance mismatch). Fat thickness, tissue curvature, and dimensions of the body, limbs, and head relative to the wavelength all affect the energy distribution [13].

RF penetration in human tissues

In considering the amount of energy absorbed by the human body, it is necessary to recognize that the percentage of incident radiation which is actually absorbed depends on frequency and the orientation of the subject relative to the field.

In human tissues, RF radiation may be absorbed, reflected or may pass through the tissue. What actually happens will depend on the body structure and the tissue interfaces involved. These interfaces are the transitions from tissue to tissue or tissue–air–tissue and are clearly complex in the human body.

The depth of RF penetration of the human body is also an important factor [14]. In the HF band, the deeper penetration is used for diathermy treatment where the deposition of heat is intended to have a beneficial effect on that part of the body considered to need treatment. The deep deposition of RF energy needs to be carefully controlled to avoid damage to tissues, which might not be noticed by the subject due to lack of sensory perception of heat in the organs concerned.

The measurement of the RF characteristics of human tissue can, for the most part, only be done with chemical simulation of tissue, since there are problems with the use of excised human tissue for this purpose. The penetration depth is usually given as the depth where the incident power density has been reduced by a factor of e_{-2}, i.e. down to about 13.5% of the incident power density.

Factors that determine energy absorbed in tissues

The magnitude and spatial distribution of EM fields within biological tissues depend on the dielectric properties of tissue (dielectric constant and conductivity), which are dominated by the water content. Therefore, tissues can be divided into those with high water content, such as eye, muscle, skin, liver, and kidney, and those with low water content, such as fat and bone.

Recently, it has been reported [15] that bone material has a higher dielectric constant and conductivity than previously published.

Other tissues that contain intermediate quantities of water, such as brain, lung, and bone marrow, have dielectric properties that lie between tissues with high and low water content. The dielectric constant and conductivity of tissues vary over a wide range and are frequency dependent.

For exposed object, the highest local SAR is usually at or near the surface, and for curved surfaces, high SAR exist at different positions.

Human body is a complex biological system, form from different layers of tissues. Each layer has its own dielectric properties. When human body exposed to a RF field, the field propagates within the multilayered object. A portion of the energy is reflected from each boundary, and a portion is transmitted into the next layer. The amount of energy transmitted and reflected depend on the dielectric properties of the tissues. Fat thickness, tissue curvature, dimensions of the body, limbs, bone, skin... all affect the energy transmitted and reflected.

From the equation (1), SAR depends on electrical conductivity, electric field intensity and the density of the tissue.

Dielectric Materials and Applications: ISyDMA'2016   Materials Research Forum LLC
Materials Research Proceedings **1** (2016) 207-213   doi: http://dx.doi.org/10.21741/2474-395X/1/52

The need for extensive data on the dielectric properties of human tissues has prompted scientists to develop Tissue-equivalent materials to simulate dielectric properties of biological tissues at frequencies of interest.

Biological tissues are inhomogeneous and show considerable variability in structure or composition and hence in dielectric properties. Such variations are natural and may be due to physiological processes or other functional requirements.

Researches tried to create and compile a database of dielectric properties of tissues [16] to be used by scientific community in solving electromagnetic interaction problems. This has been achieved through measurement in the frequency rang 10 Hz to 20 GHz and modelling the frequency dependence of the dielectric properties of over 30 body tissues to parametric expressions for inclusion in numerical solutions.

Dielectric properties and SAR

Many researchers tried to study the interaction between RF, microwaves radiation and human body [17]. The aim of their work is to see if there is any harmful effects on human by using different sources radiation. However, they face many problems especially how to model human's body with its complex layouts and dielectric properties.

Male and female body

A comparison of dielectric values in human body exposed to a specific frequency have been made [3]. For this, they choose both male and female body. SAR values increase with the increase of conductivities of human body tissues, and usually decrease with the increase of relative permittivities of human body tissues. Conductivity (σ) and relative permittivity (ε) of human tissues are the determining factors for both optimal RF communication and dosimetry. Recently, the variability of dielectric parameters of human tissues has been found.

We might take it for granted that the SAR value would increase with the increment of dielectric values. In fact, unaveraged and averaged SAR values are highly influenced by dielectric properties.

Pregnant-woman and non pregnant-woman

Researchers tried to make a realistic whole-body pregnant-woman model and specific absorption rates for pregnant-woman exposure to electromagnetic plane waves [2]. The model of pregnant woman is constituted of about 7 million voxels, and segmented into 56 different tissues and organs.

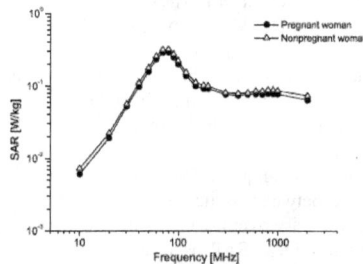

Figure 1 Comparison of whole-body averaged SAR between the pregnant-woman model and non-pregnant-woman model

The WBA-SARs of the pregnant-woman model compared with those of the non-pregnant-woman model are shown in figure 3. The difference in WBA-SAR between the pregnant and Non-pregnant models is at most 1.09 dB. The WBA-SARs of both models agree well with each other under all conditions. These results suggest that pregnancy affects the WBA-SAR of a woman due to the difference in dielectric properties.

Dielectric Materials and Applications: ISyDMA'2016   Materials Research Forum LLC
Materials Research Proceedings 1 (2016) 207-213   doi: http://dx.doi.org/10.21741/2474-395X/1/52

Children and adult

When risks are complex, or the scientific findings are not robustly quantifiable, the need for timely preventive action makes a precautionary approach an essential part of policy making. Many societies believe that this is particularly true regarding children (including the unborn child): they represent the future of the society, have the potential for longer exposure than adults have, and yet are less able to manage their own risk [1].

Researches tried to focus on children as they are representing the future of our planet, and see if they are more exposed to radiofrequency and microwaves radiations. The problem is that children are growing, and their dielectric properties of tissues are changing, as they are growing up, which make the research more difficult.

The level and distribution of radiofrequency energy absorbed in a child's head during the use of a mobile phone for example compared to those in an adult head has been a controversial issue in recent years. The methods used to determine specific absorption rate (SAR) and assess compliance with exposure standards using an adult head model [18]-[20] may not adequately account for potentially higher levels of exposure in children due to their smaller head size.

The first adult human head model from the National Library of Medicine's Visible Man model [15] was further modified by the Air Force Research Laboratory (Brooks AFB, TX) to indicate the tissue type, associated tissue density, and tissue dielectric parameters in each voxel of the model.

Table 1 Tissue Dielectric Parameters and Densities for NIT Visible Human Head Model at 900 MHz

Tissue type	Dielectric constant	Conductivity (S/m)	Density (kg/m3)
Blood	61.36	1.54	1060
Bone	20.79	0.34	1860
Bone marrow	11.27	0.23	1030
Cartilage	42.65	0.78	1100
CSF	68.64	2.41	1010
Dura	44.43	0.96	1030
Vitreous humour	68.90	1.64	1010
Fat	11.33	0.11	920
Parotid gland	45.25	0.92	1050
Gray matter	52.72	0.94	1030
Muscle	55.96	0.97	1040
Skin	41.41	0.87	1010
White matter	38.89	0.59	1030
Mucous membrane	46.08	0.84	1010
Lens	35.84	0.48	1100
Cornea	55.23	1.39	1050
Sclera	55.27	1.27	1170

The head models contain 17 different tissue types (table 1) with dielectric properties (permittivity and conductivity) and densities assigned according to Nagoya Institute of Technology (NIT)

Table 2 Calculated Results for Different-Sized Brooks Head Models Normalized to 0.6 W Output Power at 835 MHz

	Adult	10-years-old-child	5-years-old child
Peak at surface W/Kg	24,7	20,3	32,6
Peak in interior W/Kg	8,85	7,8	10,6
Peak in brain W/Kg	2,62	2,96	3,21
Whole head –average SAR W/Kg	3,93	3,43	3,44

Dielectric Materials and Applications: ISyDMA'2016   Materials Research Forum LLC
Materials Research Proceedings 1 (2016) 207-213   doi: http://dx.doi.org/10.21741/2474-395X/1/52

Table 2 shows differences between the peak one-voxel SAR values in interior of the head models. For the 5-year-old child, this value is about 20% higher, and for the 10-year-old child, it is about 12% lower compared to the adult head model. The corresponding peak one-voxel values in the brain of the child head are 23% and 14% higher for the 5- and 10-year-old child head model compared to the adult head. This can explain this difference by the shorter distance between the brain and the source in the child head model. However, the main reason is that the dielectric properties of the head are different from a person to another [21].

Discussion & conclusion

This paper has analyzed the variability of SAR in different bodies (male, female, pregnant-female and child). In the first case, All presented studies prove that the value of SAR change from male's body to female's body, due to the difference in dielectric properties between the two bodies. Then, we focused on female body, and we remarked that as the frequency change, the value of the whole body average SAR also change, which is normal. However, we noticed that there is a difference between the value of pregnant and non-pregnant female due to the different dielectric properties of their bodies. Finally, we were interested in children's head and we have noticed that the value of the whole body average SAR is higher than the value for adult. In addition, Modern children will experience a longer period of exposure to RF fields from mobile-phone use than adults will, because they started using mobile phones at an early age and are likely to continue using them. To face this problem, we advise parents to restrict the use or length of calls of their children, even if it's not enough but at least it may reduce the impact.

Most of research focus on RF and microwaves radiation, devices and their positions, and neglect the most important part, which is biological compositions of human body tissues. A good study of this part may lead to a global understanding of the problem.

References

[1] G. Bit-Babik, A. W. Guy, C-K. Chou, A. Faraone, M. Kanda, A. Gessner, J. Wang and O. Fujiwara. Simulation of Exposure and SAR Estimation for Adult and Child Heads Exposed to Radiofrequency Energy from Portable Communication Devices. RADIATION RESEARCH 163, 580–590 (2005). http://dx.doi.org/10.1667/RR3353

[2] Tomoaki Nagaoka, Toshihiro Togashi, Kazuyuki Saito,Masaharu Takahashi, Koichi Ito and Soichi Watanabe, An anatomically realistic whole-body pregnant-woman model and specific absorption rates for pregnant-woman exposure to electromagnetic plane waves from 10 MHz to 2 GHz. 2007 Phys. Med. Biol. 52 6731. http://dx.doi.org/10.1088/0031-9155/52/22/012

[3] L. S. Xu,, Max Q.-H. Meng, and Chao Hu, Effects of Dielectric Values of Human Body on Specific Absorption Rate Following 430, 800, and 1200 MHz RF Exposure to Ingestible Wireless Device. IEEE TRANSACTIONS ON INFORMATION TECHNOLOGY IN BIOMEDICINE, VOL. 14, NO. 1, JANUARY 2010.

[4] Ismail Uysal Price William DeHay Erdem Altunbas Jean-Pierre Emond R. Scott Rasmussen David Ulrich, Non-Thermal Effects of Radio Frequency Exposure on Biologic Pharmaceuticals for RFID Applications, IEEE RFID 2010.

[5] International Commission on Non-Ionizing Radiation Protection (ICNIRP), "Guidelines for Limiting Exposure to Time-varying Electric, Magnetic, and Electromagnetic Fields (Up to 300 GHz)," Health Physics 74: 494-520 (1998).

[6] J. L. Kirschvink, "Microwave absorption by magnetite: a possible mechanism for coupling nonthermal levels of radiation to biological systems," Bioelectromagnetics, vol. 17, pp. 187–194, 1996.
http://dx.doi.org/10.1002/(SICI)1521-186X(1996)17:3<187::AID-BEM4>3.0.CO;2-#

Dielectric Materials and Applications: ISyDMA'2016   Materials Research Forum LLC
Materials Research Proceedings 1 (2016) 207-213   doi: http://dx.doi.org/10.21741/2474-395X/1/52

[7] E. Marani and H. K. P. Feiraband, "Future perspectives in microwave applications in life sciences," Eur. J. Morphol., vol. 32, pp. 330–334, 1994.

[8] M. Porcelli, G. Cacciapuoti, S. Fusco, R. Massa, G. D'Ambrosio, C. Bertoldo, M. D. Rosa, and V. Zappia, "Non-thermal effects of microwaves on proteins: thermophilic enzymes as model system," FEBS Lett., vol. 402, pp. 102–106, 1997. http://dx.doi.org/10.1016/S0014-5793(96)01505-0

[9] http://www.wisenv.com/radiation/health-effects.html

[10] D. I. de Pomerai, B. Smith, A. Dawe, K.North, T. Smith, D. B. Archer, I. R. Duce, D. Jones, and E. P. M. Candido, "Microwave radiation can alter protein conformation without bulk heating," FEBS Lett., vol. 543, pp. 93–97, 2003. http://dx.doi.org/10.1016/S0014-5793(03)00413-7

[11] L.J. Challis, Mechanisms for Interaction Between RF Fields and Biological Tissue. Bioelectromagnetics Supplement 7:S98^S106 (2005). http://dx.doi.org/10.1002/bem.20119

[12] Robert F. Cleveland, Jr. Jerry L. Ulcek, Questions and Answers about Biological Effects and Potential Hazards of Radiofrequency Electromagnetic Fields Office of Engineering and Technology Federal Communications Commission Washington, D.C. 20554, OET BULLETIN 56, august 1999.

[13] C.K. Chou, H. Bassen, J. Osepchuk, Q. Balzano, R. Petersen, M. Meltz, R. Cleveland, J.C. Lin, and L. Heynick, Radio Frequency Electromagnetic Exposure: Tutorial Review on Experimental Dosimetry, Bioelectromagnetics 17:195-208 (1 996)

[14] RONALD KITCHEN, RF and Microwave Radiation Safety Handbook, 1993,2001

[15] Camelia Gabriel, Compilation Of The Dielectric Properties Of Body Tissues At Rf And Microwave Frequencies, February 1996

[16] http://niremf.ifac.cnr.it/tissprop/htmlclie/htmlclie.php

[17] Richmond, Biological effects and health implications of microwave radiation, Stephen F. Cleary department of biophysics medical college of Virginia. pp 1-261, (1971).

[18] Brian B. Beard, Wolfgang Kainz, Teruo Onishi, Takahiro Iyama, Soichi Watanabe, Osamu Fujiwara, Jianqing Wang, Giorgi Bit-Babik, Antonio Faraone, Joe Wiart, Andreas Christ, Niels Kuster, Ae-Kyoung Lee, Hugo Kroeze, Martin Siegbahn, Jafar Keshvari, Houman Abrishamkar, Winfried Simon, Dirk Manteuffel, Neviana Nikoloski, Comparisons of Computed Mobile Phone Induced SAR in the SAM Phantom to That in Anatomically Correct Models of the Human Head. IEEE TRANSACTIONS ON ELECTROMAGNETIC COMPATIBILITY, VOL. 48, NO. 2, MAY 2006

[19] Anu Goel and Asst. Prof. Richa, Evaluating the Effect of Distance on Specific Absorption Rate Values inside a Human Head Model. ISSN 2351-8014 Vol. 12 No. 1 Nov. 2014, pp. 186-189.

[20] Antonios Drossos, Veli Santomaa, Niels Kuster, The Dependence of Electromagnetic Energy Absorption Upon Human Head Tissue Composition in the Frequency Range of 300–3000 MHz. IEEE TRANSACTIONS ON MICROWAVE THEORY AND TECHNIQUES, VOL. 48, NO. 11, NOVEMBER 2000.

[21] M Mart´ınez-B´urdalo, A Sanchis, A Mart´ın and R Villar, Comparison of SAR and induced current densities in adults and children exposed to electromagnetic fields from electronic article surveillance devices. Phys. Med. Biol. 55 (2010) 1041–1055. http://dx.doi.org/10.1088/0031-9155/55/4/009

Dielectric Materials and Applications: ISyDMA'2016   Materials Research Forum LLC
Materials Research Proceedings 1 (2016) 214-217   doi: http://dx.doi.org/10.21741/2474-395X/1/53

Stimulator interface control for various electrotherapy applications

Adil SALBI*, Seddik BRI

Materials and Instrumentations (MIN), Department of Electrical Engineering, High School of Technology, ESTM - Moulay Ismail University, Meknes, Morocco

*Corresponding author: E-mail: adilsalbi@outlook.fr

Keywords: Electro-Stimulation System, Functional Rehabilitation, Muscular Contraction

Abstract. The aim of this work is to suggest a new reliable solution of electro-stimulation system for various therapeutic applications like functional rehabilitation and athletics training. This solution consists in creating a similar signal to that arriving of the central nervous system (CNS) to provoke muscular contraction of the peripheral nervous system (PNS). Thus, an electronic circuit is designed to generate a train of biphasic electrical rectangular pulses, which pulse width and frequency defined by the user. These stimuli allow muscular contraction of all types of muscular fibers by specifying the corresponding level of stimulation intensity. This stimulator system is composed of a clock, an inverter and a logical sequencer which shifts the clock signal. The stage of power permits to adapt the signals arrived on a transformer booster coupled to a command stage to regulate the amplitude of the impulse. We added a graphical user interface (GUI) to simplify control and to adapt the stimulator to various medical applications by changing stimulation parameters.

Introduction

At the time of an incomplete paralysis, certain residual functions are preserved; functional rehabilitation is then imaginable in the case of electrotherapy for the maintenance and the recovery of a maximum of functions. More than muscles damages, electrotherapy in the form of functional electric stimulation (FES) can be used in the sporting world at the time of body building training by the athletes [1, 2].

In this article, we present the design of an electric system of stimulation based on the electric generation of two signals by a circuit ordered in amplitude.

Principe of functional electro-stimulation

Electro-stimulation or electrotherapy indicates the whole uses of the electric current for the treatment of the diseases. This art aims to restore or compensate certain defective functions of the sensor motor system. It has already shown great successes as in the case of pacemaker, by controlling the rhythm beats of the heart, the cochlear implants restoring hearing, or even more recently of the implants "brain deep" aiming at removing shakings in the Parkinson's disease [3]. There exist two types of currents: the alternative current (AC), biphasic and the direct currents (DC), monophasic [4]. The biphasic signal is used to eliminate the effects of polarity, the thing which reduces the risks of cutaneous burns or mucosal in cavitary [1 – 5]

Electro-stimulation system

The basic principle of the system of functional electrical stimulation put back on the generation of more or less width impulses and specific frequency (Fig.1). The average value of the signal is always worthless, which avoids a familiarization of the muscle, an undesirable cutaneous ionization or a polarization of the metal prostheses.

Dielectric Materials and Applications: ISyDMA'2016   Materials Research Forum LLC
Materials Research Proceedings **1** (2016) 214-217   doi: http://dx.doi.org/10.21741/2474-395X/1/53

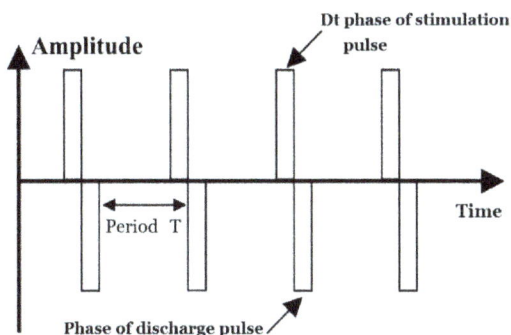

Fig.1. Shape of electro stimulation signal

The synoptic diagram of the system is presented on Fig.1 and it is composed of sequential clocks that generate stimuli and of command stage.

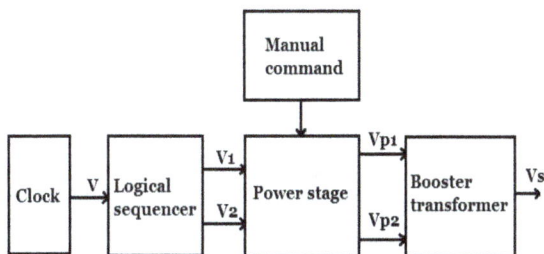

Fig.2. Block diagram of the electrical stimulator

The clock signal passes through the logical sequencer to generate two copies of the signal shifted in time by 625 μs. Then on the power stage, we have the ability to control the amplitude of the two signals simultaneously with a control signal (V_B) generated by the control stage, see following figure.

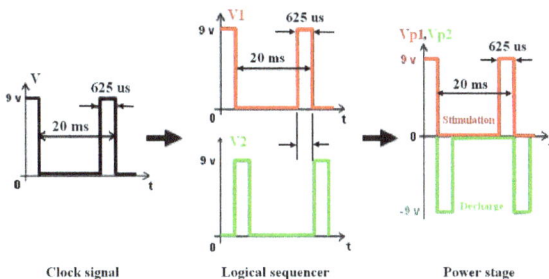

Fig.3. Process of stimulation signal generation

Dielectric Materials and Applications: ISyDMA'2016   Materials Research Forum LLC
Materials Research Proceedings **1** (2016) 214-217   doi: http://dx.doi.org/10.21741/2474-395X/1/53

The final stimulation signal is a biphasic pulse train that its amplitude is controlled by the command signal V_B. The signal period and width of impulsions are the same as clock signal (T = 20 ms, Dt= 625 us).

The amplitude signal is controlled with a digital system that can be programmed by user.

Command system and user interface

The control system is constituted, as shown in Fig. 4, of an interaction human machine (IHM) based on LabVIEW graphical interface and an acquisition card of National Instrument (DAQ NI-USB 6281).

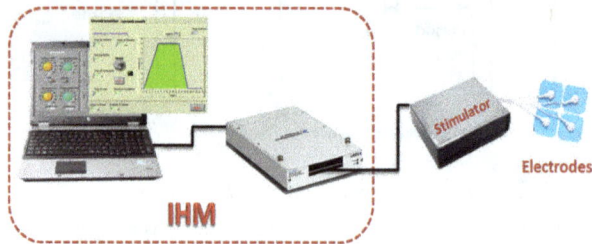

Fig.4. Command system of stimulator

The choice of the acquisition card is based on the criteria of speed, resolution, type and number of input-output pin, cost, availability, and simplicity.

The user selects a program stored or he specifies the temporal parameters of the stimulation program (T_C, T_M, T_D and T_R) using user interface (Fig. 5). Then the background program begins generating a digital signal which is converted by the DAQ to provide V_B waveform.

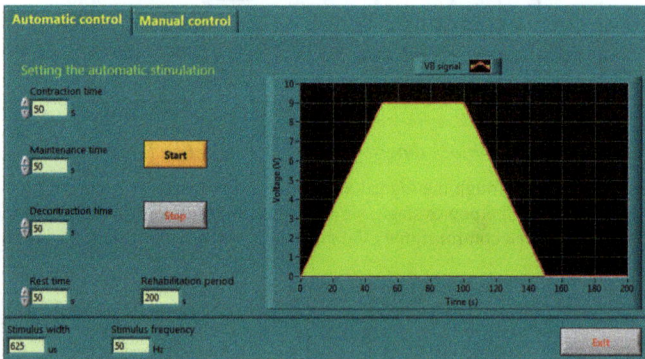

Fig.5. User interface of automatic control

The GUI is created using LabVIEW software. It consists of two tabs entitled Automatic control and Manual control. As it is presented in the front panel in the Fig. 5, under the Automatic control tab, the user draws the waveform of the control signal V_B by specifying the temporal parameters via the digital commands: contraction time, maintenance time, and relaxation time and rest time. The program period is the period of V_B control signal which equals to the sum of the periods T_C, T_M, T_D and T_R. It also indicated, at the bottom of the window, the width (625 µs) and the frequency (50 Hz) of the used electrical stimulation signal.

Dielectric Materials and Applications: ISyDMA'2016   Materials Research Forum LLC
Materials Research Proceedings 1 (2016) 214-217   doi: http://dx.doi.org/10.21741/2474-395X/1/53

Conclusion

This technique aims to restore or rehabilitate certain motor functions of the nervous system via a sequence of well-defined electrical pulses. Our work consists in conceiving an electronic system generates a sequence of pulses adjustable in amplitude and which has a null median value. This system is equipped of an automatic control of amplitude. The command signal can be programmed by user via graphical interface and make various applications.

References

[1] A. Salbi and S. Bri, "Conception of Electro-Stimulation System", International Journal of Engineering and Technology (IJET), vol. 6, 5, pp. 2136 – 2143, Oct-Nov 2014.

[2] F. Crépon, "Électrothérapie et physiothérapie: Applications en rééducation et réadaptation", Elsevier Masson, 2012.

[3] D. Thomas and A. Killésith, "La stimulation électrique fonctionnelle, appareillage et technique de facilitation", Ann. Kinésith. 5, pp. 437-447, 1978.

[4] J.-D. Techer et al., "New implantable stimulator for the FES of paralyzed muscles", Solid-State Circuits Conference, 2004. http://dx.doi.org/10.1109/ESSCIR.2004.1356716

[5] P. Decherchi et al, "Électromyostimulation et récupération fonctionnelle d'un muscle dénervé", Sci. Sports, vol. 18, 5, pp. 253 - 263, Oct. 2003. http://dx.doi.org/10.1016/S0765-1597(03)00144-8

Dielectric Materials and Applications: ISyDMA'2016
Materials Research Proceedings **1** (2016) 218-221

Materials Research Forum LLC
doi: http://dx.doi.org/10.21741/2474-395X/1/54

Rare-earth doped silica-hafnia glass ceramic for silicon solar cell efficiency enhancement

S. BELMOKHTAR[1], A. BOUAJAJ[*1], M. BRITEL[1], N. LOTFI[2], F. BELLUOMO[3],
C. ARMELLINI[4], F. ENRICHI[4], M. FERRARI[4], F. ENRICHI[5]

[1]Laboratory of Innovative Technologies, LTI, ENSA–Tangier, University Abdelmalek Essaâdi, Tangier, Morocco

[2]Dept de physique, FS – Kénitra, University Ibn Toufail, Kenitra, Morocco

[3]Meridionale Impianti SpA, Via Senatore Simonetta 26, 20867 Caponago (MB), Italy

[4]CNR-IFN, Istituto di Fotonica e Nanotecnologie, CSMFO Lab and FBK Photonics Unit, Via alla Cascata 56/C, 38123 Povo (Trento), Italy

[5]Veneto Nanotech, Laboratorio Nanofab, Via delle Industrie 5, 30125 Marghera, (Venezia), Italy

*Corresponding author: E-mail: dbouajaj@yahoo.fr

Keywords: Down Conversion, Rare Earth, Sol Gel, Transfer Efficiency

Abstract. The efficiency of solar cells may be improved by better exploitation of the solar radiation managing the photons coming from the solar spectrum to better fit the absorption band of the employed solar cells. This can be done by inserting in the front or rear of the solar cell an optically active layer doped with rare earth ions which acts as down-converter or up-converter. In this work we will focus on down conversion process using cooperative energy transfer between a Tb^{3+} ion and two Yb^{3+} ions. Down converting silica-hafnia planar waveguides ($70\%SiO_2$–$30\%HfO_2$) doped with different concentrations of Tb and Yb ([Tb+Yb]/[Si+Hf] = 5%, 7%, 9%) were deposited on a silica glass substrate by a sol gel route using the dip-coating method and finally treated at 1000°C. The evaluation of the transfer efficiency between Tb^{3+} and Yb^{3+} is obtained by comparing the luminescence decay of Tb with and without Yb co-doping ions. A maximum transfer efficiency of 55% was found for the highest rare earth doping concentration.

Introduction

According to the International Energy Agency (IEA), 86.7% of the energy consumed in worldwide comes from non-renewable resources. To deal with climate change and the depletion of fossils resources, it is urgent to develop and improve the production of renewable energy. In particular, solar energy is an alternative, the resource is inexhaustible.

One of the key to improve the solar cells performance is the adequacy of the solar spectrum with a semiconductor's band-gap [1]. The most used solar cells are based on crystalline Si (c-Si). In this context, the down-conversion would be beneficial if a photon with a wavelength shorter than approximately 500 nm (around twice the energy of the silicon band gap) could be converted into two photons with wavelengths around 1000 nm (with a energy just above the band gap of c-Si) [2,3].

Rare earth ions are commonly used as a dopant because of their several electronic transitions in the visible and infrared, and their relative insensitivity to the matrix in which they are incorporated [4,5].

The ytterbium (Yb^{3+}) ion is typically used for its emission band in the near infrared, combined with praseodymium ion (Pr^{3+}) or thulium (Tm^{3+}) or chromium ion (Cr^{3+}) or terbium ion (Tb^{3+}) for their absorption band in blue region. The rare earth chosen in this work are Tb^{3+} and Yb^{3+}. The down conversion process using cooperative energy transfer between a Tb^{3+} ion and two Yb^{3+}ions permits to cut high energy photon at wavelength shorter than 488 nm into two low energy photons around 980nm[6].

Dielectric Materials and Applications: ISyDMA'2016   Materials Research Forum LLC
Materials Research Proceedings 1 (2016) 218-221   doi: http://dx.doi.org/10.21741/2474-395X/1/54

The choice of the matrix is a crucial point to obtain an efficient down-conversion process. The materials need to be chosen to minimize non-radiative transition process from the rare-earth ions to the host matrix[6]. Sol gel-derived silica-hafnia is a reliable and flexible system that has proved to be suitable for rare earth doping and fabrication of glass-ceramic planar waveguides. In silica-hafnia glass-ceramic the rare earth ions are embedded in hafnia nanocrystals which have a cut-off frequency of about 700 cm^{-1}[7]. However, the presence of hafnia nanocrystal produces a strong reduction of the non-radiative transition process reflected by a lengthening of the measured emission lifetime.

In the current paper we present the result achieved from $70SiO_2$-$30HfO_2$ samples activated by different molar concentrations of rare earths [Tb + Yb]/[Si + Hf] = 5%, 7%, 9%, prepared by sol–gel route using the dip-coating technique. All samples were in form of glass–ceramic (GC) treated at 1000°C.

Experimental

Silica-hafnia ($70SiO_2$-$30HfO_2$) samples activated by different molar concentrations of terbium and ytterbium ions were prepared by sol–gel route using the dip-coating technique, keeping constant the rate [Yb]/[Tb] = 4, following the experimental procedure described in [2,3]. Silica-hafnia films were deposited on cleaned pure SiO_2 substrates and the final films, obtained after 20 dips, were stabilized by a treatment of 5 min in air at 900 °C. As a result of the procedure, transparent and crack-free films were obtained (G samples). To obtain GC samples, an additional heat treatment was performed in air at a temperature of 1000°C for 30 min in order to nucleate hafnia nanocrystals inside the film. $70SiO_2$-$30HfO_2$ GC planar waveguides doped with rare earth ions were thus produced. Table 1 gives the compositional and optical parameters of the obtained silica–hafnia GC planar waveguides.

The thickness of the waveguides and the refractive index at 632.8 and 543.5 nm were obtained by a m-lines apparatus (Metricon, mod2010) based on the prism coupling technique, using a Gadolinium Gallium Garnet (GGG) prism, with the setup reported in [8]. The characterization techniques used in this study are: Transmission Electron Microscopy (TEM), photoluminescence (PL) and for the lifetime analysis the spectrofluorimeter was operated in Multi Channel Scaling (MCS) mode. The details are reported in [9]

Table I Rare earth concentration, refractive index and layer thickness of the prepared silica–hafnia gc waveguides.

Sample label	[Tb] mol%	[Yb] mol%	n@543.55 nm [± 0.001]		n@632.8 nm [± 0.001]		Number of dip	Layer thickness [±0.2μm]
			TE_0	TM_0	TE_0	TM_0		
BR3GC	1	0	1,5991	1,5826	1,5849	1,5676	20	0,6487
B3GC	1	4	1,6097	1,5977	1,5960	1,5821	20	0,6676
BR4GC	1.4	0	1.5914	1.5793	1.5787	1.5650	20	0.7123
B4GC	1.4	5.6	1,6187	1,6069	1,6056	1,5922	20	0,7091
B5RGC	1.8	0	1,6109	1,5916	1,5967	1,5759	20	0,6467
B5GC	1.8	7.2	1.6097	1.5833	1.5975	1.5833	20	0.7495

Result

Fig. 1 shows the TEM images of the most doped G and GC samples, with a total rare earth content of [Tb + Yb]/[Si + Hf] = 9%. The only difference in the two samples is the final annealing temperature at 900°C or 1000°C. It is worth observing that the sample treated at lower temperature is amorphous, while the higher temperature treatment induces the precipitation of small crystallites which size is of the order of 3–4 nm. This is fully in agreement with the XRD reported in [10].

Dielectric Materials and Applications: ISyDMA'2016   Materials Research Forum LLC
Materials Research Proceedings **1** (2016) 218-221   doi: http://dx.doi.org/10.21741/2474-395X/1/54

The photoluminescence spectra of the Tb^{3+}-Yb^{3+}co-doped B5GC sample under the 270 nm excitation are shown in Figure 2. The intense emission band centered at 977 nm, with a shoulder at 1027 nm, attributed to the $^2F_{5/2} \rightarrow {}^2F_{7/2}$ transition of Yb^{3+} ions. The emission of the Yb^{3+} ion after excitation is an indication of the presence of an efficient energy transfer from Tb^{3+} to Yb^{3+} implying an effective down-conversion.

In figure 3, the decay curves of the Tb^{3+} $^5D_4 \rightarrow {}^7F_5$ emission at 543.5 nm are plotted for B5GC sample. Nearly single exponential luminescence decay is observed for the sample without Yb^{3+}. The fast luminescence decay observed for the co-doped samples is attributed to the energy transfer from the Tb^{3+}: $^5D_4$ to the Yb^{3+}: $^2F_{5/2}$ [11].

The energy transfer efficiency η_{Tb-Yb} can be obtained experimentally by dividing the integrated intensity of the decay curves of the $Tb^{3+} \rightarrow Yb^{3+}$ co-doped glass ceramics by the integrated intensity of the Tb^{3+} single doped curve [11]:

$$\eta_{Tb-Yb} = 1 - \frac{\int I_{Tb-Yb}dt}{\int I_{Tb}dt} \quad (1)$$

The effective quantum efficiency can be defined by the ratio between the number of emitted photons and the number of photons absorbed by the material. In our case, a perfect down-conversion system would have an effective quantum efficiency value of 200%, corresponding to the emission of two photons for one absorbed. The relation between the transfer efficiency and the effective quantum efficiency is linear [11] and is defined as:

$$\eta_{EQE} = \eta_{Tb-r}(1 - \eta_{Tb-Yb}) + 2\eta_{Tb-Yb} \quad (2)$$

Where, the quantum efficiency for Tb^{3+} ions η_{Tb-r} is set equal to 1. The evaluated values of energy transfer efficiency and effective quantum efficiency for the different samples are reported in Table 2.

We can see clearly that that the transfer efficiency increase with the increase of the total rare earth content, with the highest value for the [Tb+Yb] = 9% is about 55%.

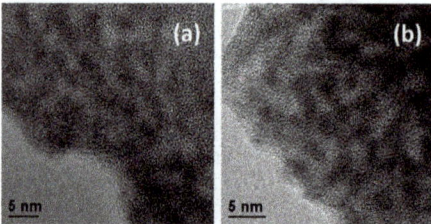

Fig 1: TEM images of $70SiO_2$–$30HfO_2$ G (a) and GC (b) waveguides

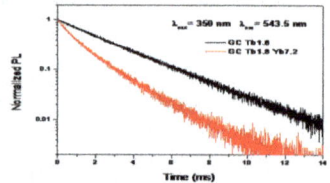

Figure 3: Decay curves of the luminescence from $^5D_4$ metastable state of Yb^{3+} for B5GC

Table 2 Transfer efficiency and effective quantum efficiency as function of (tb+yb) molar concentration for glass-ceramic samples (gc) with constant molar ratio yb/tb=4.

Composition (Tb+Yb) concentration in mol %)	5%	7%	9%
Transfer efficiency	24.6%	32.3%	54.6%
Effective quantum efficiency	124.6%	132.3%	154.6%

Conclusion

$70SiO_2$–$30HfO_2$ glass ceramics co-doped Tb^{3+}/Yb^{3+} with different molar concentrations of rare earths [Tb + Yb]/[Si + Hf] = 5%, 7%, 9% were prepared by sol-gel method and dip coating processing. TEM images show that the crystallization occurs after heat treatment at 1000 °C. Near-infrared emission at 980 nm assigned to the $^2F_{5/2} \rightarrow {}^2F_{7/2}$ transition

Dielectric Materials and Applications: ISyDMA'2016   Materials Research Forum LLC
Materials Research Proceedings 1 (2016) 218-221   doi: http://dx.doi.org/10.21741/2474-395X/1/54

of the Yb^{3+} ions was observed upon excitation at 270 nm. The energy transfer efficiencies were estimated from the decay curves of the $^5D_4$ metastable state of the Tb^{3+} ion. The Tb–Yb energy transfer efficiency increases with the increase of [Tb + Yb] concentration and the value found with B5GC sample (Tb1.8, Yb7.2) is more significant (54.6 %).

References

[1] C. Strumpel, M. McCann, G. Beaucarne, V. Arkhipov, A. Slaoui, V. Svrcek, C. del Cañizo, I. Tobias, Modifying the solar spectrum to enhance silicon solar cell efficiency – an overview of available materials, Sol. Energy Mater. Sol. Cells 91 (2007) 238–249. http://dx.doi.org/10.1016/j.solmat.2006.09.003

[2] G. Alombert-Goget, C. Armellini, A. Chiappini, A. Chiasera, M. Ferrari, S. Berneschi, M. Brenci, S. Pelli, G.C. Righini, M. Bregoli, A. Maglione, G. Pucker, G. Speranza, Frequency converter layers based on terbium and ytterbium activated HfO2 glass–ceramics, Proc. SPIE 7598 (2010). 75980P–1/9.

[3] G. Alombert-Goget, C. Armellini, S. Berneschi, A. Chiappini, A. Chiasera, M. Ferrari, S. Guddala, E. Moser, S. Pelli, D.N. Rao, G.C. Righini, Tb3+/Yb3+ coactivated Silica–Hafnia glass ceramic waveguides, Opt. Mater. 33 (2010) 227–230. http://dx.doi.org/10.1016/j.optmat.2010.09.030

[4] A. M. Srivastava, D. A. Doughty, W. W. Beers, "Photon Cascade Luminescence of Pr3+ in LaMgB5O10," J. Electrochem. Soc. 143 (12), 4113-4116 (1996). http://dx.doi.org/10.1149/1.1837346

[5] W. Strek, P. Deren, A. Bednarkiewicz, "Cooperative processes in KYb(WO4)2 crystal doped with Eu^{3+} and Tb^{3+}ions," J. Lumin. 87-89, 999-1001 (2000). http://dx.doi.org/10.1016/S0022-2313(99)00505-0

[6] B.S Richards, "Luminescent layers for enhanced silicon solar cell performance: Down conversion," Solar Energy Material & Solar Cells 90, 1189-1207 (2006). http://dx.doi.org/10.1016/j.solmat.2005.07.001

[7] D. A. Neumayer and E. Cartier, "Materials characterization of Zro2-SiO2 and HfO2-SiO2 binary oxides deposited by chemical solution deposition," J. Appl. Phys. 90, 1801-1808 (2001). http://dx.doi.org/10.1063/1.1382851

[8] S. Ronchin, A. Chiasera, M. Montagna, R. Rolli, C. Tosello, S. Pelli, G.C. Righini, R. R. Goncalves, S.J.L. Ribeiro, C.S. De Bernardi, F. Pozzi, C. Duverger, R. Belli, M. Ferrari, Erbium-activated silica–titania planar waveguides prepared by rf-sputtering, SPIE 4282 (2001) 31. http://dx.doi.org/10.1117/12.424788

[9] A. Bouajaj, S. Belmokhtar, M.R. Britel, C. Armellini, B. Boulard, F. Belluomo, A. Di Stefano, S. Polizzi, A. Lukowiak, M. Ferrari, F. Enrichi. Tb^{3+}/Yb^{3+} codoped silica–hafnia glass and glass–ceramic waveguides to improve the efficiency of photovoltaic solar cells. Optical Materials 52 (2016) 62–68. http://dx.doi.org/10.1016/j.optmat.2015.12.013

[10] S. Belmokhtar, A. Bouajaj, M. Britel, S. Normani, C. Armellini, B. Boulard, F. Enrichi, F. Belluomo, A. Di Stefano, M. Ferrari. Energy transfer from Tb^{3+} to Yb^{3+} in silica hafinia glass ceramic for photovoltaic applications. J. Mater. Environ. Sci. 7 (2) (2016) 515-518. ISSN : 2028-2508 CODEN: JMESCN.

[11] P. Vergeer, T. J. H. Vlugt, M. H. F. Kox, M. I. Den Hertog, J. P. J. M. van der Eerden, and A. Meijerink, "Quantum cutting by cooperative energy transfer in YbxY1−xPO4:Tb^{3+}," Phys. Rev. B 71, 014119-1 – 014119-11 (2005)

Dielectric Materials and Applications: ISyDMA'2016
Materials Research Proceedings 1 (2016) 222-225

Materials Research Forum LLC
doi: http://dx.doi.org/10.21741/2474-395X/1/55

Optimizing the consumption of a pottery kiln in Salé (Oulja)

Maha BAKKARI, Fatiha LEMMINI, Kamal GUERAOUI*

Department of physics, Faculty of sciences, Mohammed V University, Rabat, Morocco

*Corresponding author: E-mail: kgueraoui@yahoo.fr

Abstract. Reducing the energy consumption is one of the main challenges of the global economy. Several experimental studies have shown that the energy consumption in a pottery kiln is high and the excess consumption is due to heat losses of the combustion gas leaving the furnace through the chimney. The recovery of this heat can lead to big energy savings. To study the impact of heat losses, from the furnace, on energy consumption we developed a digital program that determines the temperatures of the kiln. Thermal losses of the kiln's walls are computed which helps to deduce the losses of the combustion gas.

Keywords: Kiln, Furnace, Energy Consumption, Pottery, Heat Losses

Introduction

The objective of this work is the study of the thermal behavior of a kiln used for baking pottery pieces, in order to optimize the energy consumption to meet Morocco's commitments under the conventions climate change.

Study the gas kiln

A. Description of the kiln

Figure 1 shows the photo of the furnace.

Fig. 1. Kiln's photo

B. Thermal balance of the kiln

The useful energy for the pottery items is given by the equation:

$$Q_u = m_a c_{pa}(T_f - T_i) + m_e c_{pe}(100 - T_i) + m_e L_{ev} + m_e c_{pv}(T_f - 100) \qquad (1)$$

This heat is composed of:
- A sensible heat to bring the clay mass (m_{ar}) from the initial temperature (T_i) at the final temperature (T_f) ;
- A latent heat (L_{ev}) to evaporate the water mass (m_e) at 100°C ;
- A sensitive heat to bring the mass of water (m_e) from 100°C to the final temperature (T_f).

The heat from the burners "Series of Energy Management "Ovens, dryers and kilns", Energy and Resources, Canada, is given by:

Dielectric Materials and Applications: ISyDMA'2016   Materials Research Forum LLC
Materials Research Proceedings 1 (2016) 222-225   doi: http://dx.doi.org/10.21741/2474-395X/1/55

$Q_C = PCI * Mass\ flow$ (2)

The burners produce a high velocity flame which gives an important heat exchange and a uniform temperature in the furnace [1].

The PCI is the heat of combustion of the combustible gas which is the enthalpy of reaction [2].

Experimentation

A. Useful and consumption Energies

The heat transmitted by the fuel turns into: - useful heat to the products and - heat losses to the environment (combustion gas leaving through the chimney and heat transmitted from the furnace walls).

Several studies have shown that the chimney loss can reach 65% "Series of Energy Management "Ovens, dryers and kilns", Energy and Resources, Canada.

Heat losses through the furnace walls are given by:

$$q_p = \frac{\Delta T}{\sum R_\lambda}$$ (3)

- ΔT is the temperature potential in the walls;
- $\sum R_k$ is the thermal resistance of the wall.

Using experimental measurements we have calculated the losses through the furnace walls (Fig.2).

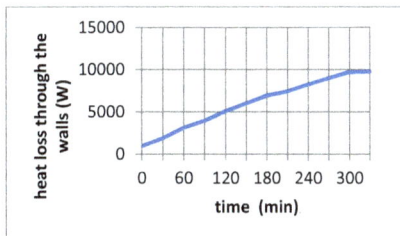

Fig.2. Wall heat loss

Table1 provides the combustion energy, the useful energy, the lost energy and their percentages.

Table1 Combustion energy, useful energy, lost energy and their percentages

Combustion energy (MJ)	Useful energy (MJ)	Lost energy (MJ)	
		Through furnace walls	through chimney
3817,145	139,41	1680,05	1997,69
100%	4%	44%	52%

To explore the possibility of energy recovery, a numerical model was developed to calculate losses through the furnace walls in order to determine the chimney losses.

Modeling the heat transfer through the furnace walls

The modeling is based on the transfer of heat through the wall of the kiln (Fig.3).

Dielectric Materials and Applications: ISyDMA'2016   Materials Research Forum LLC
Materials Research Proceedings 1 (2016) 222-225   doi: http://dx.doi.org/10.21741/2474-395X/1/55

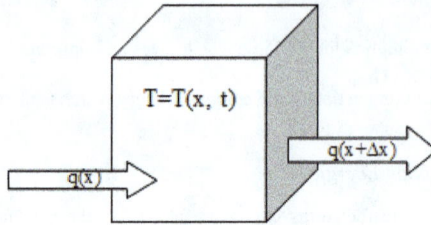

Fig. 3.Heat transfer through the wall

$$\frac{\partial}{\partial x}(k\frac{\partial T}{\partial x}) = \rho c \frac{\partial T}{\partial t} \tag{4}$$

The following tables give the values of thermal conductivity (K) in different temperatures [3]:

Table 2 Variation of thermal conductivity for ceramic fibr

T(k°)	293	323	366	373	423
k(W/m².K°)	0,04	0,041	0,05	0,054	0,066

Table 3 Variation of thermal conductivity for rok wool

T (K°)	533	856	1089	1255	1366
K (W/m².K°)	0,06	0,13	0,23	0,3	0,34

By using the moinder carree method, we obtain the following equations which give the thermal conductivity (K) as a function of temperature:
- For Ceramic fiber:

$$k(T) = 0.1070 * e^{0.00016*T(x)} \tag{5}$$

- For Rock wool:

$$k(T) = 0.0557 * \log(T(x)) - 0.2769 \tag{6}$$

The figure representing experimentally measured temperatures and those obtained by the program shows a good agreement.

Dielectric Materials and Applications: ISyDMA'2016   Materials Research Forum LLC
Materials Research Proceedings **1** (2016) 222-225   doi: http://dx.doi.org/10.21741/2474-395X/1/55

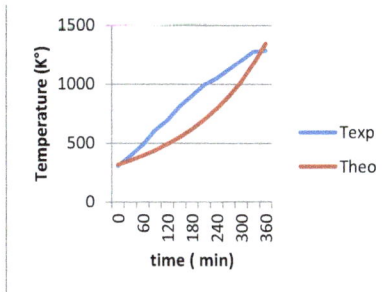

Fig. 4. Experimental temperature values (Texp) & Theoretical (Theo).

Figure 4 shows the comparison between the results obtained experimentally and the ones obtained numerically. We can note the good correlation between the two ways. This will lead us perform numerically without using expensive materiel.

Conclusion

The experimental study of the gas furnace for pottery shows that the losses through the chimney exceed 50%.

To investigate the possibility of recovering this heat supplied to the furnace by the combustion gases, a numerical model was developed.

This numerical model has been validated by comparing the experimental values of temperature inside the kiln with those computed by the model.

References

[1] L. Balli, A. Touzani,"Experimental Study and modeling of heat exchange in four to Pottery", TERMOTEHNICA, VOL. 1/2014.

[2] Industrial Chemistry, Volume 2.

[3] Morgan Thermal Ceramics Superwool plus Insulating Fiber.

Dielectric Materials and Applications: ISyDMA'2016   Materials Research Forum LLC
Materials Research Proceedings 1 (2016) 226-229   doi: http://dx.doi.org/10.21741/2474-395X/1/56

Copper oxide nanowires synthetized by facile route for electrical application

Naoual AL ARMOUZI*, Mesbah EL YAAGOUBI, Mustapha MABROUKI, Noureddine KOUIDER

Industrial Engineering Laboratory Sultan Moulay Slimane University Faculty of Science and Technology, Beni Mellal, Morocco

*Corresponding author: E-mail: nalarmouzi@gmail.com

Keywords: Coper Oxide, AFM, DRX, Nanowires, Nanoprticule

Abstract. Copper oxides have been synthesized by thermal oxidation of copper foils (Cu); the textural and structural properties of the films were characterized by atomic force microscopy (AFM) and X-ray diffraction (XRD). The effect of annealing temperature on the morphology of the nanowires is studied; it is found that the annealing temperature plays an important role in the morphology and in the crystallite size of CuO. Synthesis conditions, such as temperature and the nature of substrate, which is of importance for practical applications.

Introduction

The cupric oxide (CuO) is a P-type semiconductor with monoclinic crystal structure and 1.2eV band gaps [1]. It can be widely used in applications such as solar energy [2], sensors [3] catalysis [4] and photocatalytic [5] CuO is very promising for use in solar photovoltaic because of its excellent stability and good electrical properties.

Various types of CuO nanostructurescan be synthesized by several methods such as thermal evaporation [6] direct thermal oxidation [7], sol–gel [8] and Spray pyrolysis [9]. Among all sintering methods, the direct thermal oxidation is the simplest method and has the great potential to be adopted for large scale production of CuO nanowires.

In this work, we used a simple synthesis method for the preparation of CuO nanowires; it is the direct thermal oxidation. In comparison with other methods, CuO nanowires produced by direct thermal oxidation have excellent crystallinity.

Materials and methods

The annealed samples:

Synthesis of CuO nanowires was carried out by thermal oxidation of sheets copper (Cu) and (CuZn) of 2cm×2cm size, was first rinsed by acetone followed by de-ionized water several times. After it had been dried using an air gun, the Cu and CuZn plates was annealed in the furnace (Thermolyne Furnace 6000) at 300°C, 400°C, 500°C and 600°C for 24 h in air. CuO nanowires were contained in the black ash-like top layer formed on the Cu plate. After developing thin films of Cu, it is then conducted their structural and microstructural characterization by the X-ray diffraction (XRD-Bruker advance D8) and atomic force microscopy (AFM) Flex FM nanosurf using dynamic mode (tapping mode), with 133 Khz tip resonant frequency.

Results and discussion

A. Interpreting the results of DRX

The Ray diffraction pattern X,fig.1,shows that the both copper oxide phases CuO and Cu_2O are formed at the temperature 300°C, but the majority phase is the Cu_2O, because Cu_2O is the first phase formed after the first oxidation and the formation of CuO is slower. while the temperature is increased; more we facilitate the transformation of Cu_2O into CuO. The intensity of peak CuO increasedwhen the temperature of treatment increases from 300°C to 600°C.These observations are

Dielectric Materials and Applications: ISyDMA'2016   Materials Research Forum LLC
Materials Research Proceedings 1 (2016) 226-229   doi: http://dx.doi.org/10.21741/2474-395X/1/56

in accord with J.T. Chen and al (2008) [7]. We note a decrease in the peak intensity corresponding to the copper phase, and this decrease is due to the oxidation of copper which result the disappearance of the copper phase.

The copper oxide is monoclinic structurewith lattice parameters a = 4.6047Å, b= 3.4112Å and c= 5.1187Å, β= 99.35°.

The average crystallite size of CuO is obtained by Scherrer's relationship [10]: D= O,9λ /βcosθ

It was found between 144.9763 to 145.0269 nm.

We remark in fig.2 CuZn intesity decrease from 300°C to 600°C due to its oxidation but is not total, since still observed the CuZn phase in 600°C. And we noticed that the peak intensities of ZnO phase increases when the temperature increasesthe phases ofCuZn and ZnO are cubique and hexagonal with the lattice parameters a=b=c= 3.3347Å and α= β= ɤ= 90°and a= b= 3.2195Å, c=5.1682Årespectively. The average crystallite size of CuZn and ZnO is between 54.14 to 118.44 nm and 26.54 to 47.21 nm respectively.

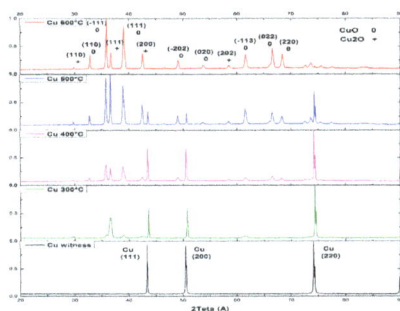

Fig. 1: XRD pattern of the copper oxide nanowires grown at different

Temperature

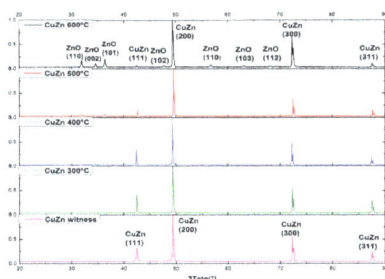

Fig. 2: XRD pattern of the copper-Zinc nanowires grown at different Temperature

B. *Interpreting the results of AFM images*

The thin film prepared showed a surface with the roughness of 10.10 and 17.165 nm at 300°C and 600°C respectively, and it was composed of big grains of CuO (Fig.3 A and Bimages in the left).The both images presented in the right shows The average roughness of the films of CuZn annealed at 300°C and 600∘C was 10.51 and 17.51 nm respectively, indicating that roughness increases with increase of annealing temperature.

Dielectric Materials and Applications: ISyDMA'2016   Materials Research Forum LLC
Materials Research Proceedings **1** (2016) 226-229   doi: http://dx.doi.org/10.21741/2474-395X/1/56

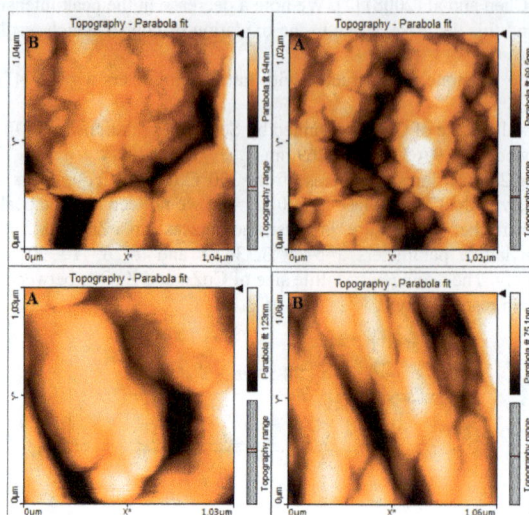

Fig. 3: AFM images for Cu A and B in the left annealing in 300°C and 600°C respectively. CuZn images A and B in the right, annealing in 300°C and 600°C respectively (scan area, 1μm×1 μm)

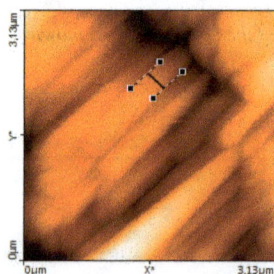

Fig. 4: AFM image for CuO nanowires (scan area, high resolution, 3 μm×3 μm)

We can see that the AFM images at high resolution revealed the formation of CuO nanowires with a diameters ranging between 127nm and 283nm and a length ranging between 720.3 nm and 1.904μm.

Conclusion

Copper oxide nanowires have been successfully prepared under different temperature from 300°C to 600°C. The prepared high purity CuO nanoparticles were investigated using combined techniques of XRD and AFM. The results show that when the temperature increse, the peak of CuO increse. Moreover the film roughness is higher.We obtain CuO nanowires only from Cu substrate.

Dielectric Materials and Applications: ISyDMA'2016   Materials Research Forum LLC
Materials Research Proceedings **1** (2016) 226-229   doi: http://dx.doi.org/10.21741/2474-395X/1/56

References

[1] Cho, Y.-S. and Y.-D. Huh. "Preparation of CuO hollow spheres by oxidation of Cu microspheres." Bulletin of the Korean Chemical Society 30(6): 1410-1412, 2009. http://dx.doi.org/10.5012/bkcs.2009.30.6.1410

[2] Dhanasekaran, V. and T. Mahalingam."Surface modifications and optical variations of ($-$ 111) lattice oriented CuO nanofilms for solar energy applications." Materials Research Bulletin 48(9): 3585-3593, 2013. http://dx.doi.org/10.1016/j.materresbull.2013.05.072

[3] S. H. Kim, A. Umar, S-W. Hwang. "Rose-like CuO nanostructures for highly sensitive glucose chemical sensor application." Ceramics International 41(8): 9468-9475, 2015. http://dx.doi.org/10.1016/j.ceramint.2015.04.003

[4] Chao Yang, Jide Wang, Feng Xiao, Xintai Su. "Microwave hydrothermal disassembly for evolution from CuO dendrites to nanosheets and their applications in catalysis and photo-catalysis." Powder Technology 264: 36-42, 2014. http://dx.doi.org/10.1016/j.powtec.2014.05.012

[5] Liu, X., Z. Li, Q. Zhang, F. Li, T. Kong. "CuO nanowires prepared via a facile solution route and their photocatalytic property." *Materials Letters*, 72 : 49-52, 2012. http://dx.doi.org/10.1016/j.matlet.2011.12.077

[6] Huang, L., et al. "Preparation of large-scale cupric oxide nanowires by thermal evaporation method." *Journal of Crystal Growth*, 260 : 130-135, 2004. http://dx.doi.org/10.1016/j.jcrysgro.2003.08.012

[7] J.T. Chen, F. Zhanga, J. Wanga, G.A. Zhanga, B.B. Miaoa,X.Y. Fana, D. Yana, P.X. Yan."CuO nanowires synthesized by thermal oxidation route." Journal of Alloys and Compounds 454(1): 268-273, 2008. http://dx.doi.org/10.1016/j.jallcom.2006.12.032

[8] Armelao, L., et al. "A sol–gel approach to nanophasic copper oxide thin films" *Thin Solid Films*, 442 :48-52, 2003. http://dx.doi.org/10.1016/S0040-6090(03)00940-4

[9] Chiang, C.-Y., C- Y. Chiang, K. Aroh, h. H. Ehrman. "Copper oxide nanoparticle made by flame spray pyrolysis for photoelectrochemical water splitting–Part I. CuO nanoparticle preparation." *international journal of hydrogen energy*, 37 : 4871-4879,2012.

[10] HADJERSI, F. "Investigation des propriétés structurales, optiques et électriques des films ITO élaborés par pulvérisation cathodique RF; Effet du recuit ", 2011.

Dielectric Materials and Applications: ISyDMA'2016
Materials Research Proceedings 1 (2016) 230-233

Materials Research Forum LLC
doi: http://dx.doi.org/10.21741/2474-395X/1/57

Adhesion of staphylococcus aureus on thin films of zinc oxide

Y. NAJIH[1], M. ELYAAGOUBI[1], N. KOUIDER[1], J. BENOURRAM[1], M. MABROUKI*[1], H. LATRACHE[2]

[1]Mechanical Engineering dept. Industrial Engineering Laboratory, Sultan MoulaySlimane University, FST, BeniMellal, Morocco

[2]Bioprocessing and Bio-interfaces Laboratory, Sultan MoulaySlimane University, FST, BeniMellal, Morocco

*Corresponding author: E-mail: mus_mabrouki@yahoo.com

Keywords: Staphylococcus Aureus, Surface Energy, Atomic Force Microscopy, Zinc Oxide, Contact Angle

Abstract. The general context of this work is to study the effect of physicochemical and morphological properties on bacterial adhesion. In this case, the adhesion of *Staphylococcus aureus* (gram positive bacteria) has been studied in thin films of zinc oxide (ZnO). The contact angle measurements were used to calculate the surface energy components of each substrate and analysis of adherent cells on the surface was made by the atomic force microscopy. The results showed that the glass surface is governed by short range forces (Lewis acid–bases forces) and that ZnO film is governed by long range forces (van der Waals forces). Images obtained by atomic force microscopy revealed that the ZnO layer has an antibacterial effect against *S. aureus*.

Introduction

The relevance of contaminated surfaces in spreading pathogenic microorganisms to foods is already well established for food processing plants [1-2]. *Staphylococcus aureus*is among the most common pathogenic bacteria isolated from surfaces in food processing plants [3], where it can adhere and form biofilms [4]. Several substratum surface properties have been shown to influence bacterial adhesion. These properties include surface charge [5], topography [6], hydrophobicity, surface chemistry and surface energy [7]. To contribute to the understanding of the phenomenon of adhesion, we determined the physicochemical properties of the solid surface. In our case we have prepared thin layers of zinc oxide (ZnO) on glass slides by the sputtering under different powers. ZnO are used as quantitative labels for biological assays [8]. The objective was to predict the adhesion of *S. aureus* on glass substrates with a ZnO film. This bacterium was chosen as a model in this study because it has thes ability to adhere and colonizethe inert surface.

Material and methods

Microorganisms, culture conditions, and sample preparation

The strain used as a model for this study was *S. aureus* ATCC 25923 Bacteria were incubated over night at 37 ° C in Luria Bertani medium containing the following (per liter of distilled water). Tryptone 10 g, 5 g of yeast extract and 10 g NaCl. After 24 h incubation, cells were harvested by centrifugation for 15 min at 8400 g and washed twice with and suspended in 0.1 M KNO_3 solution.

ZnO thin films were prepared by reactive RF-sputtering in an ALCATEL SC451 deposition system equipped with an ALCATEL ARF 601 RF generator operating at 13.56MHz witch was described in previous work [9].

Surface caracterisation

The contact angle measurements were performed using a goniometer (GBX instruments, France) by the sessile drop method witch was described in previous work [9].

Dielectric Materials and Applications: ISyDMA'2016   Materials Research Forum LLC
Materials Research Proceedings 1 (2016) 230-233   doi: http://dx.doi.org/10.21741/2474-395X/1/57

Atomic force microscopy is a scanning probe microscopy instrument that consists of a force-sensing cantilever, a piezoelectric scanner, and a photodiode detector. TheAFM characterization was carried out on a Nano-surf Flex FM in the non-contact tapping mode.

Surface Energy calculation and hydrophobicity

The components of the surface energy of a surface were determined by carrying out contact angle measurements using three probe liquids with known surface tension parameters and using Young Van Oss equation [10] :

$$(\cos \theta + 1)/2 = (\gamma_s^{LW}\gamma_L^{LW})^{1/2} / \gamma_L + (\gamma_s^+\gamma_L^-)^{1/2}/\gamma_L + (\gamma_s^-\gamma_L^+)^{1/2}/\gamma_L \qquad (1)$$

According to van Oss[10] the hydrophobicity of a given material (s) can be defined in terms of the variation of the free energy of interaction between two moieties of that material immersed in water (w):

$$\Delta G_{iwi} = -2[((\gamma_i^{LW})^{1/2} - (\gamma_i^{LW})^{1/2})^2 + 2((\gamma_i^+\gamma_i^-)^{1/2} + (\gamma_w^+\gamma_w^-)^{1/2} - (\gamma_i^+\gamma_w^-)^{1/2} - (\gamma_w^+\gamma_i^-)^{1/2})] \qquad (2)$$

Where:

θ : Contact angle between liquid and surface

ΔG_{iwi} : Free energy of interaction

γ_i: Surface energy

γ_i^{LW} : Lifshitz van der Waals component of surface energy

γ_i^+: Acid component, γ_i^-: base component

L: liquid, S: solid

Results

Surface physic-chemical analyses

The physicochemical characteristics of different ZnO substrates are calculated from the Young-van Oss equation. Table 1 summarizes the values of the surface energy components of each substrate (Lifshitz / van der Waals and acid / base component).

TABLE 1: Values of the contact angle, and surface energy components [mJ/m².].

Substratum	contact angle			surface energy component					ΔG_{iwi}
	Θ_W	Θ_F	Θ_D	γ^{LW}	γ^+	γ^-	γ^{AB}	γ^{Tot}	
ZnO 150 w	70	51.9	36.3	43	0.1	12.2	2.21	45.21	-36.61
ZnO 200 w	78.4	65	40.5	39.04	0.2	11	3.00	42.40	-37.08
ZnO 250 w	75.8	57.4	39.3	40	0.01	9.3	0.70	40.70	-44.97
Glass	34.2	31	29.3	32.7	1.9	45.3	18.5	50.1	22.48

The characteristics presented in Figures 1 summarizes the results obtained.

Dielectric Materials and Applications: ISyDMA'2016   Materials Research Forum LLC
Materials Research Proceedings 1 (2016) 230-233   doi: http://dx.doi.org/10.21741/2474-395X/1/57

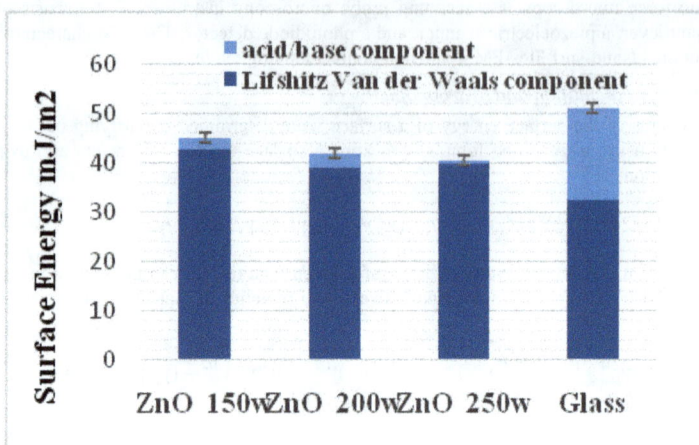

Fig. 1:Componentsofthe surface energy (Lifshitz Van der Waals / acid-base)fortheglassandforthe
varioussubstratesof the ZnO film (mJ/m^2).

Bacterial adhesion

Figure 2 and 3summarizes the comparison of the numbers of adherent cells on the two surfaces as a
function of time.

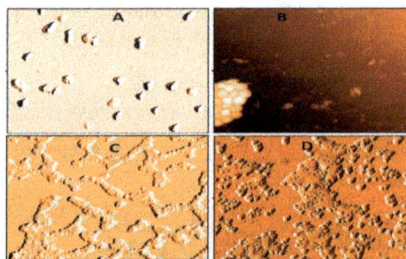

Fig. 2:AFM 2D Topography, Images of Staphylococcus aureus A : 1 hour and C : 2 hours on
the glass surface. B : 1 hour and D : 2 hours on ZnOsurface.

Discussion

The analysis by atomic force microscopy shows that the glass surface has a favorable environment
for the growth of colonies of S. aureus cells. As function of time, the cells of *S. aureus* adhere well to
the surface of glass and still colonized the entire surface. Opposed to the surface of ZnO the number
of adherent cells has decreased during the second hour of suspension.

Conclusion

ZnO thin films were deposited on glass substrate using sputtering, with different power values. The
effect of physicochemical properties on the adhesion of *Staphylococcus aureus* has been well
observed in this work. Analysis by AFM showed a difference between *S. aureus* adhesion on the

Dielectric Materials and Applications: ISyDMA'2016   Materials Research Forum LLC
Materials Research Proceedings 1 (2016) 230-233   doi: http://dx.doi.org/10.21741/2474-395X/1/57

glass surface and on the surface of ZnO films. This shows that ZnO thin films has excellent anti adhesion activity verses*S. aureus.*

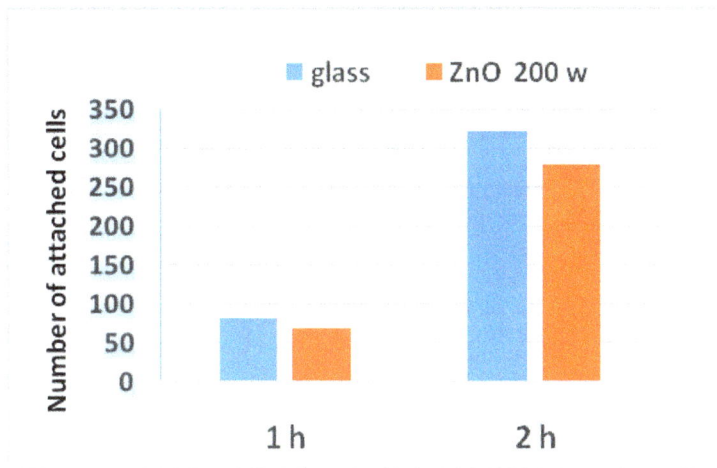

Fig. 3:Number of attached cells on the glass surface and the surface of ZnO Film

References

[1] H.D. Kusumaningru.,G. Riboldi,W.C Hazeleger, and Beummer, R.R. Int. J. Food Microbiol., vol.85, n.3, pp.227-236, 2003.

[2] N. Fuster-Valls, M. Hernández-Herrero, M. Marín-De-Mateo, and J.J. Rodríguez-Jerez, Food Control. , v.19, n.3, pp.308-314, 2008. http://dx.doi.org/10.1016/j.foodcont.2007.04.013

[3] S. Ammor, I. Chevallier,A. Lague, J. Labadile, R. Talon, andE. Dufour, Food Microbiol, v.21, n.1, pp.11-17, 2004. http://dx.doi.org/10.1016/S0740-0020(03)00051-0

[4] L. Kunigk, M.C.B.Almeida and J. Braz. Microbiol, v.32, n.1, p.38-41, 2001. http://dx.doi.org/10.1590/S1517-83822001000100009

[5] G. Harkes, J. Feijen and J. Dankert, Biomaterials,vol. 12, pp. 853, 1991. http://dx.doi.org/10.1016/0142-9612(91)90074-K

[6] A.M. Gallardo-Moreno, M.L. Gonzalez-Martin, J.M. Bruque andC. Perez-Giraldo, Colloids Surf A Physicochem. Eng. Asp, vol. 99, pp. 249, 2004.

[7] A.W.J. Van Pelt, A.H. Weerkamp, M.H.W.J.C. Uyen, H.J. Busscher, H.P. de Jong and J. Arends, Appl. Environ. Microbiol, vol. 49, pp. 1270 1985.

[8] B. Gu, C. Xu, C. Yang, S. Liu and M. Wang. Biosens. Bioelectron, vol. 26 pp. 2720 - 2723, 2011. http://dx.doi.org/10.1016/j.bios.2010.09.031

[9] Y. Najih, M. Elyaagoubi, N. Kouider,J. Bengourram and M. Mabrouki,International Journal of Emerging Technology and Advanced Engineering, , vol. 5, Issue. 8, pp. 345-35,August 2015.

[10] CJ.Van Oss, Curr. Opin. Coll. Interf, Sci. 2pp. 503–512, 1997.

Dielectric Materials and Applications: ISyDMA'2016
Materials Research Proceedings 1 (2016) 234-237

Materials Research Forum LLC
doi: http://dx.doi.org/10.21741/2474-395X/1/58

Theoretical modelling of the temperature effect in dye sensitized solar cells

A. ABOULOUARD*[1], A. JOUAITI[1], B. ELHADADI[1], M. ADAR[2]

[1]Laboratory of Sustainable Development, Faculty of Sciences and Technologies, USMS, Beni Mellal, Morocco

[2]Industrial Engineering Laboratory, Faculty of Sciences and Technologies,USMS, Beni Mellal, Morocco

*Corresponding author: E-mail: a.aboulouard@gmail.com

Keywords: Dye Sensitized Solar Cells, Simulation, Temperature

Abstract. A computer simulation and calculation method have been proposed in this work to provide the effect of temperature on dye sensitized solar cells (DSSCs).the result shows that the temperature has an effect on the open circuit voltage and efficiency.

Introduction

Over the past two decades, a lot of works have been done on the behavior of semiconductor nanoporous thin film electrodes [1,2,3]. A simple theoretical modelling of the temperature effect in dye sensitized solar cells is presented in this paper based on the differential equation which is described in [3,4,5]. Using Matlab tool a computer program has written in order to calculate the temperature effect.

The model

This study has done under a steady-state condition of an irradiated DSSCs, the injected electron in the excited dye molecules flows in the porous semiconductor (TiO$_2$) thin film, and recombines with the electrolyte at the TiO$_2$/electrolyte interface.

The processes is described by the following diffusion differential equation [3,4,5]:

$$D\frac{\partial^2 n(x)}{\partial x^2} - \frac{n(x) - n_0}{\tau} + \phi\alpha e^{-\alpha x} = 0 \tag{1}$$

Where n(x) is the excess concentration of the photo generated electrons at position within the film measured from the TiO$_2$/transparent conducting oxide (TCO) interface, n_0 is the concentration of electrons under equilibrium conditions in dark, τ is the conduction band free electrons life time, D is the diffusion coefficient of electrons, Φ is the light intensity; and α is the light absorption coefficient of the porous film.

The possibility of trapping-detrapping electrons was not taken into consideration in (1) because it is only important under nonsteady-state conditions [2]. Under short-circuit conditions, electrons are easily extracted as photocurrent and none of electrons are drawn directly on the counter electrode.

Therefore, the two boundary conditions are [3,4,5]:

$$n(0) = n_0 \tag{2}$$

$$\left(\frac{dn}{dx}\right)_{x=d} = 0 \tag{3}$$

Where d is the thin film electrode thickness.

Dielectric Materials and Applications: ISyDMA'2016   Materials Research Forum LLC
Materials Research Proceedings 1 (2016) 234-237   doi: http://dx.doi.org/10.21741/2474-395X/1/58

An optimal thickness of the electrode has used in the current study (5μm) according to Meng description [6].

The short-circuit current density J_{SC} can thus be obtained as [3,4,5]:

$$J_{SC} = \frac{q\phi L\alpha}{1-L^2\alpha^2}\left[-L\alpha+\tanh(\frac{d}{L})+\frac{L\alpha e^{-d\alpha}}{\cosh(d/L)}\right] \qquad (4)$$

Where q is the charge of an electron and L is the electron diffusion length given by [3,4,5]:

$$L = \sqrt{D\tau} \qquad (5)$$

If the DSSCs operates under a potential difference V between the Fermi level of the TiO_2 and the redox potential of the electrolyte, the density of the electrons at the TiO_2/TCO interface (x = 0) increases to n giving a new boundary condition [3,4,5] :

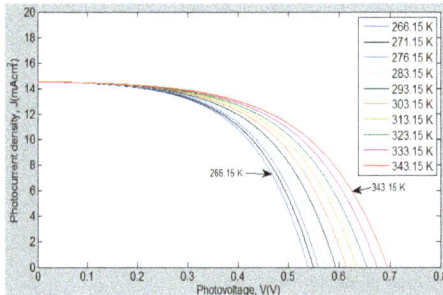

Fig. 1. Variation of the DSSCs J-V characteristics with different temperature.

$$n(0) = n \qquad (6)$$

The second boundary condition at x = d remains unchanged as shown in (3). Solving (1) yields give us the relationship between J and V [3,4,5].

$$V = \frac{kTm}{q}\ln\left[\frac{L(J_{SC}-J)}{qDn_0\tanh(\frac{d}{L})}+1\right] \qquad (7)$$

Where k is the Boltzmann constant and m is the ideality factor.

The efficiency η of DSSCs is defined by:

$$\eta = \frac{J_{MP}V_{MP}}{P_{in}} \qquad (8)$$

Where J_{MP}, V_{MP} are successively the current and the voltage at the point of maximum power, their values have determined from the J-V characteristics.

P_{in} is the light intensity which equals to 100 mWcm^{-2} under a steady-state condition of irradiated DSSCs.

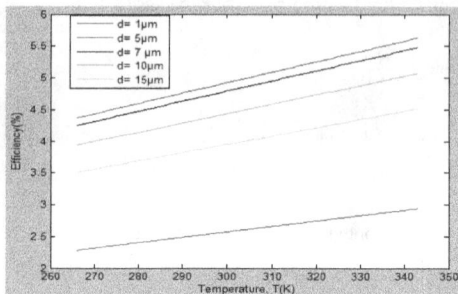

Fig. 2. The effect of temperature on efficiency with different thickness of the electrode.

TABLE I *Input values used in parametric analyses*

Parameter	Value	References
Electron concentration in a dark condition n_0 (cm^{-3})	10^{16}	[7,8]
Light intensity $\emptyset$ (cm^{-2}s^{-1})	10^{17}	[3,9]
Light absorption coefficient α(cm^{-1})	5000	[3,9]
Electron lifetime τ (ms)	10	[3,10]
Ideality factor m	4.5	[3,9]
Electron diffusion coefficient (cm^{2}s^{-1})	$5*10^{-4}$	[9,10]
Thin film electrode thickness d (µm)	5	[6]

Simulation results

The J-V characteristics of the DSSCs under different temperature are plotted in Fig.1.

As shown in Fig. 1 the temperature increases according to the open circuit voltage, due to electrons movement from the conduction band to the valence band of TiO$_2$.

The temperature effect on efficiency is illustrated in Fig. 2, on different thickness of the electrode. The figure shows an increase in efficiency of DSSCs due the temperature arise. As result the optimal electrode thickness has found to be 5µm; as an improvement of Meng's result [6].

References

[1] B. M. O'Regan, and M. Grätzel, "A low-cost, high-efficiency solar cell based on dye-sensitized colloidal TiO2 films", Nature, vol. 353, pp. 737-740,1991. http://dx.doi.org/10.1038/353737a0

[2] J. Nelson, in: A.J. Bard, M. Stratmann (Eds.), Encyclopedia of Electrochemistry, (Wiley-VCH, Weinheim, 6, 432 (2002).

Dielectric Materials and Applications: ISyDMA'2016   Materials Research Forum LLC
Materials Research Proceedings 1 (2016) 234-237   doi: http://dx.doi.org/10.21741/2474-395X/1/58

[3] R. Gomez, and P. Salvador, "Photovoltage Dependence on Film Thickness and Type of Illumination in Nanoporous Thin Film Electrodes According to a Simple Diffusion Model", Sol. Energy Mater. Sol. Cells vol. 88, pp. 377–388 ,2005. http://dx.doi.org/10.1016/j.solmat.2004.11.008

[4] S. Sodergren, A. Hagfeldt, J. Olsson ,and S. E. Lindquist,"Theoretical Models for the Action Spectrum and the Current-Voltage Characteristics of Microporous Semiconductor-Films in Photoelectrochemical Cells'', J. Phys. Chem. vol. 98, pp. 5552–5556,1999. http://dx.doi.org/10.1021/j100072a023

[5] M. Ni, M. K. H. Leung, D. Y. C. Leung, and K. Sumathy, ''An Analytical Study of Porosity Effect on Dye-Sensitized Solar Cell Performance,'' Sol. Energy Mater. Sol. Cells, vol. 90, pp. 1331–1344, 2006. http://dx.doi.org/10.1016/j.solmat.2005.08.006

[6] N. Meng, M. K. H. Leung, and D. Y. C. Leung,"Theoretical Modelling of the Electrode Thickness on Maximum Power Point of Dye-Sensitized Solar Cell", The Canadian Journal of Chemical Engineering, vol. 86, pp. 35-42, 2008. http://dx.doi.org/10.1002/cjce.20015

[7] G Rothenberger,D. Fitzmaurice, and M. Gratzel, ''Optical Electrochemistry. 3. Spectroscopy of Conduction-Band Electrons in Transparent Metal-Oxide Semiconductor-Films- Optical Determination of the Flat-Band Potential of Colloidal Titanium-Dioxide Films,'' J. Phys. Chem, vol. 96, pp. 5983–5986, 1992. http://dx.doi.org/10.1021/j100193a062

[8] J. Ferber, and J. Luther, ''Modeling of Photovoltage and Photocurrent in Dye-Sensitized Titanium Dioxide Solar Cells,'' J. Phys. Chem. vol. B 105, pp. 4895–4903, 2001.

[9] J. Lee, J.G. M. Coia, and N. S. Lewis, ''Current Density Versus Potential Characteristics of Dye-Sensitized Nanostructured Semiconductor Photoelectrodes. 2. Simulation,'' J. Phys. Chem. vol. B 108, pp. 5282–5293, 2004.

[10] L. Dloczik, O. Ileperuma, I. Lauermann, L. M. Peter, E. A. Ponomarev, G. Redmond, N. J. Shaw, and I. Uhlendorf, ''Dynamic Response of Dye-Sensitized Nanocrystalline Solar Cells: Characterization by Intensity Modulated Photocurrent Spectroscopy'', J. Phys. Chem. vol. B 101, pp.10281–10289,1997.

Dielectric Materials and Applications: ISyDMA'2016   Materials Research Forum LLC
Materials Research Proceedings 1 (2016) 238-242   doi: http://dx.doi.org/10.21741/2474-395X/1/59

Studies on the dielectric properties of olive pomace grains reinforced polymer matrix composites

L. KREIT[*1], Z. SAMIR[1], M.E. ACHOUR[1], A. OUERIAGLI[2], A. OUTZOURHIT[2], L.C. COSTA[3], M. MABROUKI[4]

[1]LASTID Laboratory, Sciences Faculty, Ibn Tofail University, Kenitra, Morocco

[2]LPSCM Laboratory, FSS, Cadi Ayyad University, Marrakech, Morocco

[3]I3N and Physics Department, University of Aveiro, Aveiro, Portugal

[4]Laboratory LGI, FST, Sultan Moulay Slimane University, Beni Mellal, Morocco

*Corresponding author: E-mail: kreit.lamyaa@gmail.com

Keywords: Biocomposites, Dielectric Relaxation, Kramers-Kronig Relation

Abstract. In this work, the polyester polymer matrix filled with olive-pomace grains was investigated by means of dielectric spectroscopy in the frequency range 10 KHz –1MHz and temperature from 300 K to 400 K. A numerical Kramers-Kronig transformation allowed the calculation of dielectric losses from the measured dielectric constant. The obtained relaxation processes were attributed to the orientation polarization imputed to the presence of polar water molecules in the pomace grains and to the accumulation of charges at the pomace grains/polyester interfaces. The analysis of the temperature dependence of the relaxation time using the Arrhenius equation shows the existence of two activation energies which are attributed to contributions of different relaxation processes.

Introduction

Over the last few years, natural fibers have become the main reinforcement for high performance composite materials [1]. The uses of lignocellulosic fibers have ranged from the construction industry to the automotive industry. The main attraction of bio-fiber reinforced composites lie in their low density and high strength. The most interesting aspect about natural fibers is their positive environmental impacts. As they come from a natural resource, they are a completely biodegradable and nonabrasive material and can be easily eliminated after the degradation of the polymer. The high sound attenuation of lignocellulosic composites is another added advantage. Vegetable fibers also provide a relativity reactive surface, which can be used for grafting specific groups and inducing new functional entities. In addition, the recycling by combustion of polysaccharide filled composites is easier in comparison with inorganic fillers [2].

Relaxation phenomena is one of the most studied topics in materials sciences and the spectra of the real and imaginary complex permittivity are used extensively to present this kind of dielectric phenomenon. In our work, the analysis of relaxation phenomenon may be rendered difficult because of dc-conductivity contribution that hides any dielectric relaxation process [3]. In order to overcome this shortcoming caused by high electrical conductivity, an alternative method, consisting of using the representation of the ohmic conduction free based on the derivation has been introduced in this paper. Furthermore, using the Arrhenius equation, the analysis of the temperature dependence of the relaxation time shows the existence of two activation energies which are attributed to contributions of relaxation processes. The mechanisms that induce relaxation processes are not yet clearly understood. So, it is appropriate to investigate how the dielectric properties of such bio-fiber reinforced composites are affected by the additional polarization effects caused by the presence of nanoparticles.

Dielectric Materials and Applications: ISyDMA'2016   Materials Research Forum LLC
Materials Research Proceedings 1 (2016) 238-242   doi: http://dx.doi.org/10.21741/2474-395X/1/59

Experimental details

Materials

We used the natural olive-pomace grains of 63 μm size, loaded in a polyester polymer matrix, forming pellets with 13 mm diameter, and thickness of about 3 mm. The matrix used in this work was unsaturated polyester resin 154TB, including 31 wt% of styrene monomer, requiring 30 minutes for gelation, at constant T=300 K, and was acquired from Cray Valley/Total, USA. The pomace/polyester nanocomposite has been prepared by mixing 5.87g of liquid polyester resin and 0.2% weight of cobalt octanone, as reaction activator. Pomace particles were introduced in different fractions, before adding 1% of hardener to make each mixture cohesive. The methyl ethyl ketone peroxide (MEKP) is used to harden the mixtures and to facilitate fabrication process. Each olive pomace/Polyester composite was mixed at room temperature for 5 minutes, promoting the gelation and after it was poured into the mold. After two hours the samples were unmolded, taking 24 hours for having a complete polymerization.

Dielectric measurements

The complex permittivity function, $\varepsilon^* = \varepsilon' - i\varepsilon''$, was measured by using a Hewlett Packard Network Analyzer (Model 4192A, USA). The dielectric constant ε' and the loss factor ε'' of the samples were calculated from the admittance $Y^* = G + iB = iC_o\omega\varepsilon^*$ of the equivalent circuit leading to ε' $=2hB/\varepsilon_o d^2 \pi^2 F$, where B is the susceptance, G the conductance, F the frequency, ε_o the vacuum dielectric constant, and h and d are the thickness and the diameter of the sample, respectively. The measurements were performed in the frequency range 10 KHz –1 MHz under isothermal conditions for temperatures ranging between 300 and 400 K.

Results and discussion

The Figure 1 shows the real and the imaginary parts of the complex permittivity of the olive pomace/Polyester composite at a temperature of 300 K, as a function of the frequency and for different concentrations of the olive-pomace. The both real and imaginary parts increase with the olive pomace and display a dispersive behavior versus frequency. The analysis of relaxation processes is often hampered due to the ohmic conduction losses that obscure any relaxation process [4]. This problem was observed in this composite. However, it is still possible to work around this issue by using the Kramers-Kronig relation [5]. This technique eliminates the ohmic conduction from the measured loss spectra by using an elegant way based on inversion technique, i.e. on calculating the imaginary part from the real part [Eq. (1)][6-8]:

$$\varepsilon''_{deriv} = -\frac{\pi}{2}\frac{\partial \varepsilon'(\omega)}{2\,\partial\ln\omega} \approx \varepsilon'' \tag{1}$$

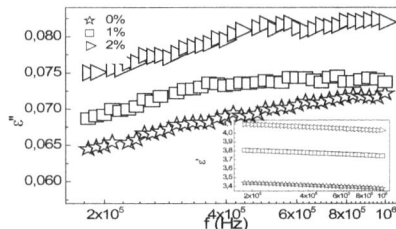

Fig. 1. real part (inset) and imaginary part, of the complex permittivity for different concentrations of the olive-pomace, at a temperature of 300 K.

As one can see in Figure 2 and Figure 3, this inversion technique allowed us to obtain conductivity free dielectric loss peaks, which are a clear signature of the relaxation in the composites. This

Dielectric Materials and Applications: ISyDMA'2016   Materials Research Forum LLC
Materials Research Proceedings 1 (2016) 238-242   doi: http://dx.doi.org/10.21741/2474-395X/1/59

relaxation is associated to polarization of the water molecules linked to cellulosic olive pomace grains which formed a monomolecular layer wrapping the external surface of the grains. Water molecules are known to be tightly bound the surface hydroxyl group of cellulose olive pomace grains. The analysis of the obtained results shows the influence of the temperature and concentration of olive-pomace grains on the dielectric properties of the composite [9].

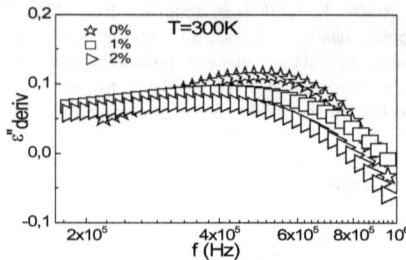

Fig. 2 .Variation with frequency of the ohmic conduction free dielectric loss, ε''_{deriv}, for concentrations of the olive-pomace 0.0 , 1.0 and 2.0% , at a temperature of 300 K.

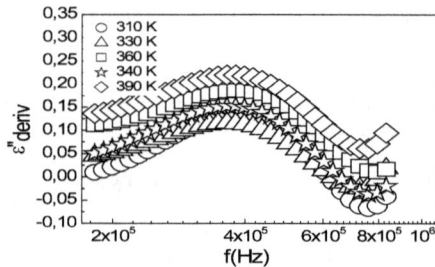

Fig. 3. Ohmic conduction free, $\varepsilon''deriv$, at various temperatures for a concentration of the olive-pomace (ϕ=2.0 %).

In order to further elucidate the dielectric relaxation in the olive pomace/Polyester, it is important to estimate the activation energy associated with the relaxation process which could be obtained from the temperature dependence frequency, according to the relation [Eq. (2)]:

$$f_{max} = f_0 e^{-\frac{E_a}{K_B T}} \tag{2}$$

with E_a, K_B and T being the activation energy, the Boltzmann's constant and the absolute temperature respectively. It should be noted that the activation energy was calculated from the slopes of the straight lines of $\ln(f_{max})$ versus (1000/T) for the concentrations of olive-pomace grains of 0.0, 1.0%, 2.0% and 3.0% represented in Figure 4.

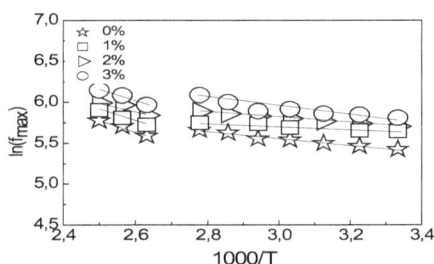

Fig.4. Ln(f_{max}) vs. 1000/T for different concentrations of olive-pomace .The solid lines are the least square linear fits to Eq. (3)

The values of activation energies are given in Table 1.

Table1. Activation energies for pomace grains/polyesters composites having 0.0 and 3.0 vol. percentages of pomace grains.

Fraction Volume	Temperature range	
	300 K-370 K	380 K-400 K
0.0%	0.0019 eV	0.0053 eV
1.0%	0.0014 eV	0.0042 eV
2.0%	0.0009 eV	0.0050 eV
3.0%	0.0019 eV	0.0048 eV

As seen in Table 1, there are two temperature regions of the maximal frequency in this composite: 300 - 370 K and 380 - 400 K. This behavior can be taken as indication that the olive pomace grains have different energies and structural configurations. As shown in figure 4 there is one straight line for the maximal frequency versus temperature plot at the given concentration are applicable in each temperature region. It is very clear from this figure that there is a strong dispersion of maximal frequency with concentration.

Conclusion

In the present work investigation regarding the dielectric behavior of polyester matrix and composites based on pomace grains, have been studied over the temperature range from 300 to 400K and over the frequency range from 10 KHz to 1 MHz. The results were analyzed using the Kramers-Kronig transformation which allows the calculation of dielectric losses from the dielectric permittivity measured. In addition to the relaxation associated to the glass transition of the polyester resin matrix and ionic relaxation caused by the mobility of dielectric charges, the presence of pomace grains in the composite gives rise to other relaxation associated to water polarization. The analysis of the temperature dependence of the relaxation time using the Arrhenius equation shows the existence of two activation energies which are attributed to contributions of different relaxation processes.

Acknowledgment

The authors acknowledge FCT-CNRST bilateral cooperation, and FEDER by funds through the COMPETE 2020 Program and National Funds through FCT - Portuguese Foundation for Science and Technology under the project UID/CTM/50025/2013

Dielectric Materials and Applications: ISyDMA'2016   Materials Research Forum LLC
Materials Research Proceedings **1** (2016) 238-242   doi: http://dx.doi.org/10.21741/2474-395X/1/59

References

[1] A. Hodzic, R. Shanks, Natural Fibre Composites: Materials, Processes and Properties; Woodhead Publishing Series in Composites Science and Engineering, Elsevier Science, 2014.

[2] I. Ben Amor, H. Rekik, H.Kaddami, M.Raihane, M.Arous and A.Kallel, "Studies of dielectric relaxation in natural fiber-polymer composites", Journal of Electrostatics 67, 2009, 717-722. http://dx.doi.org/10.1016/j.elstat.2009.06.004

[3] Z. Samir, Y. el Merabet, M.P.F. Graça, S. Soreto Teixeira, M. E. Achour, L. C. Costa, "Dielectric behavior of carbon nanotube particles filled polyester polymer composites", J. Comp. Materials, 2016 (in press). http://dx.doi.org/10.1177/0021998316665682

[4] M. Wuubbenhorst, J. v. Turnhout, "Analysis of complex dielectric spectra. I. One-dimensional derivative techniques and threedimensional modeling", Journal of Non Crystalline Solids, 2002,305, 40–49. http://dx.doi.org/10.1016/S0022-3093(02)01086-4

[5] JR. Macdonald, WR. Kenan, Impedance spectroscopy- Emphasizing solid materials and systems. John Wiley and Sons, New York, Interscience 1987; 368.

[6] P. Steeman, J. Vanturnhout, "Fine structure in the parameters of dielectric and viscoelastic relaxations". Macromolecules 1994, 27(19),5421–5427. http://dx.doi.org/10.1021/ma00097a023

[7] A. Molak, M. Paluch, S. Pawlus, J. Klimontko, Z. Ujma, I. Gruszka, "Electric modulus approach to the analysis of electric relaxation in highly conducting (Na0.75Bi0.25)(Mn0.25Nb0.75)O3 ceramics", Journal of Physics 2005, 38(9), 1450–1460. http://dx.doi.org/10.1088/0022-3727/38/9/019

[8] H. Mingjuan, Z. Kongshuang, "Effect of volume fraction and temperature on dielectric relaxation spectroscopy of suspensions of PS/PANI composite microspheres", The Journal of Physical Chemistry C 2008, 112, 19412–19422. http://dx.doi.org/10.1021/jp803530m

[9] Tagmouti, S.E.Bouzit, L.C.Costa, M.P .F .Garça and A.Outzourit "Impedance Spectroscopy of Nanofluids based on Multiwall Carbon Nanotubes", Spectroscopy Letters, 48:10, 761-7666, 2015. http://dx.doi.org/10.1080/00387010.2015.1034874

Dielectric Materials and Applications: ISyDMA'2016
Materials Research Proceedings **1** (2016) 243-246

Materials Research Forum LLC
doi: http://dx.doi.org/10.21741/2474-3951/2/60

Experimental verification on enhancing electric and dielectric phenomena of transformer nanofluids

O. E. GOUDA*[1], A. THABET[2]

[1]Electrical Power & Machine Engineering Dept., Faculty of Engineering, Cairo University, Giza, Egypt

[2]Nanotechnology Research Center, Elect. Engineering Dept., Faculty of Energy Engineering, Aswan University, Aswan, Egypt

*Corresponding author: E-mail: Prof_ossama11@yahoo.com

Keywords: Electrical Conductivity, Breakdown Voltage, Nanoparticles, Nanofluids, Diala D, Transformer Oil

Abstract. Transformer oil-based Nano-fluids with nanoparticles suspensions have substantially positive or negative voltage levels with respect to inclusion types and their concentrations in pure transformer oil. This paper has been studied the effects of powder nanoparticles for enhancing electrical and physical properties of pure transformer oils. Also, it has been dispersed low concentrations of Nano-sized Zinc oxide spherical particles in Diala D insulating oil for enhancing their physical and dielectric properties. Therefore, this paper presents performance of transformer oil nanofluids as alternative pure oil insulation for enhancing power transformer characterization and verifying the experimental results of dielectric strength and electrical properties of dielectric oil nanofluids medium based on nanotechnology techniques.

Introduction

Power transformer oil, as an insulating component is one of the dominant factors in the life-time of a unit due to its sensitive to surrounding conditions and routine operation. Therefore, it is important to study physical and electrical characteristics of oil and evaluate its status. New developments and production techniques make power transformer oil based nanofluids that changing oil transformer characterization. The oil conductivity at a certain frequency, which can be calculated from the results of dielectric response, is an important factor for the quality of the oil. There are several different methods to measure the conductivity of oil [1]. Predicting the electrical conductivity of power transformer oil based nanofluids is of the most importance in both academic and industrial research. Maxwell formula and recent formulas are well established techniques for the evaluation of effective electrical conductivity of power transformer oil based nanofluids [2, 3]. According to the theory of colloid chemistry, in nanofluids there are charges on the surface of nanoparticles, and opposite charges would be attracted around the particles [4], namely the electrical double layer (EDL). The surface charge of nanoparticles, together with the ion cloud that constitute the EDL, would actively increase the electrical conductivity of nanofluids [5].

Theoretical model and experimental setup

Theoretical model

Effective electrical conductivity of transformer oil based nanofluids on dependence of volumetric concentration and temperature has been predicted by using theoretical models [2, 3]. Fig. 1 shows a schematic diagram for electrical double layer and zeta potential in nanofluids. However, zeta potential is the value that can be related to the stability of colloidal dispersions and the degree of repulsion between adjacent.

Dielectric Materials and Applications: ISyDMA'2016   Materials Research Forum LLC
Materials Research Proceedings 1 (2016) 243-246   doi: http://dx.doi.org/10.21741/2474-395X/1/60

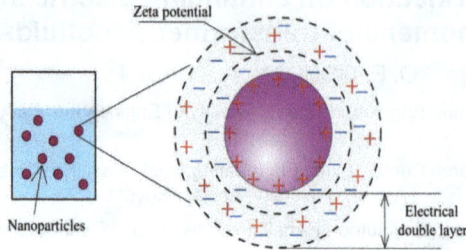

Fig. 1 Schematic for electrical double layer in nanofluids

There is a motion of dispersed nanoparticles relative to a fluid under the influence of a spatially uniform electric field, electrophoresis phenomenon, that is caused by the presence of a charged interface between the particle surface and the surrounding fluid. Thus, adding the electrophoresis conductivity to Maxwell can be expressed as follows,

$$\sigma_{eff} = \sigma_M + \sigma_E = \sigma_f \left(1 + \frac{3\varphi}{(\alpha+2)/(\alpha-1)-\varphi}\right) + \frac{2\varphi\varepsilon_f^2\zeta^2}{\eta_o r^2} e^{\lambda(T-T_o)} \tag{1}$$

Where, ϵ_f is permittivity of transformer oil, r is radius of nanoparticles, ζ is zeta potential of nanoparticles inside the transformer oil, ηo is initial dynamic viscosity at T_o temperature, λ is ratio of viscosity index ($\alpha = \sigma_p/\sigma_f$),

Experimental setup

Oil tester device used in testing specifications are: single phase AC 180 V- 250 V, 50 Hz, voltage output: 0-100 kV AC, 0-80 kV AC, 0-60 kV. Diala D oil has been selected as electrical insulating oil since it offers inherent natural resistance to oil degradation and is suitable for long life applications as transformers, grid and industrial transformers up to maximum load. Additives of nanoparticles (Dia.: 50 nm) to the base Diala D oil transformer have been fabricated by using mixing, and ultrasonic processes for selected Zinc oxide (ZnO) nanoparticles.

Results and discussion

Characterization of power transformer oil based on nanofluids has been illustrated the effective electrical conductivity performance with respect to physical properties, dimension and concentrations of nanoparticles. Also, the breakdown voltage of new nanofluids oil transformer are measured using two copper electrodes in uniform electric field.

Analytical analysis

According to the effect of dielectric constant of transformer oil, as shown in Fig. 2, the effective electrical conductivity of power transformer oil based nanofluids has been increased by increasing the volume percentages of ZnO.

Moreover, Fig. 3 shows the effective electrical conductivity of power transformer oil based nanofluids that have been increased by increasing volume percentages of ZnO, and so; increasing viscosity index increases the effective electrical conductivity of power transformer oil.

Experimental analysis

Fig. 4 shows breakdown voltages for pure Diala D oil transformer and number of tests by using stirrer and without using stirrer in uniform electric field. On the other hand, Fig. 5 shows increasing percentage of zinc Oxide (ZnO) nanoparticles in the Diala D oil transformer up to 1% by weight increases dielectric strength of the industrial materials respectively.

Dielectric Materials and Applications: ISyDMA'2016   Materials Research Forum LLC
Materials Research Proceedings **1** (2016) 243-246   doi: http://dx.doi.org/10.21741/2474-395X/1/60

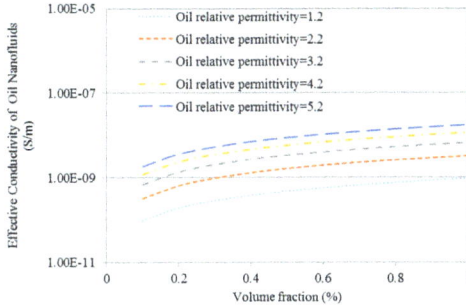

Fig. 2 Effect of concentration of ZnO nanoparticles on effective conductivity of nanofluids

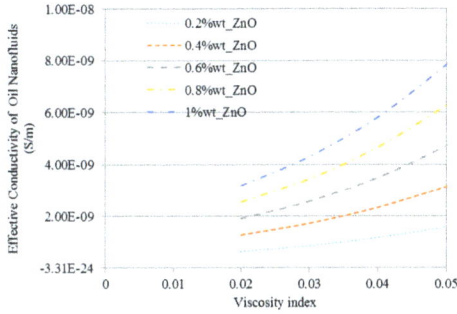

Fig. 3 Effect of viscosity index on effective conductivity of transformer oil nanofluids

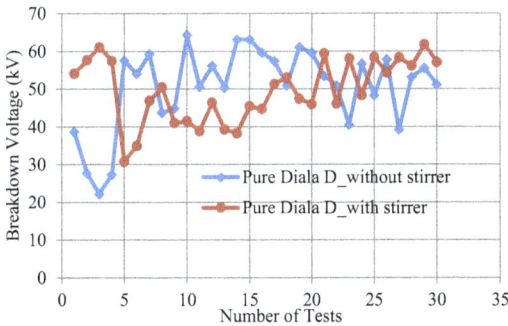

Fig. 4 Breakdown voltages in pure Diala D under uniform electric fields

Dielectric Materials and Applications: ISyDMA'2016   Materials Research Forum LLC
Materials Research Proceedings **1** (2016) 243-246   doi: http://dx.doi.org/10.21741/2474-395X/1/60

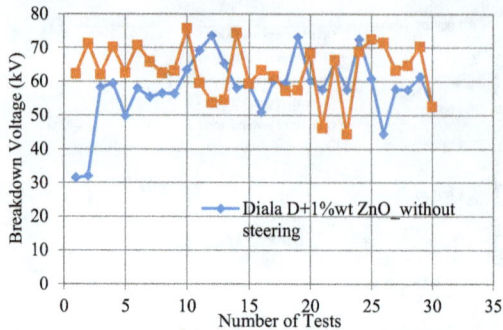

Fig. 5 Breakdown voltages in (Diala D +1% wt ZnO) under uniform electric fields

Conclusions

Volumetric percentages of Zinc oxide nanoparticles have lightly raising the effective electrical conductivity of power transformer oil nanofluids. Furthermore, increasing the electric applied field increases the magnitude of deposited charge on the nanoparticles as it increasing the space where the electrons can be deposited. Thus, it has been investigated experimentally that ZnO/DialaD nanofluids have higher electrical and dielectric properties than pure insulating liquids by adding Zinc oxide nanoparticles.

References

[1] Yuan Zhou, Miao Hao,George Chen , Gordon Wilson,Paul Jarman, "Frequency-dependence of conductivity of new mineral oil studied by dielectric spectroscopy", International Conference on High Voltage Engineering and Application, Shanghai, China, September 17-20, 2012.

[2] J.C. Maxwell, 3rd ed, "A Treatise on Electricity and Magnetism," vol. 1, Clarendon, Oxford, 1904.

[3] J. Miao, M. Dong, L.-P. Shen, "A modified Electrical conductivity model for insulating oil-based nanofluids", IEEE International Conference on Condition Monitoring and Diagnosis, Bali, Indonesia, pp. 23-27 September 2012. http://dx.doi.org/10.1109/cmd.2012.6416357

[4] Shen Z, Zhao ZG, Kang WL, "Colloid and Surfacr Chemistry," Bei Jing, Chemistry industry press, 2011.

[5] M. R. Meshkatoddini, "Aging Study and Lifetime Estimation of Transformer Mineral Oil" American Journal of Engineering and Applied Sciences Vol.1, No.4, pp. 384-388, 2008. http://dx.doi.org/10.3844/ajeassp.2008.384.388

Dielectric Materials and Applications: ISyDMA'2016
Materials Research Proceedings **1** (2016) 247-248

Materials Research Forum LLC
doi: http://dx.doi.org/10.21741/2474-395X/1/61

Influence of various types of dielectric epoxylite resin on field electron emission properties from the nano and micro scale tungsten and carbon fiber tips

S.S. ALNAWASREH, A.M. AL-QUDAH, M.A. MADANAT, E.S. BANI ALI,
A.M. ALMASRI, MARWAN S. MOUSA*

Department of Physics, Mu'tah University, Al-Karak, Jordan

*Corresponding author: E-mail: marwansmousa@yahoo.com & mmousa@mutah.edu.jo

Keywords: Field Electron Emission, Composite Microemitters, Field Electron Microscope, Fowler-Nordheim

Abstract. This investigation deals with the process of electron emission from composite micro-emitters consisting of a tungsten or carbon fiber core that has been coated with two types of dielectric materials: 1- Clark Electromedical Instruments Epoxylite resin (CEI- resin), 2- Epidian 6 (based on Bisphenol A) (E6-resin). Various properties of these emitters were measured, including the current-voltage (I-V) characteristics that are presented as Fowler-Nordheim (FN) plots, in addition to the corresponding electron emission images. Measurements were carried out under ultra-high vacuum (UHV) conditions with a base pressure of 10^{-8} mbar. Field electron microscope (FEM) standardized with 10 mm tip (cathode) – screen (anode) distance was employed to characterize the electron emitters.

Introduction

Composite emitters, consisting of clean tungsten and carbon fiber tips with known profile, which were coated by dielectric materials has been reported [1-3], as potential electron source with improved characteristics. Such research is very important target for applied electronics. The first explanation of what appeared to be a field electron emission initiated electric discharge was given by Winkler [4, 5]. Cold field electron emission (CFE) is an electron emission regime in which most electrons escape by tunneling

from states below the emitter Fermi level [5,6]. The objective of the investigation reported here is to study the impact of dielectric materials' coatings: 1- Clark Electromedical Instruments Epoxylite resin (CEI- resin), 2- Epidian 6 (based on Bisphenol A) (E6-resin) on the characteristics of tungsten and carbon fiber micro-emitters. Comparison includes the current-voltage (I-V) characteristics presented in the form of Fowler-Nordheim (F-N) plots, in addition to the emission current stability (electron emission images).

Results

Tungsten and carbon fiber nano and microemitters coated with two types of dielectric materials, have been investigated. The results are compared with those obtained from clean uncoated tungsten and carbon fiber emitters. Fig.1 and Fig.2, show the improvement in the I-V characteristics of field emission from tungsten and carbon fiber emitters after coating by Clark Electromedical Instruments Epoxylite resin (CEI-resin). By comparing the effects of different coating materials, the following effects were found. Clark Electromedical Instruments Epoxylite resin improves the characteristics of the tungsten emitters more than those of the carbon fibers. Likewise, Epidian 6 epoxy resin (based on Bisphenol A) improved (although to a less degree) the characteristics of the tungsten emitters more than those of the carbon-fiber emitter.

Dielectric Materials and Applications: ISyDMA'2016   Materials Research Forum LLC
Materials Research Proceedings 1 (2016) 247-248   doi: http://dx.doi.org/10.21741/2474-395X/1/61

(a)

(a)

(b)

(b)

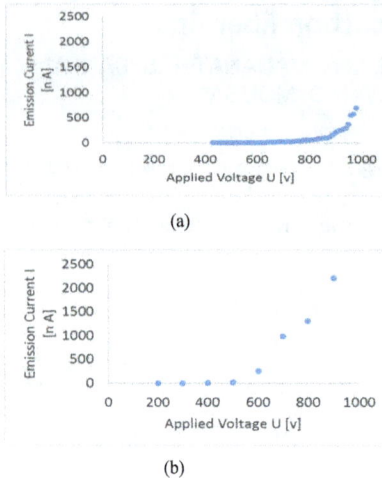

Fig. 1: (a) The I-V characteristics of a clean tungsten tip, taken as voltage increases, (b) I-V characteristics for composite tungsten/CEI-resin emitter during the second increasing-voltage cycle.

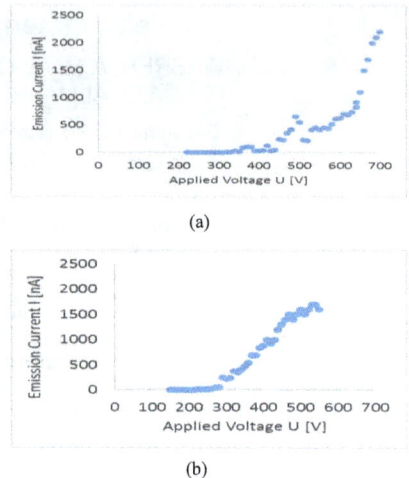

Fig. 2: (a) The I-V characteristics of a clean carbon fiber tip, taken as voltage increases, (b) I-V characteristics of composite carbon-fiber/CEI-resin, during the first decreasing voltage cycle.

References

[1] Al-Qudah, A. A., Mousa, M. S., Fischer, A. (2015), Effect of insulating layer on the field electron emission performance of nano-apex metallic emitters. IOP Conf. Series: Materials Science and Engineering. 92, 1-8, 012021. http://dx.doi.org/10.1088/1757-899x/92/1/012021

[2] Moran Meza, J. A., Lubin, C., Thoyer, F., Villegas Rosales, K. A., Guitarra Espinoza, A. A., Martin, F., and Cousty, J. (2015) Fabrication of ultra-sharp tips from carbon fiber for scanning tunneling microscopy investigations of epitaxial graphene on 6H-SiC(0001) surface. Carbon. 86, 363-370. http://dx.doi.org/10.1016/j.carbon.2015.01.050

[3] Mousa, M. S. (2007) Influence of a dielectric coating on field electron emission from micro-point electron sources. Surf. Interface Anal. 39, 102-110. http://dx.doi.org/10.1002/sia.2470

[4] Winkler, J. H., Gedanken von den Eigenschaften, Wirkungen und Ursachen der Electrizität nebst Beschreibung zweier electrischer Maschinen. (Verlag B. Ch.Breitkopf, Leipzig, 1744).

[5] Alnawasreh, S., Mousa, M. S., and Al-Rabadi, A. N. (2015) Investigating the effects of sample conditioning on nano-apex carbon fiber tips for efficient field electron emission. Jo. J. Phys. 8, (3), 95-101.

[6] Latham, R. V., Mousa, M. S. (1986) Hot electron emission from composite metal-insulator micropoint cathodes, J. Phys. D: Appl. Phys. 19, 699-713. http://dx.doi.org/10.1088/0022-3727/19/4/021

Dielectric Materials and Applications: ISyDMA'2016   Materials Research Forum LLC
Materials Research Proceedings **1** (2016) 249-252   doi: http://dx.doi.org/10.21741/2474-395X/1/62

TEM and XRD characterizations of epitaxial silicon layer fabricated on double layer porous silicon

S. GOUDER[*1], L. TEBESSI[1], R. MAHAMDI[2], S. ESCOUBAS[3], L. FAVRE[3], M. AOUASSA[3], A. RONDA[3], I. BERBEZIER[3]

[1]University Tébessa, (12000) Algeria

[2]LEA, Electronics Department, Batna 2 University, Batna, (05000) Algeria

[3]IM2NP Aix-Marseille, UMR CNRS n°7334, Faculté des Sciences St-Jérome-Case142, 13397 Marseille Cedex 20 France

*Corresponding author: E-mail: soraya.gouder@gmail.com

Keywords: Double Layer, Porous Silicon, Molecular Beam Epitaxy, Transmission Electron Microscopy, High Resolution X-Ray Diffraction

Abstract. Single crystal Silicon (Si) layers have been deposited by molecular beam epitaxy on double-layer porous silicon (PSi). We show that a top thin layer with a low porosity is used as a seed layer for epitaxial growth. While, the underlying higher porosity layer is used as an easily detectable etch stop layer. The morphology and structure of epitaxial Si layer grown on the double-layer PSi are investigated by high resolution X-ray diffraction and transmission electron microscopy. The results show that, an epitaxial Si layer with a low defect density can be grown. Epitaxial growth of thin crystalline layers on double-layer PSi can provide opportunities for silicon-on-insulator applications and Si-based solar cells provided that the epitaxial layer has a sufficient crystallographic quality.

Introduction

During the last decades, epitaxial growth of silicon (Si) on porous silicon (PSi) has been widely investigated. It can be applied to form Si-based solar cells [1] and Silicon On Insulator (SOI) substrates via the full isolation by porous oxidized silicon (FIPOS) process [2] and epitaxial layer transfer (ELTRAN) [3]. For all these applications, defect free epitaxial layers are needed. Our approach in this work consists in using a double-layer PSi on a Si wafer, namely a low porosity upper layer on a buried high porosity layer. Prior to the epitaxial growth, a double layer PSi is annealed at high temperature (1100°), to get a good seeding layer for epitaxy. However, the underlying high porosity forms a separation layer, which is mechanically weak and can be used as an easily detectable etch stop material.

Experimental

The double PSi structures are formed by electrochemical etching of a (001)-oriented B-doped Si wafer dipped in hydrofluoric acid (HF). An increase of the current density during etching causes stacked porous layers to form with correspondingly varied porosities. PSi layers are annealed under H atmosphere at high temperature T (1100°C) for 60 seconds. The Si layers were grown on double-layer PSi substrates by molecular beam epitaxy (MBE). After cleaning, samples were introduced directly into the ultra high vacuum chamber to avoid the contamination of PSi surface, where they are outgassed at 400°C for 15min.

The morphology and microstructure of the samples were investigated by transmission electron microscopy (TEM) and high resolution x-ray diffraction (HR-XRD). For TEM cross-section examination, a JEOL2010F microscope with 200 kV acceleration voltage was used. The crystalline structure was investigated by HR-XRD with an X'Pert PRO MRD_Diffractometer. Both HR-XRD pattern and reciprocal space mappings are performed, respectively for the Si (004) and (224)

Dielectric Materials and Applications: ISyDMA'2016   Materials Research Forum LLC
Materials Research Proceedings 1 (2016) 249-252   doi: http://dx.doi.org/10.21741/2474-395X/1/62

reflexions. The (004) HR-XRD pattern allows the determination of the vertical lattice parameter. The asymmetrical (224) reciprocal space mapping allow the determination of the strain relaxation.

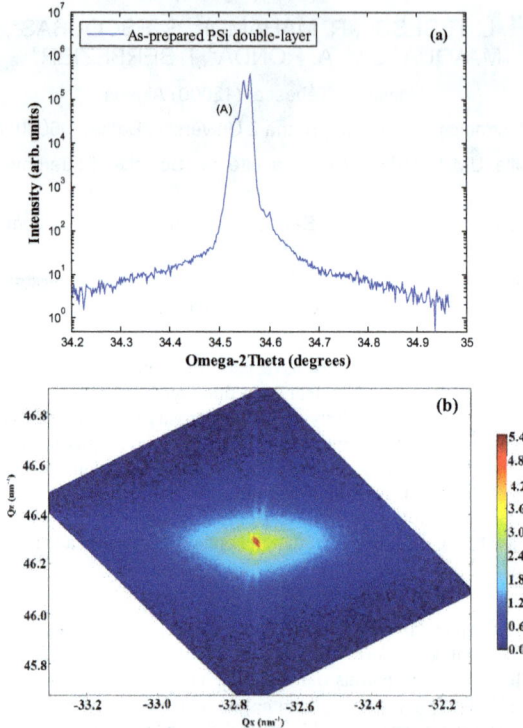

Fig 1: (a) Omega-2Theta (004) scans of the as-prepared PSi double-layer, (b) (224) reciprocal space mapping of the annealed PSi double-layer.

Results and discussion
The double PSi structure is composed of a thin (200nm) PSi layer with a low porosity 20%, this starting layer shoes a spongy microstructure with small pore channels following arbitrary directions with lots of interconnections. It is deposited on a thick (7μm) highly 35% porous PSi layer. Annealing in hydrogen atmosphere influences the structure of the double PSi layer greatly. The open porosity of the starting layer has changed to a closed one and we find isolated, mostly facetted pores, similar to those in the monolayer systems [4]. The structure of the annealed separation layer consists in very large, elongated pores with an average diameter of ~ 300 nm. TEM image evidences the flatness of the top surface which is confirmed by AFM measurements of the root mean square (RMS) roughness with RMS=8.4Å. An effective way of analysing the stress and detecting the lattice strain between the layers in double-layer structure is the use of the X-ray diffraction. Figure. 1a depicts the HR-XRD spectra acquired from a PSi double layer. It shows different peaks representative of the different lattice parameters in the growth direction perpendicular to the wafer surface. Beside this double peak which has already been observed in our recent paper [4], an additional broad peak is

Dielectric Materials and Applications: ISyDMA'2016   Materials Research Forum LLC
Materials Research Proceedings 1 (2016) 249-252   doi: http://dx.doi.org/10.21741/2474-395X/1/62

visible on the left side (A). This one is ascribed to the layer with the lower porosity. Before annealing, the second PSi layer with the higher porosity is more expanded than that with the lower porosity and the lattice spacing is greater. But, after annealing, the second PSi layer with the higher porosity contracts more than that with the lower porosity. The tensile strain observed in PSi was explained by the desorption of –OH species during annealing [5], which causes the swelling of the lattice structure. The (224) reciprocal space map shows that the double layer PSi structure is fully strained (Fig. 1b).

In a second step, Si was deposited on double layer PSi by MBE with a thickness of 300 nm at 550°C growth temperature, in order to avoid a dramatic sintering of the pores. Cross-sectional TEM image (Fig. 2) confirmed the quality of the grown layers, but also, revealed a low defect density which are mainly dislocations originating at the double-layer PSi-epitaxial film interface. Moreover, the corresponding transmission electron diffraction (TED) pattern obtained on selected area indicates a polycrystalline layer of Si (inset Fig. 2). In addition, HR-XRD has been applied for complementary investigation of microstructure and morphology for the experimental Si on double layers PSi. After the epitaxial growth, the lattice shrinks and the corresponding peaks in the XRD spectrum shift to the high angle side relative to the substrate peak.

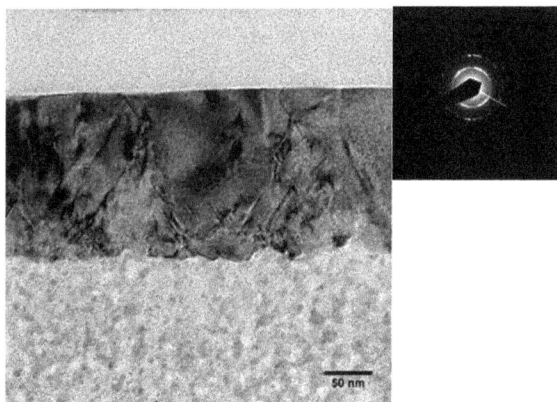

Fig 2: TEM cross section image of the epitaxial Si layer grown on the double-layer PSi; in the inset is presented a TED diffraction pattern indicates a polycrystalline layer of the grown layers.

Conclusion

In summary, a double-layer PSi with a lower porosity layer and an underlying higher porosity was used as the substrate for Si epitaxy. It was found that, the sintered PSi double-layer serves as a seed layer for the epitaxial growth of the monocrystalline Si film, while the lower PSi layer remains unchanged for the etching stop stage. It was also found that, epitaxial layer with a low defect density can be grown on such a dual layer. Despite this, the grown layers are relatively flat and continuous. Furthermore, Si thin films epitaxially deposited by MBE on annealed double-layer PSi could be integrated in Silicon On Insulator (SOI) technology and Si-based solar cells.

Dielectric Materials and Applications: ISyDMA'2016   Materials Research Forum LLC
Materials Research Proceedings **1** (2016) 249-252   doi: http://dx.doi.org/10.21741/2474-395X/1/62

Acknowledgments

The authors would like to thank CP2M, where TEM study was performed. (Multidisciplinary Centre of Electron Microscopy and Microanalysis: University of Aix Marseille, Faculty of sciences of Saint Jérôme, Avenue Normandie Niemen, 13397 MARSEILLE).

References

[1] V. Y. Yerekhov, and I. I. Melnyk, "Porous silicon in solar cell structures: a review of achievements and modern directions of further use," renew. Sustainable Energy, vol. 3, pp. 291-322, December 1999. http://dx.doi.org/10.1016/S1364-0321(99)00005-2

[2] K. Imal, "FIPOS (Full Isolation by Porous Oxidized Silicon) Technology and its applications to LSI'S" Electron Devices, IEEE Transactions, Vol. 31, PP. 297-302, March 1994.

[3] T. Yonehare, K. Sakagushi, and N. Sato, "Epitaxial layer transfer by bond and etch back of porous Si," Appl. Phys. Lett, vol. 64, pp. 2108-2110, January 1994. http://dx.doi.org/10.1063/1.111698

[4] S. Gouder, R. Mahamdi, M. Aouassa, S. Escoubas, L. Favre, A. Ronda, and I. Berbezier. "Investigation of microstructure and morphology for the Ge on Porous Silicon/Si substrate hetero-structure obtained by Molecular Beam Epitaxy". Thin solid Films, vol. 550, pp 233-238, 2014. http://dx.doi.org/10.1016/j.tsf.2013.10.183

[5] I. Berbezier, J.M. Martin, C. Bernardi, J. Derrien, "EELS investigation of luminescent nanoporous p-type silicon," Appl. Surf. Sci. Vol. 102, pp. 417-422, 1996. http://dx.doi.org/10.1016/0169-4332(96)00090-6

Dielectric Materials and Applications: ISyDMA'2016
Materials Research Proceedings 1 (2016) 253-255

Materials Research Forum LLC
doi: http://dx.doi.org/10.21741/2474-395X/1/63

Lawsonite thin film as radiative cooling minerals for dew harvesting

M. BENLATTAR*[1], M. MAZROUI[1], A. MOUHSEN[2], E.M. OUALIM[3]

[1]Hassan II University, Faculty of Sciences Ben M'sik, Department of Physics, 7955 Sidi Othman Casablanca, Marocco

[2]Hassan I University, Science and technology Faculty of Settat, Settat, Morocco

[3]Hassan I University, High School of Technology of Berrechid, BP 218 Berrechid, Morocco

* Corresponding author. E-mail: m.benlattar@um5s.net.ma, benlattar1975@gmail.com

Keywords: Condensation, Radiative Cooling, Lawsonite, Harvesting Dew, Dew Yield

Abstract. Radiative cooling is the phenomenon responsible for dew formation on plants. Harvesting humidity from the air has two different pathways: fog and dew, but harvesting dew has not been sufficiently exploited. In this paper, we describe the resources for radiative cooling as well as lawsonite mineral for exploiting this natural phenomenon. Further, this paper describes the development of dew harvesting systems for use in the semi-arid mirleft south Morocco. Numerical simulations using the energy balance equation were performed. Harvesting dew can be used as a renewable complementary source of water both for drinking and agriculture in specific arid or semi-arid areas. In order to form global estimates of dew collection potential via a dew formation model, we combined meteorological parameters with radiative properties of a specific collector sheet (natural lawsonite, $CaAl_2Si_2O_7(OH)_2 \cdot H_2O$ deposited on glass) to enhance the dew yield. The daily yields show that significant amounts of dew water can be collected.

Introduction

Since1905, scientists has tried to collect dew to obtain water, using the so-called Zibold condensers [1-3]. Recently, many studies concerning natural condensation of water vapor are oriented towards agriculture (plants and animals), the source of drinking water and the study of the soil cooling [4-8]. We attempt to create an efficient and cheap dew plate condenser has not yet been exploited. In this modeling study we focus on the harvesting dew onto lawsonite radiant mineral as a dew condenser, and investigate the potential for its collection. The planar dew condenser was set at an angle of 30° with respect to horizontal. Numerical simulations using the energy balance equation to identify the meteorological factors which determine the degree of cooling, and to assess their effect on harvesting dew were performed. These meteorological parameters were found to be ambient temperature (T_{amb}), cloud cover (N), wind speed (V_w), soil heat flux (G), and relative humidity (Hr). The temperature of the collector sheets (Ts) dew forms on is also important. Dew formation is the result of nocturnal radiation. Practically, Dew is a natural phenomenon that occurs under particular meteorological conditions and on a dew plate condenser with high radiative cooling properties specially designed for this purpose. Dew forms when the temperature of a surface (collector sheet or dew condenser) cools below the dew point temperature (Td) of the surrounding air so that water vapor contained in this air condenses on the collector sheets. The cooling effect of a collector sheet is caused by a radiation loss.

Dew model description

In implementing the model that describes the harvesting dew, we followed the approach presented by Nikolayev et al., Wahlren, 2001, Jacobs et al. and O. Clus [9]. The heat energy balance as given by the following equation:

$$\left(\frac{dT_c}{dt}\right)(MC_c + m_w C_w) = q_{IR} + q_{cd} + q_{conv} + q_{cond} \tag{1}$$

where T_c, M and C_c are the dew condenser's temperature, mass and specific heat capacity, respectively. The dew condenser's mass is given by $M = \rho_c S_s \tau_c$, where ρ_s, S_c and τ_c are its density, surface area (square meter) and thickness. C_w and m_w are the specific heat capacity and mass of water, representing the cumulative mass of dew water that has condensed onto the collector sheet.

Results and discussion

Figure 1 illustrates the sensitivity of the modelled dew formation to changes in the dew condenser thickness at different values of dew condenser temperature T_c. The effect of dew condenser temperature is more complex: increasing the dew condenser temperature T_c reduces the collected dew, whereas decreasing the dew condenser temperature increases convective heating. Collected dew declines linearly with dew condenser thickness. It has a big influence on collected dew for both cases: T_c decreases or increases.

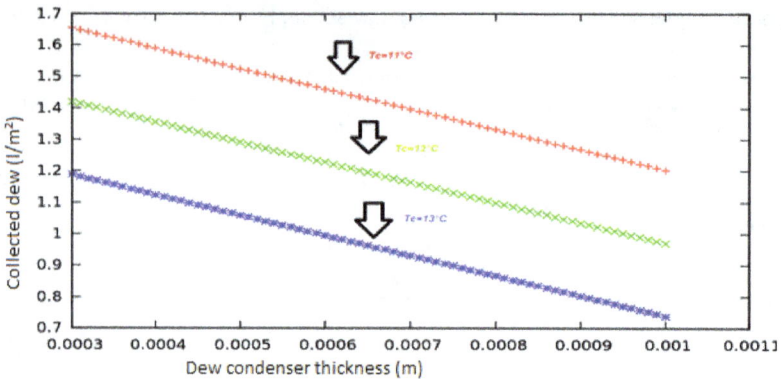

Fig. 1. Effect of dew condenser thickness on collected dew at different values of condenser temperature.

Conclusion

The numerical simulations for dew formation on lawsonite collector sheet were investigated by implementing a dew collection model bases on solving the energy balance equations. We observe that dew collection yield depends also on the dew condenser thickness and dew condenser temperature. The result obtained is that the lawsonite sheet condenser collected between 0.5 and $0.165 L/m^2$/night of water. The lawsonite thin film has high emissivity across 8-13µm; which indicates that it can be used as good radiative cooling and dew water condenser mineral.

References

[1] F. I. Zibold 1905 Significance of underground dew for water-supply in Feodosia-city Collect. Foresty Trans. 3 (1905) 387–41.

[2] M. A. Knapen Dispositif intérieur du puits aérien Knapen Extrait des mémoires de la société des ingénieurs civils de France (Bulletin de Janvier-Fevrier) (Paris: Imprimerie Chaix) 1929.

[3] L. Chaptal La captation de la vapeur d'eau atmosphérique, Ann. Agronomiques 2 (1932) 540–55.

[4] A. F. G. Jacobs, B.G. Heusinkveld, S. M., Berkowicz, Dew deposition and drying in a desert system: a simple simulation model. J. Arid Environ. 42 (1999) 211-222.
http://dx.doi.org/10.1006/jare.1999.0523

[5] W. J. Liu, J. M. Zeng, Wang, H. M. Li, W. P. Duang, on the relationship between forests and occult precipitation (dew and fog precipitation). J. Nat. Resour. 16 (2001) 571- 75.

[6] X.Y. Li, Effect of gravel and sand mulches on dew deposition in the semiarid region of China. J. Hydrol. 260 (2002) 151-160. http://dx.doi.org/10.1016/S0022-1694(01)00605-9

[7] R. Marek, J. Straub, Analysis of the evaporation coefficient and the condensation coefficient of water. Int. J. Heat Mass Tran. 44 (2001) 39-53. http://dx.doi.org/10.1016/S0017-9310(00)00086-7

[8] V.S. Nikolayev, D. Beysense, A. Gioda, I. Milimouk, E. katiouchine, J.P. Morel, Water recovery fromdew, Journal of Hydrology, 182 (1996) 19-35. http://dx.doi.org/10.1016/0022-1694(95)02939-7

[9]O. CLUS «Condenseurs radiatifs de la vapeur d'eau atmosphérique (rosée) comme source alternative d'eau douce»; UNIVERSITE DE CORSE PASQUALE PAOLI FACULTE DES SCIENCES ET TECHNIQUES, Thèse de Doctorat Spécialité (Physique énergétique génie des procèdes,2007.

Dielectric Materials and Applications: ISyDMA'2016   Materials Research Forum LLC
Materials Research Proceedings 1 (2016) 256-260   doi: http://dx.doi.org/10.21741/2474-395X/1/64

Dilute nitride GaInAsN: a promising material for concentrator photovoltaic cells

Tarik BOURAGBA*[1], Martine MIHAILOVC[2], Joël LEYMARIE[2]

[1]Dept. Engineering sciences department, EIGSICA, Casablanca, Morocco

[2]Optical spectroscopy of solids department, LASMEA, Aubière cedex, France

*Corresponding author: E-mail: tarik.bouragba@eigsica.ma

Keywords: Component 10 Band kp Model, Dilute Nitride Materials, GaInAsN, GaInNAsN, Photoluminescence, Solar Cells, TDOA

Abstract. Nitride alloys GaInAsN and GaInNAsSb are an exciting new generation of semiconductors based on combined GaN (Wutzite crystal) with GaAs (Sb) (Zincblend crystal). Nowadays it is possible to achieve good qualities of dilute nitride films with 1 eV band gap taking advantage from lattice-matched to gallium arsenide and Germanium. In this paper, we provide explanation to some anomalous nitride phenomena such blueshift of the Photoluminescence (PL) energy after annealing and the Stoke's shift observed between absorption and emission spectra, band gap energies of both GaInAsN and GaAs:N

Introduction

The III–V–N semiconductor solid solutions have attracted the world attention because of their interesting physical properties and potential applications in optoelectronic devices and solar cells applications.

The most important feature of GaInAsN quaternary solid solutions is that they can be grown with lattice match to a GaAs or InP substrate with a small amount of nitrogen incorporation, thus it lead to avoid the growth problems encountered in strained layers, such as the ternary GaAsN and GaPN semiconductors.

Its known that narrow band gap materials such GaInAsN improves drastically tunnel current J_{max} that important for concentrator solar cells application.

Also, several improvements have been performed using multi-junction III-V solar cells, not only for space applications but also terrestrial concentrator solar cells. The highest photovoltaic conversion efficiency has been recorded so far is 46% [1-2]. In order to further improve the performance of solar cells while reducing the cost by using dilute nitride and dilute antimonide materials, which can be grown lattice matched onto gallium arsenide and germanium substrates.

Our aim in this paper is to calculate separately the lattice parameter and band gap energy of the (GaInAsN) quaternary semiconductor alloy. The absorption energy was calculated and compared to PL emission energy in order to extract the Stoke's shift, the impact of annealing post-treatment was also shown and discussed by using photoluminescence spectroscopy.

Calculation method

10 band k.p model was used to evaluate the band gap energies and to predict the interband transitions on InGaAsN [3].

This model developed by Lindsay and O'reilly [4].This formalism (tight binding) sp3s * based on the calculation of the interaction energies between orbitals in tetrahedral atomic bond is used to describe the band structure of GaInAsN. The modified 10-band kp model, which is an extension of the conventional 8-band kp model [5], was adopted to establish the band dispersion relation of the material Ga (In) AsN..

Two main reasons are the behind the development of this model:
• Provide a better description of the E + band.

• Optimize BAC model

The Hamiltonian is quantized with respect to the z axis growth perpendicular to the growth plane. So for kx = ky = 0, the Hamiltonian 10 × 10 decouples two independent 5×5Hamiltonian:

$$H = \begin{pmatrix} E_N & V_{NM} & 0 & 0 & 0 \\ V_{NM}^* & E_M & 0 & \sqrt{2}U & U \\ 0 & 0 & E_{HH} & 0 & 0 \\ 0 & \sqrt{2}U^* & 0 & E_{LH} & Q \\ 0 & U^* & 0 & Q^* & E_{SO} \end{pmatrix}$$

The index N, M, HH, LH et SO were respectively related to Nitrogen level, GaInAs, Heavy hole, light hole and spin-orbite bands where V_{NM} is the coupling parameter that describe the interaction between Nitrogen level and the conduction band of GaInAs.

Results and discussions

Results

In the present works, one set of samples were studied, it is composed by three samples of $In_{0.4}$ $Ga_{0.6}As_{1-y}N/GaAs$ quantum wells deposited on GaAs substrates by molecular beam epitaxy. The nominal thicknesses for both wells and the barriers are respectively 7 nm and 80 nm. The nominal nitrogen concentrations are 0.9% sample S564, 1.3% sample S563, and 1.8% sample S571.

Photoluminescence (PL) and thermally detected of optical absorption (TDOA) have been carried out on these structures. This later is an original technique and well suited for the study of absorption from QWs. This sensitive method is based on the detection of phonons emitted via non-radiative processes, which occur during the de-excitation of a system after optical absorption.

As showed in the figure 1, as-grown InGaAsN thin films generally exhibit poor luminescence efficiencies.

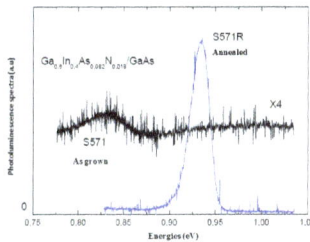

Fig.1. Photoluminescence spectra recorded on GaInAsN/GaAs sample before and after annealing

Post-growth annealing removes drastically defects from the as-grown crystal; it improves the photoluminescence PL intensity by inducing an homogenization of In content and lead to the diffusion of N atoms outside the well. Consequently, it an undesirable blueshift in the emission peak wavelength is produced [6].

Using the 10 band k.p model, it can be possible to calculate the absorption energies; we found that our results are in good agreement with the experimental data. We could explain that the reduction in PL linewidth and the observed Stokes shift is related to the homogenization of the In distribution.

The table 1 below summarizes the optical absorption obtained from TDOA spectra on the three studied samples (see fig 2).

Dielectric Materials and Applications: ISyDMA'2016   Materials Research Forum LLC
Materials Research Proceedings **1** (2016) 256-260   doi: http://dx.doi.org/10.21741/2474-395X/1/64

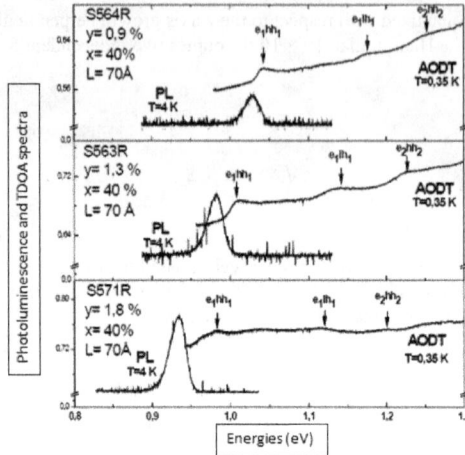

Fig.2. PL and TDOA spectra performed on three GaInAsN/GaAs samples

Table 1 Summary of TDOA energies and Stoke's shift values extracted from the experimental results

Samples	TDOA energies values			Stoke's shift
	e_1hh_1	e_1lh_1	e_2hh_2	
S564r (x_{In}=40% Y_N=0.9% L=70Å)	1.048	1.183	1.263	14
S563r (x_{In}=40% Y_N=1.3% L=70Å)	1.018	1.148	1.273	27
S571r (x_{In}=40% Y_N=1.8% L=70Å)	0.989	1.120	1.207	47

We have also recorded absorption spectra performed on three GaAs:N samples non-intentionally doped barriers (see fig 3), recorded values are summarized in table 2.

We can clearly see the evolution of the narrow pics of GaAs:N as shown on fig3, It is also seen that the absorption edge for all N-containing samples is red-shifted while N content is increasing.

Dielectric Materials and Applications: ISyDMA'2016   Materials Research Forum LLC
Materials Research Proceedings **1** (2016) 256-260   doi: http://dx.doi.org/10.21741/2474-395X/1/64

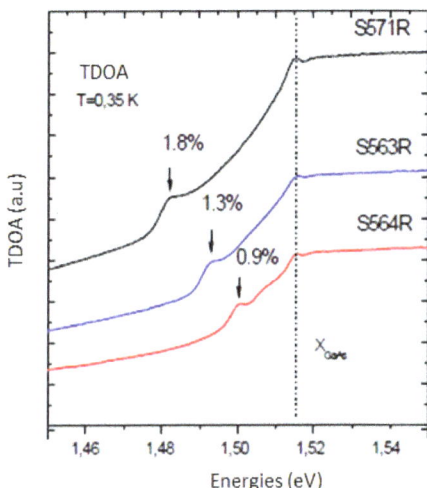

Fig.3. TDOA spectra near the narrow pic for GaAs;N barrier

Table.2. TDOA energies of GaAs:N compared to calculated value using 10 band k.p model

Samples	Experimental vs. Calculated energies of narrow pics	
	Experimental value	Calculated value
S564r	1500	1498
S563r	1493	1492
S571r	1483	1480

Discussions

As for annealed samples, in the framework of our model, changes in the nitrogen environment (short range order) result in changes of the coupling parameter V_{NM} whereas In concentration. On the other hand, thickness of the square well, and N concentration are kept equal to the values found for the as-grown sample (No nitrogen diffusion was considered). Consequently, the V_{NM} values have been used as fitted value and we observed that the coupling parameters values decrease while N content is increasing.

Concerning, the shift energy between absorption and emission (called Stokes shift). It appears that this later is even higher than the electron-phonon interactions are strong. Nitrides materials exhibit low band gap, and a shift between the main photoluminescence pic (PL) and absorption (TDOA).

However, this offset is important to low temperature and vanishes at higher temperature. We can confirm that carrier-phonon interactions are greater at high temperatures; it is likely that the shift observed is just a carrier thermalization at low temperature [5-7].

Dielectric Materials and Applications: ISyDMA'2016   Materials Research Forum LLC
Materials Research Proceedings **1** (2016) 256-260   doi: http://dx.doi.org/10.21741/2474-395X/1/64

Conclusion

Dilute nitride - GaInAsN - 0.7 μm thick layers are grown on GaAs substrates by low-temperature MBE. These heterostructures are investigated by using both spectroscopy techniques thermally detected of optical absorption and photoluminescence spectroscopy.

Using 10 band k.p model allowed us to fit well the experimental results obtained from our set of GaInAsN samples and stokes shift energies has been extracted and well explained.

Both TDOA and PL spectra have been used to follow the interband transitions from the quantum wells structure.

The absorption edge for all N-containing samples is red shifted compared to the GaAs one, this shift was well explained by using 10 band k.p model and the huge bowing introduced once nitrogen content is growing.

References

[1] New world record for solar cell efficiency at 46%,

[2] PRC.Kent, Semicond. Sci. Technol 17 (2002) p851-p859.
http://dx.doi.org/10.1088/0268-1242/17/8/314

[3] T. Bouragba, M. Mihailovic, H. Carrère, P. Disseix, A. Vasson, J. Leymarie, E. Bedel, A. Arnoult, and C.Fontaine, IEE Proc: Optoelectron.151, 309, 2004.
http://dx.doi.org/10.1049/ip-opt:20040929

[4] A. Lindsay , E.P. O'Reilly, Solid State Comm 112, p443, (1999).
http://dx.doi.org/10.1016/S0038-1098(99)00361-0

[5] E.O.Kane, J. Phys. Chem. Solids, 1, 249 (1957).
http://dx.doi.org/10.1016/0022-3697(57)90013-6

[6] M. Albercht, Applied Physics Letters V81, Number15, p2719-p2721 (2002).
http://dx.doi.org/10.1063/1.1509122

[7] L. Grenouillet, Thesis (2001)

Dielectric Materials and Applications: ISyDMA'2016
Materials Research Proceedings 1 (2016) 261-265

Materials Research Forum LLC
doi: http://dx.doi.org/10.21741/2474-395X/1/65

Study of ferrite by resonant nuclear scattering nickel-iron: the influence of directional order

Abelkrim BENLALLI

Radiation Physics Laboratory, Physics Department, University Badji Mokhtar-Annaba, Annaba
Algeria

Corresponding author: E-mail: a_benlalli@yahoo.fr

Keywords: Ferrite, Resonant Nuclear Scattering, Iron Ore, Spinel Phase, Directional Order

Abstract. The Resonant Nuclear Scattering has revealed the directional order that forms in the ferrites due to anisotropic distribution of Fe^{2+} and Fe^{3+} to the local direction of magnetization. We undertook to conduct a study on ferrites of composition:

$$Fe^{3x}_{2(1+x)} \, Fe^{2+}_{0.2-2.7x} \, Ni_{0.8-0.3x} \, \Diamond_x \, O_4$$

$\Diamond$ = vacancies, $0 < x \le 0.07$
The spectra were made on two kinds of samples: quenched and slowly cooled at different temperatures. The results show that it is possible to possible to distinguish at least three sites according to the values of x: an octahedral site (O) occupied by the Fe^{3+} ion, which will be designated by $(Fe^{3+})_O$, a tetrahedral site (T) or $(Fe^{2+})_T$, and a tetrahedral site (T) or $(Fe^{3+})_T$. The variation of intensity of the site $(Fe^{3+})_O$ according to x, and the differences of variation according to the cooling conditions suggest the existence of a partial order among the nearest neighbours of the site $(Fe^{3+})_O$ of an ion $(Fe^{3+})_T$, which will make good account of the directional order.

Introduction

The Nickel-Iron ferrites at room temperature, have respective crystal groups: R $\bar{3}$ and R $\bar{3}$m [1], [2], Fe^{2+} and Ni^{2+} octahedral sites occupy the like. A lowering of symmetry to 240 K has been observed [3], but it does not seem to affect significantly the trigonal symmetry crystal field acting on the Fe^{2+} ion with axial magnetic properties remain at low temperature [4]. Ferrites containing Fe^{2+} ions such as Fe-Ni ferrite can submit electronic relaxation phenomena causing the Fe^{2+} ions in privileged positions in the subnet-specific octahedral sites relative to the direction of the spontaneous magnetization. Rearrangements resulting lead to orientation structures or directional orders that we propose to study by Resonant Nuclear Scattering. Migration of electrons between Fe^{2+} and Fe^{3+} gives maximum relaxation at a temperature below 160 K. Migration of ions by means of gaps give two maxima: one at 277 K we call A, the other will be called B between 500 and 600 K. A relaxation phenomenon of migration is a short distance, the second B is a phenomenon long distance. The quenching effect of suppressing the directional order B in a sample. Finally a slow cooling leads to a sample with a local order across the field. The Resonant Nuclear Scattering spectra has been made on ferrites characterized X-ray diffractometer Philips X'Pert MRD. [5] [6], an operating voltage of 40 kV and a current of 30 mA to horizontal geometry, in a configuration $\Omega = 2\theta$. A wavelength of $\lambda = 1.54056$ Å is used to obtain data in the range of $20° < 2\theta < 120°$, with an angular pitch of $0.04°$ for the one hand minimize the background noise, and other by collecting all the information contained in the shape, width and lineshape. Only the spinel phase has been detected. The composition of the samples was determined by chemical analysis and obeys the formula:

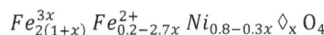

$$Fe^{3x}_{2(1+x)} \, Fe^{2+}_{0.2-2.7x} \, Ni_{0.8-0.3x} \, \Diamond_x \, O_4$$

$\Diamond$ = vacancies, $0 < x \leq 0.07$
The samples underwent the following treatments:

► Slowly Cooled samples, $(SCS)_x$ for :

x = 0.003; 0.017 and 0.05

► Samples Quenched from 800°C, $(SQ)_x$ for :

x = 0.005, 0.026 and 0.055

The Resonant Nuclear Scattering spectra were obtained at room temperature where the relaxation A manifests, at 173 K, A and B relaxations are frozen. The experimental spectra have been programmed by superposition of Lorentzian-like elementary sextuplets and by minimizing the difference between the sum of the squares of the experimental and calculated blows; for an effective thickness of the sample approximately fixed. The programming automatic spectra determines the 3 hyperfine parameters and the intensity of each sextet (the total number of sites that have been agreed in advance) and the experimental line width we assumed common to all sites. The isomer shifts δ are given relative to metallic iron which was used as a standard speed of the spectrometer.

Results
Resonant Nuclear Scattering spectra were performed on ferrites characterized by X-ray Debye-Scherrer camera with iron K_α radiation: only the spinel phase was detected, the composition of the samples was determined by scan chemical and obeys the formula

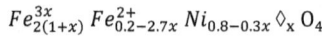

$$Fe^{3x}_{2(1+x)}\, Fe^{2+}_{0.2-2.7x}\, Ni_{0.8-0.3x}\, \Diamond_x\, O_4$$

($\Diamond$: vacancies) and x < 0.074.

The values of the three hyperfine parameters H, δ, Δ does not vary at a given temperature as a function of x. Their values according to temperature are given in Tables I, II and III.

On the other hand the intensities according to the total amount of iron present in the sample varies with x as follows:

1 - The intensities of sites $(Fe^{3+})_T$ and $(Fe^{2+})_T$ vary with x independently of the treatment undergone by the material whatever the temperature, while this is not the case for the other intensities of tetrahedral sites to 300 K. (see fig. 2). These behavioral differences disappear at 173 K (see Fig. 3).

2- If the change in the intensity of the site (Fe3+)O with x is not affected by the temperature change, that of the site (Fe3+)T is more important than 173 K to 300 K.

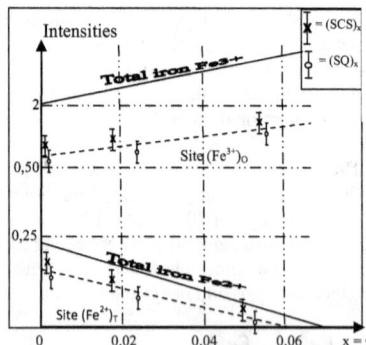

Fig. 1: Intensities (normalized to the total quantity of iron) of sites

$(Fe^{3+})_o$ and $(Fe^{2+})_T$ at 300 K.

3 - The decrease of the intensity with x in site $(Fe^{3+})_T$ seems independent of the temperature , while its increase with x is transformed to 173 K in a decrease reflecting change affecting the site $(Fe^{3+})_O$.

Table I Hyperfine parameter Values δ different sites at different temperatures

Site	δ (mm/s)		
	80 K	173 K	300 K
$(Fe^{2+})_T$	0.29 ± 0.02	0.26 ± 0.02	0.22 ± 0.02
$(Fe^{3+})_T$	0.48 ± 0.02	0.44 ± 0.02	0.40 ± 0.02
$(Fe^{3+})_O$	0.90 ± 0.02	0.81 ± 0.07	0.84 ± 0.05

Table II Hyperfine parameter Values Δ different sites at different temperatures

Site	Δ (mm/s)		
	80 K	173 K	300 K
$(Fe^{2+})_T$	-0.01 ± 0.02	-0.02 ± 0.02	-0.01 ± 0.02
$(Fe^{3+})_T$	-0.01 ± 0.01	-0.02 ± 0.02	-0.01 ± 0.01
$(Fe^{3+})_O$	0.30 ± 0.20	0.20 ± 0.15	0.15 ± 0.15

Table III Hyperfine parameter Values H different sites at different temperatures

Site	H (T)		
	80 K	173 K	300 K
$(Fe^{2+})_T$	512 ± 1.05	507 ± 1.80	497 ± 2.00
$(Fe^{3+})_T$	549 ± 1.06	545 ± 1.80	530 ± 2.00
$(Fe^{3+})_O$	525 ± 2.04	512 ± 3.00	489 ± 3.00

Interpretation

The sextuplets samples analyzed can identify without too much ambiguity from the values of isomer shift and hyperfine field [7], respectively, as from Fe^{3+} ions in tetrahedral and octahedral sites and Fe^{2+} ions in tetrahedral site. Such an allocation for sites O and T is reinforced by the variations similar levels of Fe^{3+} and Fe^{2+} observed according of x in Figure 1. It should be noted that all the Fe^{2+} ions do not participate in site T, and that even all of Fe^{3+} ions not participating in site O. It should be noted in fact that Fe^{3+} ions in site O are not fully represented, the intensity of to site is lower than 1 whatever the temperature, and a linear function of decreasing x.

As it is admitted that vacancies can not to form into sites tetrahedral in samples cooled slowly [8], we would could to move towards the presence of Fe2+ in octahedral sites (O), but this hypothesis is in contradiction with results earlier [7].

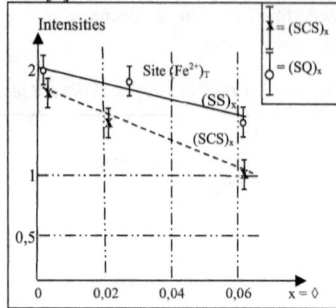

Fig. 2: Intensities (normalized to the total quantity of iron) of sites $(Fe^{2+})_T$ at 300 K

Discussion

The vacancies tend to block the exchange Fe^{3+}, Fe^{2+} in tetrahedral site (T) at temperature 173 K and in proportion to the rate vacancies. Furthermore at 77 K there is a significant decrease in the intensity of exchange between ions of sites (O) and (T), which corresponds to a partial blockage of the electronic exchange between Fe^{2+} and Fe^{3+}. The decrease in the intensity of the site $(Fe^{3+})T$ with x whatever the temperature, seems to indicate that it occurs at the octahedral sites of increasing ionic rearrangements with content in vacancies at the temperature 300 K.

These rearrangements are also function of the constitution of the material, from which we can remark that it is the quenched material that admits the fewest rearrangements, in view of the phenomenon of rearrangement existing between 300 and 400°C was cancelled remains that only to room temperature.

It is these observations that encourage to assume that the site $(Fe^{3+})T$ would come from Fe^{3+} ions in octahedral sites whose nearest neighbours $(Fe^{2+})T$ would present a partial order such that the characteristics hyperfine the indistinguishable from an exchange site $Fe^{3+} \leftrightarrow Fe^{2+}$.

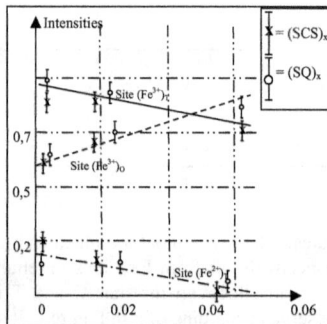

Fig. 3: Intensities (normalized to the total quantity of iron) of sites $(Fe^{3+})_o$, $(Fe^{3+})_T$ and $(Fe^{2+})_T$ at 173 K.

Conclusion

We have could highlight by Resonant Nuclear Scattering the electronic exchange $Fe^{2+} \leftrightarrow Fe^{3+}$ by its partial blockade obtained at 80 K. It is expected to make a spectrum at about 4 K where the electronic exchange to be completely blocked. The rate of increasing gaps favours the rearrangement ionic therefore directional order. In addition we observe by Resonant Nuclear Scattering a decrease of electronic exchange $Fe^{3+} \leftrightarrow Fe^{2+}$ content in when the vacancies increases for temperatures lower than 300 K.

This is in accordance with results previously obtained on magnetite [9], and with other results obtained by some authors on irradiated ferrites [10].

Appendix

SCS = Slowly Cooled samples
SQ =
δ = Isomer shift
Δ = Quadrupol Splitting
H = hyperfine field
Indices T and O respectively represent :
Tetrahedral and Octahedral

References

[1] Hamilton, W.C., Acta Crystallogr. Vol 54, pp 353-361, 1998.

[2] Hassel, O. et Salvesen, J. R., Z. Phys. Chem. Vol 216, pp 345-352, 2002

[3] A. Ramdani, R.Gerardin., J.Solid State Chem vol 65 pp 309-321, 1986.
http://dx.doi.org/10.1016/0022-4596(86)90103-9

[4] Ohtsuka, T., J. Phys. Soc. Japan vol 64, pp 1245-1254, 2000.

[5] B. E. Warren, X-ray Diffraction, Addison-Wesley, New York 1970.

[6] J. I. Langford, J. Appl. Cryst. Vol 11, pp 10-14, 1978.
http://dx.doi.org/10.1107/S0021889878012601

[7] Linnet, J. W., Rahman, M.M., J. Phys. & Chem. Solids. Vol 93- pp 7562-7568, 1989.

[8] P.Gisse, M.Sidir, V.L.Lorman., J. Physique II. vol 7 pp 1817-1826, 997.

[9] C.J.Howard & H.T.Stokes., Acta Cryst. Vol B60 pp 674-684, 2004.
http://dx.doi.org/10.1107/S0108768104019901

[10] Q.Shen, S.Kycia & I.Dobrianov., Acta Cryst. Vol A56, pp 268-279, 2000.
http://dx.doi.org/10.1107/S0108767300000246

Dielectric Materials and Applications: ISyDMA'2016
Materials Research Proceedings 1 (2016) 266-269

Materials Research Forum LLC
doi: http://dx.doi.org/10.21741/2474-395X/1/66

Some physical properties of the glasses within the Li$_2$O-Li$_2$WO$_4$-TiO$_2$-P$_2$O$_5$ system

H. ES-SOUFI[1], L. BIH*[1], B. MANOUN[2], D. MEZZANE[3], P. LAZOR[4]

[1]EquipePhysico-Chimie de la Matière Condensée (PCMC), Département de Chimie, Faculté de Sciences de Meknès, Morocco

[2]Laboratoire des Sciences des Matériaux, des Milieux et de la Modélisation (LS3M) Univ. Hassan 1er, 26000, Morocco

[3]Laboratoire de la Matière Condensée et Nanostructure (LMCN) Faculté des Sciences et Techniques Guéliz, Marrakech, Morocco

[4]Department of Earth Sciences, Uppsala University, SE-752 36 Uppsala, Sweden

*Corresponding author: E-mail: lbih@hotmail.com

Keywords: Phosphate,Tungstate, Glasses, Glass Transition Temperature, Conductivity

Abstract. Phosphate glasses of the compositions 20Li$_2$O-(50-x)Li$_2$WO$_4$-xTiO$_2$-30P$_2$O$_5$ (0≤x≤15, mol%) were prepared by the melt quenching method. The amorphous nature of these glasses was confirmed by the XRD diffraction. Their characteristic temperatures were determined by DSC analysis. Impedance spectroscopy is used to determine their electrical conductivity as a function of temperature. The obtained results show that the conductivity as a function of temperature follows an Arrhenius-law.

Introduction

Phosphate glasses have been studied in recent years because of their possible technological applications. Phosphate glasses are important materials in technologically comparing with the conventional oxide glasses. They possess some superior physical properties such as high thermal expansion coefficients, low melting temperature, low softening temperatures and smaller liquids viscosity than silicate glasses. Another important feature of these glasses is their ability to incorporate large amounts of transition metal without reduction of glass forming ability. These properties make them useful candidates for fast ion conducting material and other important applications such as laser hosts, nuclear waste glasses, glass-to-metal seals and biocompatible materials (1). Due to the fact that some phosphate glasses showed alkali ion conductivities even at room temperature they are being studied extensively over the last decade in order to be applied in energy storage devices and solid-state batteries (2-4). Recently (5), lithium phosphortungstate glasses belonging to the system Li$_2$O-WO$_3$-Nb$_2$O$_5$-P$_2$O$_5$ glasses have shown that their electric transport is mixed ionic-electronic conductivity and may be used as positive electrodes in solid-state batteries. Recent studies by us (6) showed that the durability of phosphate glasses can be increased by the addition of TiO$_2$, which stabilizes the phosphate glass network.

The aim of this study is to understand the possible role of TiO$_2$ in thermal stabilizing the glassy network and its contribution to the electrical conductivity of the glasses.

Experimantal procedure

Synthesis route

Phosphate glasses containing TiO$_2$ of the compositions20Li$_2$O-(50-x)Li$_2$WO$_4$-xTiO$_2$-30P$_2$O$_5$ (x=0, 5, 8, 10, 15 mol%) have been prepared by melt-quenching route from the raw materials: Li$_2$CO$_3$, Li$_2$WO$_4$, TiO$_2$and NH$_4$H$_2$PO$_4$. They are weighted according to the coefficient stoichiometriesthen, they are groundin agate mortar. Themixture in alumina crucibleis placed into electrical furnace (Nabertherm P330)and heated successively at 300°C about 12 hours and at 500°C foran hour.The

Dielectric Materials and Applications: ISyDMA'2016   Materials Research Forum LLC
Materials Research Proceedings 1 (2016) 266-269   doi: http://dx.doi.org/10.21741/2474-395X/1/66

temperature of the furnace is increased at 900±10°C during an hour; the melting is agitated for homogenization. After that, it cooled in air. Immediately after quenching, the obtained glasses wereannealed at 300°Cfor one hour, and then, slowly cooled inside the furnace to minimize the internal stresses resulting from quenching.

Technique of characterisation

XRD patterns of the glasses have been carried outat room temperature by using a Phillips D5000 diffractometer equipped by an anticathodeCu (K_λ= 1,5405 Å). The patterns were scanned through steps of 0.01°, between 10° and 100° with a fixedtime counting of 16 s.

The thermal stability of the studied glasses was studied by the Differential ScanningCalorimetry (DSC). DSC runs were carried out for ground glass batches of about 40 mg in nitrogen atmosphere at a heating rate 5°C/min using instrument DSC131Evo.

The electrical conductivity was determined from the impedance /admittance spectroscopic method. The spectra were carried out on a Hewlett Packard Model 4284A precision LCR Meter in the frequency range [20-10^6] Hz with temperatures changing from 25 to 533 K.

Results and discussion

X-ray diffraction of the glasses

By the procedure described above, homogeneous glasses were prepared. The presence of a glassy state was checked by X-ray diffraction. XRD patterns for different glass samples are presented in Fig. 1. These patterns confirm the glassy nature of the samples with broad peaks around 20–30° (2 theta values).

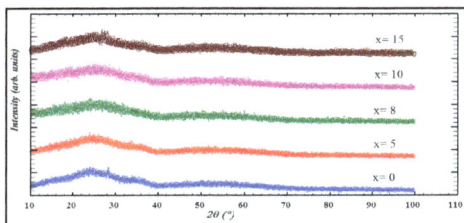

Figure 1:XRD patterns ofthe $20Li_2O$-$(50-x)Li_2WO_4$-$xTiO_2$-$30P_2O_5$glasses

Thermal analysis

The glass transition temperature, T_g, and the values of crystallization temperature,T_c, were determined from the DSC curves. The results are summarized in Table 1. It's concluded that the glass transition temperature T_g increases with increasing of TiO_2 content. The observed increase in T_g indicates the increase of bonding strength in the glass structure with increasing of TiO_2 content. We suspect that the nature of bonding is responsible for the variation of T_g.

Table 1 Tg and Tc temperaturesfor$20Li_2O$-$(50-x)Li_2WO_4$-$xTiO_2$-$30P_2O_5$ (x=0-15mol%) glasses.

$xTiO_2$(% mol)	T_g (±5°C)	T_c (±5°C)
0	407	503
5	425	532
8	435	---
10	445	---
15	462	---

Dielectric Materials and Applications: ISyDMA'2016   Materials Research Forum LLC
Materials Research Proceedings 1 (2016) 266-269   doi: http://dx.doi.org/10.21741/2474-395X/1/66

Impedance Spectroscopy analysis.

The conductivity (σ_{dc})is obtained from the relation $\sigma_{dc}= (1/Z_0)*(e/s)$, where Z_0 is the intercept on the real axis of the zero phase angle extrapolation of the highest frequency curve and (e/s) is the sample geometrical ratio.

The reciprocal temperature dependence of the dc conductivity for the different compositions is shown in the Fig. 2. The figure shows that the plots are straight lines at high temperatures indicating that the dc conductivity obeys the Arrhenius relation $\sigma_{dc}=\sigma_0\exp(-E_a/kT)$, where σ_0 is the pre-exponential factor, $E_a$the activation energy and k is the Boltzmann constant. In all the temperature range, the temperature dependence of the conductivity is not linear. Therefore, two activation energies, according to low or high temperature, could be determined for each glass.

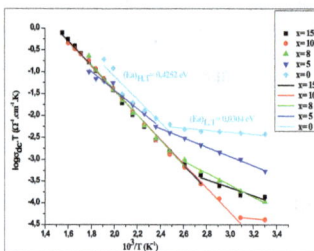

Figure 2: Temperature dependence of log (σ_{dc}.T) for the glasses

The values of the activation energies are given in table 1. The determined energy at low temperatures is very low and reveals that the conductivity is probably due to electronic conduction. However, at high temperature the conductivity is predominantly ionic.

Table2 Activation energy at low and high temperature ranges for the glasses.

x(%mol)	$(E_a)_{H.T}$ (eV)	$(E_a)_{L.T}$ (eV)
0	0,4252	0,0304
5	0,4286	0,0216
8	0,5133	0,0255
10	0,5633	0,0503
15	0,5389	0,0378

Conclusion

The glasses under study were elaborated by the solid-melt quenching route. They were characterized by thermal analysis and impedance spectroscopy. They showed mixed electronic-ionic conduction.

References

[1] P.Y. Shih, "Thermal, chemical and structural characteristics of erbium-doped sodium phosphate glasses", Mater.Chem. Phys. 84 (2004) 151. http://dx.doi.org/10.1016/j.matchemphys.2003.11.016

[2] F. Borsa, D.R. Torgeson, S.W. Martin, H.K. Patel, "Relaxation and fluctuations in glassy fast-ion conductors: Wide-frequency-range NMR and conductivity measurements"Phys. Rev. B 46 (1992) 795. http://dx.doi.org/10.1103/PhysRevB.46.795

[3] B.V.R. Chowdari, K.F. Mok, J.M. Xie, R. Gopalakrisnan, "Electrical and structural studies of lithium fluorophosphate glasses", Solid State Ionics 76 (1995)189. http://dx.doi.org/10.1016/0167-2738(94)00280-6

[4]   R. Winter, K. Siegmund, P. Heitjans, "Nuclear magnetic and conductivity relaxations by Li diffusion in glassy and crystalline LiA1Si$_4$lO", J. Non-Cryst.Solids 212 (1997) 215. http://dx.doi.org/10.1016/S0022-3093(96)00654-0

[5]   M.A. Valente, L. Bih, M.F.P. Graça, "Dielectric analysis of tungsten–phosphoniobate 20A2O–30WO3 –10Nb2O5 –40P2O5 (A=Li, Na) glass–ceramics", J. Non-Cryst. Solids, 357 (2011) 55. http://dx.doi.org/10.1016/j.jnoncrysol.2010.10.007

[6]   H. Sinouh, L. Bih, M. Azrour, A. El Bouari, S. Benmokhtar, B. Manoun, B. Belhorma, T. Baudin, P. Berthet, D. Solas,and R. Haumont,"Effect of TiO2 and SrO additions on some physical properties of 33Na2O–xSrO–xTiO2–(50- 2x)B2O3–17P2O5 glasses",Journal of Thermal Analysis and Calorimetry, Volume 111, 2013, pp. 401-408. http://dx.doi.org/10.1007/s10973-012-2394-3

Dielectric Materials and Applications: ISyDMA'2016   Materials Research Forum LLC
Materials Research Proceedings 1 (2016) 270-274   doi: http://dx.doi.org/10.21741/2474-395X/1/67

Effect of porosity and incident angle on the optical emission radiations of silicon and porous silicon under Kr$^+$ ion beam bombardment

H. TARGAOUI[1], S. MARGOUM[1], A. EL BOUJLAIDI[1], R. BENDAOUD[1], I. ASHRAF[2], K. BERRADA*[1]

[1] ESIAM - Department of Physics, Faculty of Sciences Semlalia - Cadi Ayyad University, Marrakech, Morocco

[2] Department of Physics - University Quaid-Azzam, Islamabad, Pakistan

*Corresponding author: E-mail: berrada@uca.ma

Keywords: Silicon, Porous Silicon, Optical Emission Spectroscopy, Sputtering

Abstract. This study is focused on the optical emission spectroscopy of silicon, porous silicon (PS) and their oxides. It allows us to understand the effect of the angle and porosity on optical radiations emitted when substrates of silicon, PS and their oxides are bombarded at low-pressure less than 10^{-7} torr by Kr+ ions at 5 keV. The comparison of the spectra shows that the spectral lines attributed to Si(I) have a similar behavior and the maximum of the signal is obtained for $\theta=70°$. These lines show a significant exaltation when the density of porosity of similar samples is larger. This exaltation depends on the desactivation of the excited states and it may reach a quite important factor with intensity saturation for maximum silicon surface porosity. The intensity in a reverse behavior for the line Si(III) at $\lambda=254.4nm$ and associated to Si^{++} matched for the first time in our results. However, the signal amplification is caused by an enrichment of oxygen on the porous surfaces, which become oxidized more easily when the porosity density increased.

Introduction

Porous silicon is long known for its potential applications in the field of micro- and nano-electronics. With the discovery of its visible luminescence properties, it has become an interesting material in the field of biosensor.

Optical emission induced by excited sputtered atoms under Kr$^+$ ion bombardment, from a solid surface, has been the subject of several investigations in recent years [1-2]. The emitted light strongly depends on the condition of the surface under ion bombardment and therefore, constitutes basics of surface chemical analysis.

The ejected particles present at the surface can be identified by their characteristic line emissions [3-5]. This identification has been shown by the measurement of their wavelength.

In this paper, optical emissions from sputtered particles, following 5 keV Kr$^+$, ions bombardment of pure silicon and PS targets have been studied by using Optical Emission Spectroscopy (ASSO). The light intensities are measured as a function angle of incident and porosity of PS targets. For silicon, it has already been studies that the intensity of spectral lines changes abruptly with time, by interruption of the incident beam or change in ion beam current or with the change in oxygen pressure [6-9].

The interest of this study is particularly important for our team, since we use these materials as substrate in the development of composite PS-Electroactive polymer for chemical sensors or components for optoelectronics [10-13]. The study of the morphology of these materials is carried out by Scanning Electron Microscopy (SEM), which made it possible to observe the porosity of these substrates [14].

Dielectric Materials and Applications: ISyDMA'2016   Materials Research Forum LLC
Materials Research Proceedings 1 (2016) 270-274   doi: http://dx.doi.org/10.21741/2474-395X/1/67

In this experimental work we tried to highlight the reasons along with modification which depend on one hand on the density of porosity and on the other hand on the angle of incident to the surface of the sample.

Experimental

The presented work is based on the study of three samples, silicon and PS with two different densities, 40 mA/cm^2 and 80 mA/cm^2. The PS samples are obtained by an electrochemical anodization of silicon. The anodization cell can employs platinum cathode and silicon wafer anode immersed in hydrogen fluoride electrolyte diluted with 20%. The corrosion of anode is produces by running electrical current of various densities, through the cell. The porosity of the sample is perfectly controllable according to the experimental conditions [13].

The experimental setup used is the same as described in our earlier work [1-2,5]. The schematic diagram of the experimental setup is shown in figure 1.

The formation of Kr$^+$ ions is done by electron impact on Kr (of 99.998% purity) in a plasma source. This results in the generation of few microamperes current of 5 keV Kr$^+$ ions. The ionic beam is guided to the target mounted inside an ultra-high vacuum chamber, with residual pressure of less than 10^{-7} torr.

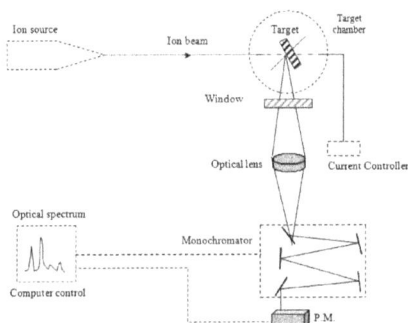

Figure 1: Schematic diagram of experimental setup

Prior to the start of actual experiment, the Si and PS targets were cleaned in-situ by Kr$^+$ ion beam. The ion beam current was measured on the sample or by using Faraday cup placed behind the target, in the direction of ion beams. The beam current densities are kept around 1μA/mm^2, for the explored range of incident angels from 20° to 90°, with respect to the normal at the surface.

The light emission was analyzed through a R320 monochromator, equipped with a 1800 groves/mm holographic grating using 400 μm slit width and a Hammatsu 4220P photomultiplier which is very sensitive to the explored wavelengths range. A micro-computer controller uses the prism program for detection process and storage of data. The selected wavelength range varied from 210 to 260nm.

Results and discussions

The light emission caused by the ion-surface interaction is technically referred as the beam-induced light emission (BLE) or sputtered ion photon spectroscopy (SIPS).

In this paper, optical emissions from sputtered particles, following 5 keV Kr$^+$, ions bombardment of pure silicon and PS targets have been studied by using Optical Emission Spectroscopy as explained in [3].

Figures 2 (a) and (b) show the emission spectra of silicon and porous silicon for the wavelength range from 200-300 nm.

Dielectric Materials and Applications: ISyDMA'2016   Materials Research Forum LLC
Materials Research Proceedings 1 (2016) 270-274   doi: http://dx.doi.org/10.21741/2474-395X/1/67

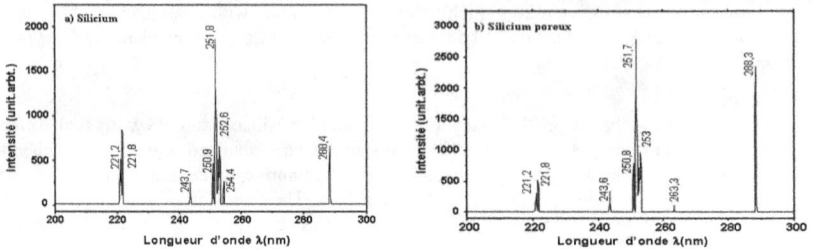

Figure 2: Luminescence spectra for the range 200-300 nm of (a) Silicon and (b) Porous silicon

Under our experimental conditions, no spectral lines have been found out side this range. The behavior of spectral line intensities obtained with an angle of 70° with respect to the normal to the surfaces of pure Si and porous Si, has been depicted in figures (2a) and (2b), respectively.

Both graphs show almost similar behavior as shown in figure 3. The intensity of spectral lines increases in the beginning, being maximum around 70° of incident angle and then decreases exponentially with the decrease of angle. However the intensity of spectral lines is quite large for porous silicon with respect to pure silicon. This indicates the enhancement of the intensities with the porosity of the silicon. In order to elaborate more the effect of angle of incidence and the porosity of the silicon we have made three different sets of measurements, pure silicon, porous silicon, with density of porosity 40 mA/cm^2 and 80 mA/cm^2, for minimum angle of incident 20° as described by Targaoui [14]. All these graphs have shown clearly the enhancement of spectral lines intensities with the increase in the density of porosity. It shows the increase in the intensity of spectral lines from 350 for pure silicon, to 400 for porous silicon with density of porosity 80 mA/cm^2, even for minimum angle of incident of 20° [14].

The variation of spectral lines intensities for the incident angle from 10° to 80° are represented in figures (3a) (3b) and (3c).

The shape of the curve obtained for the pure silicon sample is similar to other curves obtained on other porous silicon samples with different porosity densities. A "soft" signal increase with the angle up to a maximum value followed by a sharp decrease for higher angles of 70°. The porous silicon intensity increases with the density of porosity, which once again confirms the link between porosity and oxidation.

We can conclude from these results that the presence of pores creates a critical effect on the optical radiation emitted when the target is bombarded with ions Kr$^+$ kinetic energy of 5 keV and at an angle of incidence θ=70 °.

Luminescence spectra between 210 nm and 260 nm have a set of transmission lines Si(I) corresponding to the silicon transitions. One ion emission line Si(III) is associated with the wavelength at λ=254.4 nm.

For spectral lines located at λ = 221.66 nm; 243.51 nm; 250.6 nm; 251.61 nm and 252.85 nm, we have observed an increase value of their intensity depending on their porosity density (for 80mA/cm^2 of current density).

Dielectric Materials and Applications: ISyDMA'2016   Materials Research Forum LLC
Materials Research Proceedings 1 (2016) 270-274   doi: http://dx.doi.org/10.21741/2474-395X/1/67

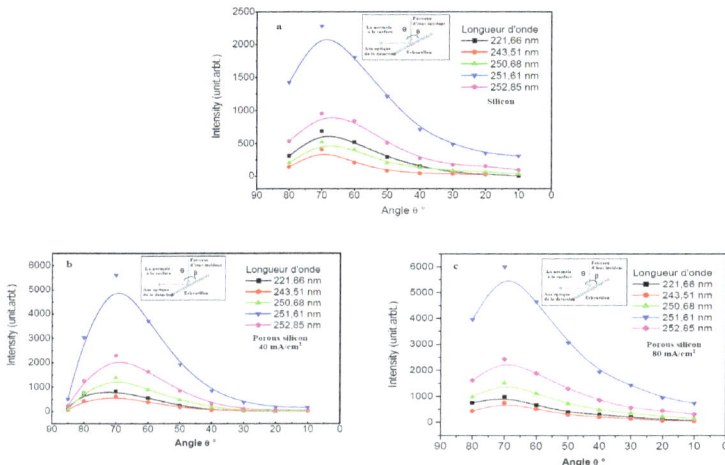

Figure 3: Optical emission intensity at various porosity (a) Silicon, (b) PS 80% and (c) PS 40%

Beyond this density on the porous silicon we obtain an amplified signal that varies very slightly. In fact, using this density value, we can reach more than 80% porous surface on the Si, and consequently the surface becomes very sensitive to oxidation. For the emission line associated to the wavelength λ=254.4 nm that has been attributed to Si(III), the intensity has shown an inverse behavior as described in figure 4 (negative dependence). The signal intensity of Si(III) decreased depending on the porosity density for an angle of incidence of θ=70°.

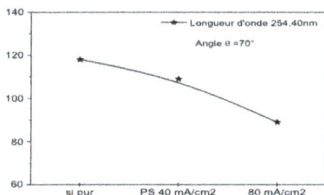

Figure 4: Evolution of intensity of λ=254.4 nm with density of porosity

Conclusion

We have presented a comparative study to see the effect of the angle and porosity on optical radiation emitted by the silicon and the porous silicon at various current densities under Kr+ ion bombardment of 5 keV.

A range of incident angles from 20° to 80° degrees, with respect to the normal to the surface, have been explored. The emission light from these surfaces, as a result of ion bombardment, has been studied over a range of wavelength from 210-260 nm. Observed lines are attributed to neutral Si(I) and ionized silicon Si(II), on explored range. The spectral lines show almost similar behavior with different incidence angles and all have maximum of signal at 70°. The comparison of the various recorded spectra indicates that the intensities of the spectral lines shows a clear exaltation according to the density of porosity in the porous silicon, this exaltation can reach a factor of 3.5 compared to pure silicon, on the contrary we noted a decrease of the intensity of the spectral line Si(II) 254,4 nm which was observed in porous silicon spectra for a first time.

Dielectric Materials and Applications: ISyDMA'2016   Materials Research Forum LLC
Materials Research Proceedings 1 (2016) 270-274   doi: http://dx.doi.org/10.21741/2474-395X/1/67

We try through this experimental study to bring the essential causes with these modifications, which depend on the one hand on the density of porosity and on the other hand of the angle of attack in relation to surface of the samples.

References

[1] Berrada, K., Targaoui, H., Kaddouri, A., Louarn, G., Froyer, G., Outzourhit, A., Optical emissions of Al-Mg alloys under krypton ions bombardment. Physical and Chemical News, 2004, Vol.17, p.49-53.

[2] El Boujlaidi, A., El Fqih, M. A., Hammoum, K., Aouchiche, H., Kaddouri, A., Continuum radiation emitted from transition metals under ion bombardment. The European Physical Journal D, 2012, Vol. 66, N°10, p.1-6. http://dx.doi.org/10.1140/epjd/e2012-30347-2

[3] Fournier, P. G., Fournier, J., Bellaoui, B., Optical emissions from sputtering species formed by the fast ion bombardment of clean and oxygenated Mg surfaces. Nuclear Instruments and Methods in Physics Research Section B: Beam Interactions with Materials and Atoms, 1992, Vol.67, N°1-4, p.604-609. http://dx.doi.org/10.1016/0168-583X(92)95882-R

[4] Agarwal, P., Bhattacharyya, S. R., Ghose, D., Transient effect in light emission during oxygen ion bombardment of a beryllium–copper target. Applied surface science, 1998, 133 (3), p.166-170. http://dx.doi.org/10.1016/S0169-4332(98)00200-1

[5] El Fqih, M., Faké, A., Berrada, K., Kaddouri., A. Effet de l'oxygène sur les radiations optiques émises lors de la pulvérisation de l'aluminium par un faisceau d'ions. Afrique Science: Revue Internationale des Sciences et Technologie, 2008. Vol. 4, N°3.

[6] Bellaoui, B., Thèse de Doctorat, Université de Paris Sud, Orsay (1996).

[7] Louarn, G., Berrada, K., Errien, N., Ait El Fqih, M., Froyer, G., Kaddouri, A., Vibrational and optical emission spectoscopies of porous and oxydized porous silicon. Physical and Chemical News, 2005, N°21, p.5-10

[8] Suchańska, M., Ion-induced photon emission of metals. Progress in surface science, 1997, Vol.54, N°2, p. 165-209. http://dx.doi.org/10.1016/S0079-6816(97)00004-X

[9] Van Der Weg, W. F. & Bierman, D. J., Excitation of Cu atoms by Ar ions and subsequent radiationless deexcitation of scattered particles near a Cu surface. Physica, 1969, Vol. 44, N° 2, p.206-218. http://dx.doi.org/10.1016/0031-8914(69)90222-5

[10] Martin, P. J., Bayly, A. R., Macdonald, R. J., De-excitation processes near the surface of ion bombarded SiO_2 and Si. Surface Science, 1976, Vol.60, N°2, p.349-364. http://dx.doi.org/10.1016/0039-6028(76)90321-6

[11] Kaddouri, A., Spectroscopie d'émission des produits de pulvérisation de solides soumis à un bombardement ionique, Thèse de $3^{\text{ème}}$ cycle, Université Paris 11 (1989).

[12] O. Varenne, Thèse de Doctorat, Université de Paris Sud, Orsay (2000).

[13] Bhattacharyya, S. R., Brinkmann, U., Hippler, R., Investigation of relative sputtering yields during ionoluminescence of Si. Applied surface science, 1999, Vol.150, no 1, p. 107-114. http://dx.doi.org/10.1016/S0169-4332(99)00229-9

[14] H. Targaoui, Thèse de Doctorat, Université Cadi Ayyad, Marrakech (327/2011).

Dielectric Materials and Applications: ISyDMA'2016
Materials Research Proceedings 1 (2016) 275-278

Materials Research Forum LLC
doi: http://dx.doi.org/10.21741/2474-395X/1/68

Studies of SnS thin films grown by SILAR method

Y. QACHAOU*[1], A. RAIDOU[1], K. NOUNEH[1], M. LHARCH[1], A. QACHAOU[1], M. FAHOUME[1], L. LAANAB[2]

[1]L.P.M.C., Faculté des Sciences, Université Ibn Tofail, BP.133-14000 Kénitra, Morocco

[2]L.C.S., Faculté des Sciences, Université Mohammed V, BP.1014 Rabat, Morocco

*Corresponding author: qachaou.yussef@gmail.com

Keywords: SILAR, SnS, Tin Sulfide, Thin Films

Abstract. Tin sulfide is a promising optoelectronic material which has a particular interest due to its absorption coefficient, a direct bandgap 1.7 eV and its non-toxic components. SnS thin films were deposited on glass substrates by successive ionic layer adsorption and reaction (SILAR) method. The cationic and anionic solutions are $SnCl_2.2H_2O$ and $Na_2S.9H_2O$ respectively, were used as precursors materials. The structure, film composition, morphology, and optical properties were investigated by using X-ray diffraction, energy dispersive X-ray analysis, Scanning electron microscopy (EDX-SEM) and spectrophotometer. X-ray diffraction (XRD) patterns indicated that the deposited SnS thin films have orthorhombic crystal structure. Uniform deposition of the material over the entire glass substrate was showed by Scanning electron microscopy (SEM). The optical band gap energy was found to be 1.7 eV.

Introduction

To obtain solar cells cost-effective thin films for the wide-scale production of solar energy, the absorbing semiconductor material used in the device has to meet many requirements. First, the components must be inexpensive and nontoxic. Second, to achieve high-energy conversion efficiency, the material must have appropriate optical properties. In this side, tin monosulfide SnS meet some of these criteria, and thus is a promising candidate as an absorbent material in photovoltaic applications [1]. Its components (tin and sulfur) are inexpensive, abundant in nature and respectful of the environment. In addition, the binary compound SnS presents an orthorhombic structure [2] with the following lattice parameters a = 3.99 Å, b = 4.32 Å and c = 11.2 Å [3]. Thus, SnS has appropriate optical properties for PV application; it has an optical band gap indirect from 1.0 to 1.1 eV direct from 1.3 to 1.5 eV and a high absorption coefficient (> 10^4 cm^{-1}) above the absorption edge [4]. Thin films of tin monosulfide can be made by physical and chemical method. Such as thermal evaporation [5], RF sputtering [6], (CBD) Chemical bath deposition [7], spray pyrolysis [8], electrodeposition [9] and SILAR method [10].

Materials and Methods

A. Elaboration: Preparation of SnS

Before the deposition, the size of glass substrates ($26 \times 76 \times 1$ mm^3; normal glass, Product Code: STB14-7105) has been carefully cleaned and degreased. The cationic and anionic precursor solutions, 0.2M $SnCl_2$ + H_2O and 0.1M Na_2S+9H_2O, were dissolved in 100 ml of distilled water. Each cycle of SILAR method contains: immersion of the substrate in a cationic precursor solution ($SnCl_2$, $2H_2O$) for 20 s, tin ions were adsorbed on the surface. Then, the substrate is rinsed in distilled water for 10 s. Next, the substrate is immersed in the anionic solution (Na_2S, $9H_2O$) for 20 s, Sulfide ions react with those of tin adsorbed on the active centers of the substrate. Again, the substrate is rinsed in distilled water for 10 s to remove the ions weakly related to the substrate [10]. So, the first cycle of growth by SILAR method is finished. We have repeated this cycle for 40 times. The experiment was performed at room temperature (27°C), without stirring solution. In process, the preparatory important parameters are concentrations, pH, temperature and immersion time.

Dielectric Materials and Applications: ISyDMA'2016   Materials Research Forum LLC
Materials Research Proceedings 1 (2016) 275-278   doi: http://dx.doi.org/10.21741/2474-395X/1/68

B. Characterization:

The analysis by X-ray diffraction (XRD) was performed using an XPERT-3 diffractometer with a radiation (Cu-Kα, λ =1,5406 Å) in the range of 2θ=10°-60°. The surface morphology and study of films composition were made, using scanning electron microscopy (EDX-SEM). The band gap energy was investigated using the Perkin Elmert Instrument Lambda 900 UV / Vis / NIR spectrophotometer.

Resultats and Discussion

A. Structural and Compositional Studies

Fig.1 shows the XRD pattern of the SnS films deposited on glass substrates, it can be observed the presence of SnS peaks (012) ; (110) ; (013) ; (104) ; (022) the clarity of the major peaks indicate a good crystallinity. The comparison of "d" values in observed XRD patterns with those from the standard (JCPDS card # 1-984) confirms the formation of SnS phase having orthorhombic (Pmcn (62)) crystal structure without a secondary phase. The lattice parameters values "a" "b" and "c" for orthorhombic structure calculated for the deposited film are found to be in good agreement with the reported values   (a = 4,07 Å ; b = 4,13 Å ; c = 11,45 Å).

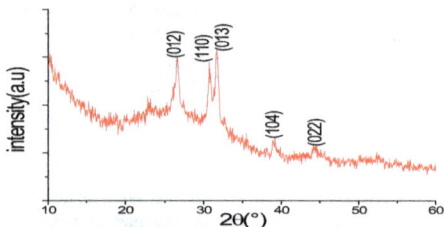

Fig. 1. X-ray diffraction pattern of SnS film.

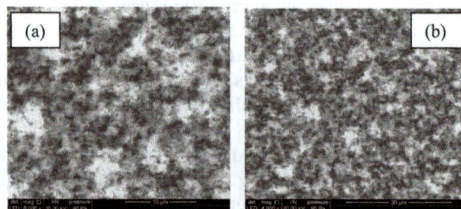

Fig. 2. SEM images (a) and (b) of SnS thin film grown on glass at different magnification.

B. Morphological Studies

The morphological features of SnS sample for different parameter deposition are investigated by SEM image. The Fig.2 (a) and (b) show the SEM of SnS thin film deposited on a glass substrate by using SILAR technique, at different magnification. The substrate is covered with the tin sulfide film, having nearly equal size distribution of grains.

   EDX studies, in Fig.3, were done at different location of thin film and the Sn to S ratio obtained was nearly equal 1, indicating there is no change in composition on the surface of the fabrication film. The corresponding S:Sn ratio was observed to be 52.05:47.95 respectively. Thus, the film as deposited was slightly riche in sulphur component.

C. Optical Studies

The optical band gap of SnS film, based on the UV-Vis-NIR spectrophotometer, varies from 1.3 to 1.8 eV, depending on the deposit method, deposition temperature and type of substrates [11]. The optical absorption theory gives the relation between the photon energy (hv) and the absorption coefficient (α) as follow [12]:

$$\alpha h v = A(h v - E_g)^n \quad (1)$$

Where the value of n is 2 and 1/2 for direct and indirect transition respectively, A is a constant, hv is the photon energy and E_g the band gap energy. This equation determines band gap (E_g), by extrapolating the linear part of $(\alpha h v)^2 = f(h v)$ graphic to intersects the hv axis, gives the band gap value for the direct transitions as shown in Fig. 4. The linear part of the plot gave a good approximation of the bandgap energy of the direct bandgap material, was found to be 1.7 eV [13].

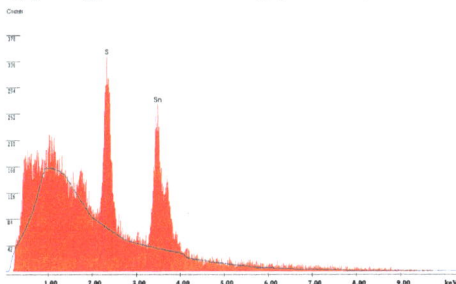

Fig. 3. EDX spectrum of SnS film deposited by SILAR on glass

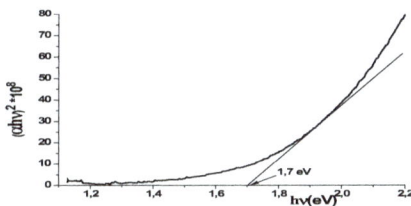

Fig. 4. Plot of (αhv)^2 vs. (hv) optical band gap of SnS film.

Conclusion

SILAR method, was used to deposit SnS thin films on glass substrates, the quality of films depends on the preparative parameters. The XRD studies showed that our thin film SnS has an orthorhombic structure, SEM image revealed nearly equal size distribution of grains, EDX indicated nearly stoichiometry between elements with S:Sn ≈ 1.08, and the optical band gap was obtained as 1.7 eV by using the UV-Vis-NIR spectrophotometer.

References

[1] N. Koteeswara Reddy, "Growth-Temperature Dependent Physical Properties of SnS Nanocrystalline Thin Films," ECS J. Solid State Sci. Tech., vol. 2, pp. 259-263, 2013. http://dx.doi.org/10.1149/2.006306jss

[2] W. Albers, C. Haas, H. J. Vink, and J. D. Wasscher, "Investigations on Sn S," J. App. Phys., vol. 32, pp. 2220-2225, October 1961. http://dx.doi.org/10.1063/1.1777047

[3] M. Kul, "Electrodeposited SnS film for photovoltaic applications," Vacuum, Vol. 107, pp. 213–218, September 2014. http://dx.doi.org/10.1016/j.vacuum.2014.02.005

Dielectric Materials and Applications: ISyDMA'2016   Materials Research Forum LLC
Materials Research Proceedings 1 (2016) 275-278   doi: http://dx.doi.org/10.21741/2474-395X/1/68

[4] G. Zhang, Z. Fu, Y. Wang, and H. Wang, "Facile synthesis of hierarchical SnS nanostructures and their visible light photocatalytic properties," Adv. Powd. Techn., vol. 26, pp. 1183-1190, July 2015. http://dx.doi.org/10.1016/j.apt.2015.05.014

[5] F. Jamali-Sheini, M. Cheraghizade, and R. Yousefi, "Impact of growth temperature on the properties of SnS film prepared by thermal evaporation and its photovoltaic performance," Cur. App. Phys., vol. 15, pp. 897-901, August 2015. http://dx.doi.org/10.1016/j.cap.2015.03.026

[6] K. Hartman, J.L. Johnson, M.I. Bertoni, D. Recht, M.J. Aziz, M.A. Scarpulla, and T. Buonassisi, "SnS thin-films by RF sputtering at room temperature," Thi. Sol. Fil., vol. 519, pp. 7421-7424, 31 August 2011. http://dx.doi.org/10.1016/j.tsf.2010.12.186

[7] E. Guneri, C. Ulutas, F. Kirmizigul, G. Altindemir, F. Gode, and C. Gumus, "Effect of deposition time on structural, electrical, and optical properties of SnS thin films deposited by chemical bath deposition," App. Surf. Sci., vol. 257, pp. 1189-1195, 1 December 2010. http://dx.doi.org/10.1016/j.apsusc.2010.07.104

[8] M. Patel, I. Mukhopadhyay, and A. Ray, "Molar optimization of spray pyrolyzed SnS thin films for photoelectrochemical applications," J. of All. and Comp., vol. 619, pp. 458-463, 15 January 2015. http://dx.doi.org/10.1016/j.jallcom.2014.08.207

[9] R. Mariappan, T. Mahalingam, and V. Ponnuswamy, "Preparation and characterization of electrodeposited SnS Thin films," Optik, vol. 122, pp. 2216-2219, December 2011. http://dx.doi.org/10.1016/j.ijleo.2011.01.015

[10] B. Ghosh, M. Das, P. Banerjee, and S. Das, "Fabrication and optical properties of SnS thin films by SILAR method, " App. Surf. Sci., vol. 254, pp. 6436-6440, 15 August 2008. http://dx.doi.org/10.1016/j.apsusc.2008.04.008

[11] T. Minemura, K. Miyauchi, K. Noguchi, K. Ohtsuka, H. Nakanishi, and M. Sugiyama, "Preparation of SnS Thin Films by MOCVD Method Using Single Source Precursor, Bis(3-mercapto-1-propanethiolato) Sn(II), " Bull. Korean Chem. Soc., vol. 33, pp. 3383-3386, 2012. http://dx.doi.org/10.5012/bkcs.2012.33.10.3383

[12] G.H. Tariq, K. Hutchings, G. Asghar, D.W. Lane, and M. Anis-UR-Rehman, "study of annealing effects on the physical properties of evaporated sns thin films for photovoltaic applications," J. of Ovo. Res., vol. 10, p.p.247-256. December 2014.

[13] J. S. Suryawanshi, S. R. Gosavi and R. S. Patil, "Fabrication and Characterization of SnS Thin Films Prepared by SILAR Method," J. of Nano Sci. & Tech., vol. 5, p.p. 31-38, March 2015.

Dielectric Materials and Applications: ISyDMA'2016
Materials Research Proceedings 1 (2016) 279-283

Materials Research Forum LLC
doi: http://dx.doi.org/10.21741/2474-395X/1/60

Study of transport phenomenon in ternary alloys AlGaN, InGaN and InGaN

Nadia BACHIR*, N.E. CHABANE SARI

University of Abou-bekr Belkaid. P.O. Box 119., Materials and Renewable Energy Research Unit, Tlemcen, Algeria

*Corresponding author: E-mail: nadia_bachir @hotmail.com

Keywords: AlGaN, InGaN and AlGaN, Steady, Monte Carlo Simulation Method

Abstract. Recent advances in growth and understanding of the physics of III-nitride semiconductor caused a vertiginous expansion of their field of applications. These materials are very useful for the application of ternary and quaternary alloys. They offer a wide variety of compositions to vary their electronic properties. Also, they are of great importance especially in the fields of optoelectronics where they find a very wide scope, since they are commonly used in green LEDs, blue and ultraviolet as well as laser diodes and ultraviolet detectors. In this work, we are interested in the study of transport phenomena in ternary alloys nitride, particularly AlGaN, InGaN and AlGaN, using the Monte Carlo simulation method.

Introduction

Like all III-V compounds family, compounds III-nitrides were very early synthesized and studied. The first monocrystals GaN high surface were epitaxial on sapphire substrates by vapour method halides (EPVH). These thick layers ($\geq 100\mu m$) have permitted to obtain since 1971, the main optical properties, especially a good estimate of the energy band-gap and the order of the three valence bands. The large gap of GaN (3.4eV at 300K) and the possibility of alloys with other nitrides (InN and AlN) have quickly raised great hopes for these materials as transmitter or detectors of visible light or ultraviolet light (UV) [1]. In this work, we study the different interactions experienced by the electrons during their movement in the alloy between the three nitrides GaN, InN and AlN. This alloys ($Al_xGa_{1-x}N$, $In_xGa_{1-x}N$ and $In_xAl_{1-x}N$) is often used in containment barrier optoelectronic structures based on nitrides. Knowledge of its properties and control its growth objectives are imperative for developing new components, particularly in the far UV. In the first approximation, the lattice constants can be deduced from those of GaN and AlN by linear interpolation; it is the Vegard law. Its effective masses, its mechanical coefficients and those of Varshni, can often be deduced by linear interpolation of GaN, AlN and InN coefficients [2].

$$a_{In_xGa_{1-x}N} = x \cdot a_{InN} + (1 - x) \cdot a_{GaN} \tag{1}$$

$$a_{Al_xIn_{1-x}N} = x \cdot a_{AlN} + (1 - x) \cdot a_{InN} \tag{2}$$

$$a_{Al_xGa_{1-x}N} = x \cdot a_{AlN} + (1 - x) \cdot a_{GaN} \tag{3}$$

a_{GaN}=4.50 A°, a_{AlN}=4.38A°, a_{InN}=4.98 A° [3], [4]

To study the electronic transport, it is essential to know the gap energy of the compound studied. GaN and AlN have gaps greater than those of authors III-V materials, for a smaller lattice constant [5].

The change in the energy band gap of the alloy depending on the composition is not linear but quadratic.

Dielectric Materials and Applications: ISyDMA'2016   Materials Research Forum LLC
Materials Research Proceedings 1 (2016) 279-283   doi: http://dx.doi.org/10.21741/2474-395X/1/69

$$E^g_{Al_xGa_{1-x}N} = x.E^g_{AlN} + (1-+x).E^g_{GaN} - x.(1-x).b \tag{4}$$

$$E^g_{In_xGa_{1-x}N} = x.E^g_{InN} + (1-+x).E^g_{GaN} - x.(1-x).b \tag{5}$$

$$E^g_{In_xAl_{1-x}N} = x.E^g_{InN} + (1-+x).E^g_{AlN} - x.(1-x).b \tag{6}$$

Eg(GaN)= 3.3eV, Eg(AlN)= 6 eV, Eg(InN)= 0.7 eV [3], [4]

The Bowing b for the three nitrides is b_{AlGaN}=0.688, b_{InGaN}=1.416, b_{InAlN}=3.477 [6]
    The electrons can undergo during their movement to two types of interactions, elastic interactions and inelastic interactions.
    The elastic interactions do not influence essentially on the behavior of the electron that it is not deviated from its trajectory. They have nevertheless been introduced for a description as quantitative as possible to the electron dynamics. Among these interactions there are piezoelectric interactions and acoustic interactions. Moreover, among the inelastic interactions are the absorption and emission intervalley interactions, intravalley interactions and optical polar interactions.

Implementation of the simulation method
In this work, we study the electric transport I the three alloys AlGaN, InGaN and InAlN using the Monte Carlo method of simulation. This method offers the possibility to reproduce the various microscopic phenomena residing in semiconductor materials. It is very important to study the transport properties of representative particles of the various layers of material over time. It is to monitor the behavior of each electron subjected to an electric field in real space and in the wave vectors space.
    The method software allows performing two basic functions. The first is devoted to calculate probabilities from the usual expressions. The second function is for determining the instantaneous quantities defined on a set of electrons (energy, speed, position) by the so-called "Self Scattering procedure [7] for which the free flight times are distributed to each electron. The implementation of this method was initially adopted by Kurosawa [8] and improved by other authors [9]. It is based on a drawing of lots process of interactions sustained by the free carriers during their movement in the compound, from probability laws. It consists to follow the behavior of each electron in real space and in wave-vectors space [10].
Consider a The general procedure for implementing the software is composed of three basic steps:
- Reading the data file of studied material parameters.
- Implementation of the software.
- Provision of output files containing the probabilities values of various interactions and the average quantities values (energy, speed, population valleys ...).
Over time, the electrons in the conduction band will have a behavior that results from the action of external applied electric field and their various interactions in the crystal lattice. In our case, the dispersal mechanisms that we consider are the elastic and inelastic dispersions that are the subject of this study.

Results and discussions
We study the electron transport within bulk cubic $In_xAl_{1-x}N$, $In_xGa_{1-x}N$ and $Al_xGa_{1-x}N$ compound, the doping concentration being set to $10^{17}cm^{-3}$; we use the method of Monte Carlo simulation. The scattering mechanisms included in our simulation are: acoustic phonon, polar and no-polar optical phonon (equivalent and nonequivalent intervalley), ionized impurities, and piezoelectric scattering. Especially, we calculate the electron drift velocity versus applied electric field for different molar fractions of indium (x = 0, x = 0.2, x = 0.5, x = 0.6, x = 0.8, and x = 1). The results are illustrated by Figures 1 to 3.

Dielectric Materials and Applications: ISyDMA'2016   Materials Research Forum LLC
Materials Research Proceedings 1 (2016) 279-283   doi: http://dx.doi.org/10.21741/2474-395X/1/69

In the three alloys, for small values of electric field, the polar optical interactions and acoustic interactions are very important and therefore the speed is very low, when the electric field increases, the effect of these interactions is very low and the speed increases until it reaches a peak for the critical fields for which the electrons have sufficient energy to occupy the upper valleys and an equilibrium is restored. In the upper valley, the electrons mass becomes larger and intervalley interactions decrease the speed of electrons.

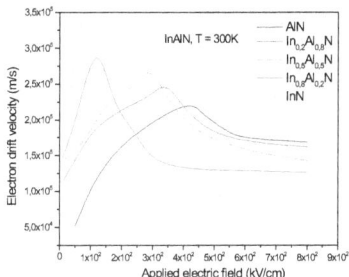

Fig.1: The electron drift velocity versus the applied electric field within $In_xAl_{1-x}N$ for different mole fractions x, at T = 300K.

With the increase of the indium molar fraction, the speed increases first and then gradually decreases, this can be explained as follows:

Both of the energy between Γ and L valleys, and the effective mass of electrons in the central valley, decrease. The electrons then have a high mobility and a high speed until the applied field reaches a critical value; they reach their maximum velocity. At this point, they are transferred to the upper valley where their population and their effective mass, increase; the intervalley collisions increase. Due to all this, their velocity begins to decrease until saturation.

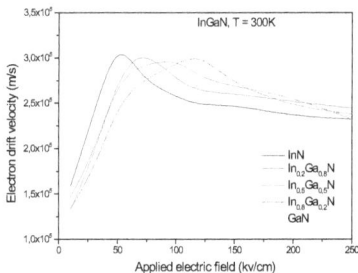

Fig.2: The electron drift velocity versus the applied electric field within $In_xGa_{1-x}N$ for different mole fractions x, at T = 300K.

With the increase of the In molar fraction, the speed increases first and then gradually decreases, this can be explained as follows:

The energy between Γ and L valleys, and effective mass of electrons in the central valley, decrease; the electrons then have a high mobility and a high speed until the applied field reaches a critical value for which they reach their maximum velocity. At this point, they are transferred to the

Dielectric Materials and Applications: ISyDMA'2016   Materials Research Forum LLC
Materials Research Proceedings 1 (2016) 279-283   doi: http://dx.doi.org/10.21741/2474-395X/1/69

upper valley where their population and their effective mass, increase; the intervalley collisions increase. Due to all this, their velocity begins to decrease until saturation.

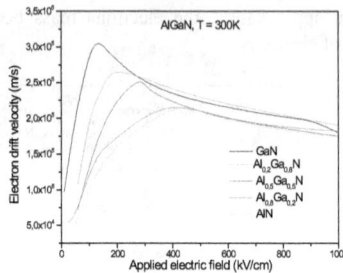

Fig.3: The electron drift velocity versus the applied electric field within $Al_xGa_{1-x}N$ for different mole fractions x, at T = 300K.

Increasing the Al mole fraction in the AlxGa1-xN gives a growth of the effective mass and the gap energy which separates the Γ and L valleys. The growth of the electron mass, results in a decrease of electrons mobility and drift velocity. When the mole fraction increases, the energy that separates the Γ and L valleys also grows and therefore, it is an increase in the critical electric field where the speed reaches its peak.

In the three alloys, for small values of electric field, the polar optical interactions and acoustic interactions are very important and therefore the speed is very low, when the electric field increases, the effect of these interactions is very low and the speed increases until it reaches a peak for the critical fields for which the electrons have sufficient energy to occupy the upper valleys and an equilibrium is restored. In the upper valley, the electrons mass becomes larger and intervalley interactions decrease the speed of electrons.

When we compare the three results
• For InAlN, the overshoot is around 2.6×10^5m/s for a critical field of about 284kV/cm. The saturation velocity is approximately equal to 1.4×10^5m/s for a field equal to 800kV/cm.
• For AlGaN, the overshoot is around 2.5×10^5m/s for a critical field of about 282kV/cm. The saturation velocity is approximately equal to 1.75×10^5m/s for a field equal to 1000kV/cm.
• For InGaN, the overshoot is around 2.98×10^5m/s for a critical field of about 99kV/cm. The saturation velocity is approximately equal to 2.3×10^5m/s for a field equal to 500kV/cm. Within InGaN alloy, the electron drift velocity is higher but the saturation field is relatively small than of InAlN.
Within AlGaN, the electron drift velocity is smaller compared with that of InAlN, but the saturation field is much higher.
    The InAlN alloy brings the benefits: the addition of In molar fraction grows the electron drift velocity, and the addition of Al molar fraction grows saturation electric field.

Conclusion
InN, AlN, GaN and their alloys, constitute a major research field of electronics in the solid state for analog microwave applications. $Al_xGa_{1-x}N$ alloys with large concentrations of Al have several advantages including their high voltage thanks to their large gap allowing higher output impedance, their high saturation speed, their high linearity, their resistance to electromagnetic pulses, their chemical stability, and their stability at high temperature. However; for $In_xAl_{1-x}N$ and $In_xGa_{1-x}N$

alloys with large concentration of In, the gap is relatively small but the electrons can reach very high speeds for low electric fields. Cubic alloys have a strong potential for optoelectronics, particularly in the far ultraviolet. However, it will be necessary to determine clearly the nature of the gap for high concentrations of aluminum. In addition, the epitaxial layer quality is still quite low compared to those of hexagonal nitrides. The development of optoelectronic applications, based on cubic nitrides will therefore by a better control of growth which is relatively difficult because of the nature of the metastable cubic phase.

References

[1] Rosen G, Materiaux pour l'Optoelectronique, Traite EGEM serie Optoelectronique, tome 7, (Hermes Science Publications), Paris, 2003

[2] Fabrice Enjalbert, Etude des hétérostructures semi-conductrices III-nitrures et application au laser UV pompé par cathode à micropointes, these de doctorat, l'U. Grenoble 1, 2004.

[3] *Semiconductors: Data Handbook*, 3rd Ed., edited by O. Madelung_Springer, Berlin, 2004.

[4] I. Vurgaftman and J. R. Meyer, Journal of Applied Phyisics **94**, 3675; 2003. http://dx.doi.org/10.1063/1.1600519

[5] I. Vurgaftman, J. R. Meyer, et L. R. Ram-Mohan, Band parameters for III–V compound semiconductors and their alloys, Journal of Applied Physics. 89, 5815; 2001. http://dx.doi.org/10.1063/1.1368156

[6] Fei Wang, Shu-Shen Li, and Jian-Bai Xia, H. X. Jiang and J. Y. Lin, Jingbo Li and Su-Huai Wei, Effects of the wave function localization in AlInGaN quaternary alloys, APPLIED PHYSICS LETTERS, 2007. http://dx.doi.org/10.1063/1.2769958

[7] K. Kubota, Y. Kobayashi, and K. Fujimoto, J. Appl. Phys. 66, 2984 -1989

[8] J. Zimmermann, etude par la methode de Monte Carlo des phenomenes de transport electronique dans le Silicium de type N en regimes stationnaire et non stationnaire. Application a la simulation des composants submicroniques, These de doctorat d'etat, U.de Lille 1,1980.

[9] T. Kurosawa, Journal of the physical society of Japan, supplement 21; 1966; 424

[10] P. J. Price, IBM Journal of Research and Development, Vol 17 ; 1973; 39. http://dx.doi.org/10.1147/rd.171.0039

Dielectric Materials and Applications: ISyDMA'2016   Materials Research Forum LLC
Materials Research Proceedings 1 (2016) 284-287   doi: http://dx.doi.org/10.21741/2474-395X/1/70

Low-frequency dispersion behavior of impedance spectra of Li$_{0.867}$Nb$_{0.988}$Ni$_{0.098}$O$_3$ ceramic

A. EL BACHIRI*[1], M. EL HASNAOUI[2], F. BENNANI[1]

[1]LPMC Laboratory, Physics Department, Faculty of Sciences, Ibn-Tofail University, BP 133, Kenitra, Morocco

[2]LASTID Laboratory, Physics Department, Faculty of Sciences, Ibn-Tofail University, BP 133, Kenitra, Morocco

*Corresponding author: E-mail: bachiriabdo2009@hotmail.fr

Keywords: Ceramic Material, Solid-State, Dielectric Relaxation, Impedance Spectroscopy

Abstract. Ceramic material sample of Li$_{0.867}$Nb$_{0.988}$Ni$_{0.098}$O$_3$ was prepared by solid-state reaction method and their impedance spectra were measured in the frequency range from 1Hz to 10MHz and temperature range from 400 to 800°C. The obtained data were analyzed by means of electric impedance formalism. Two dielectric relaxation processes were revealed in the frequency range and temperature interval of the measurements. One of these relaxations appearing in law frequencies was associated with grain boundaries and/or the interfacial polarization. Whereas the other appear at high frequencies is consistent with the grain effect.

Introduction

Research on ceramics made with Lithium niobate (LN) have drawn immense interest for integrated optics because of its unique electro-optic, acousto-optic and nonlinear optical properties [1, 2]. LN has a nonstoichiometric composition, with a high concentration of intrinsic defects. The presence of these defects makes it easier to incorporate impurities added intentionally for a desired application. The necessary charge compensation for the extrinsic defects caused by the impurity atoms can get balanced or disturbed by the intrinsic defects. These intrinsic defects are compensated by added extrinsic defects such as Mg, Zn, In, Sc, etc [3, 4]. The electrical properties of polycrystalline LiNbO$_3$ pure or doped have been less studied, although some related works are found in the literature.[5-7], This studies present the contribution of the complex impedance spectroscopy to the investigation of the properties of dielectric behavior of polycrystalline LiNbO$_3$ doped with 10% Ni, in the temperature range 400–800 °C. Impedance spectroscopy is a powerful tool for the study of the dielctrical and electrical properties of ionic, electronic, and mixed conductor ceramics. It has already been applied successfully in the investigation of ferroelectric materials such as BaTiO$_3$ ceramic [8] and LiTaO$_3$ [9]. In this study we focus our interest on the analysis of the responses of impedance spectra of the ceramic sample Li$_{0.867}$Nb$_{0.988}$Ni$_{0.098}$O$_3$.

Experimental tools

The polycrystalline compound of Li$_{0.867}$Nb$_{0.988}$Ni$_{0.098}$O$_3$ sample has been synthesized by solid-state reaction technique at high temperature, using high-purity constituents which are Lithium carbonate (Li$_2$CO$_3$), Niobium oxide (Nb$_2$O$_5$) and Nickel oxide (NiO). These constituents were mixed in the required amount according to the following chemical reaction:

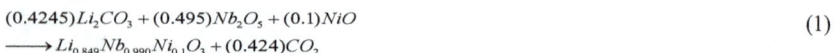

$$(0.4245)Li_2CO_3 + (0.495)Nb_2O_5 + (0.1)NiO$$
$$\longrightarrow Li_{0.849}Nb_{0.990}Ni_{0.1}O_3 + (0.424)CO_2$$

(1)

The dielectric measurements were carried out in the frequency range from 1 Hz to 10MHz and temperature range from 400 to 800°C by using an impedance analyser Schlumberger Solatron 1260.

Dielectric Materials and Applications: ISyDMA'2016   Materials Research Forum LLC
Materials Research Proceedings **1** (2016) 284-287   doi: http://dx.doi.org/10.21741/2474-395X/1/71

More details about the experimental and measurement techniques are described in our previous works [7, 9].

Results and discussion
A. Variation of the real and imaginary parts of impedance.

Impedance spectroscopy has been shown to be a useful tool to understand the transport mechanisms in different ceramic sensor device materials. We have used this technique to investigate the conduction mechanism and to understand the charge transport in the $Li_{0.867}Nb_{0.988}Ni_{0.098}O_3$ structure. Generally, the complex impedance $Z(\omega)$ can be given by the following expression: $Z^*(\omega)=Z'(\omega)-jZ''(\omega)$.

Fig. 1 shows the variation of the real (Z') and imaginary (Z'') parts of impedance versus frequency for different temperatures. The real part characterize by a constant region after decrease with increasing the frequency. The imaginary parts reach a maximum peak at two values of frequency, which indicates the double relaxation process in the system (relaxation frequencies F_1 and F_2).

B. Complex impedance analysis

The impedance data, when represented in the Nyquist diagram Z'' versus Z', with each point corresponding to different probing frequency, can give a number of semicircles, each corresponding to a particular process involved in the sample, in this representation, grain and grain boundary contributions are easily identified and the electrical properties of the bulk material can be studied separately from grain boundary interference. Moreover, the Nyquist representation allows the determination of the relaxation frequencies of the material (F_1 and F_2) which are at the apex of the semicircles.

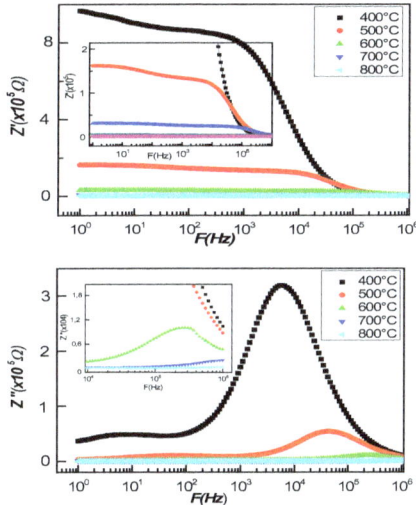

Fig. 1. Isothermal runs of the real (Z') and imaginary (Z'') part of the complex impedance versus frequency at different temperatures of $Li_{0.867}Nb_{0.988}Ni_{0.098}O_3$ ceramic.

Fig. 2 shows the Nyquist plots for studied sample at 500 °C. For this sample, there is, two arcs are clearly observed over the entire measured frequency range studied. According to the result of Bennani *et al.* [9], the first arc seems contains several contributions, such as, the contribution of

Dielectric Materials and Applications: ISyDMA'2016   Materials Research Forum LLC
Materials Research Proceedings 1 (2016) 284-287   doi: http://dx.doi.org/10.21741/2474-395X/1/71

grain boundaries, electrode-electrolyte interface and may be to the spontaneous polarization whereas the second arc (arc 2), can be assigned to the contribution of grains. The patterns are characterized by the presence of asymmetric and depressed semicircular arcs whose center does not lie on Z' axis. The behavior of impedance spectrum is suggestive of the temperature dependent hopping type of mechanism for electric conduction (charge transport) in the system and non-Debye type dielectric relaxation. According to result of Bhaumik *et al* [10] these results may be related to the defect structure in Lithium niobate [3, 4].

To simulate the electrical behavior represented experimental data in Fig. 2, we can use a circuit which in our case is presented in insert of Fig. 2. The CPE is the constant phase element [11], represented by an empirical impedance of the type $Z=A\omega^{\alpha}$. A is a constant and α is a Cole-Cole parameter. Two particular cases are observed for $\alpha = 0$ (resistor) and $\alpha = 1$ (capacitor). The fitted values of the experimental data at 500°C are enlisted in Table 1, they confirmed that we are in the non-Debye case. These results are in good agreement with those obtained by Kumari *et al.* [12]. They also confirm the presence of non-Debye type of relaxation phenomenon in the same class of ceramic materials.

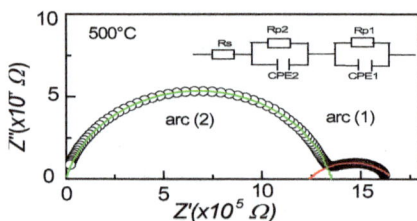

Fig. 2. Nyquist plot for $Li_{0.867}Nb_{0.988}Ni_{0.098}O_3$ ceramic at 500°C, Solid curves represent the best fit of data using the Cole-Cole model (Insert figure represent the equivalent circuit to simulate the dielectric response of impedance spectra of ceramic sample).

Table. 1. Parameters of the equivalent circuit for $Li_{0.867}Nb_{0.988}Ni_{0.098}O_3$ ceramic at 500°C.

	Rs(Ω)	Rp(Ω)	CPE(F)
Arc 1	124040	42757	$7,2\times10^{-7}$
Arc 2	50	135070	1.7×10^{-10}

Conclusion

The analysis of impedance spectra of Ni:LN ceramic was studied at different temperature. This study exhibits two typical dielectric relaxations processes which characterise the electrical transport mechanisms governing this ceramic material.

References

[1] A. Dhar, N. Singh, R. K. Singh, *J.Phys Chem Solids,* 74, 146-151(2013). http://dx.doi.org/10.1016/j.jpcs.2012.08.011

[2] V. Tatyana, W. Manfred, "Lithium *Niobate Defect Photorefraction and Ferroelectric Switching"*, 1th ed, Springer, Berlin (2008).

[3] R. Mouras, M. D. Fontana, P. Bourson, and A. V. Postnikov, *J Phys. Condens Matter* 12, 5053-5059 (2000). http://dx.doi.org/10.1088/0953-8984/12/23/313

[4] D. A. Bryan, R. Gerson, and H. E. Tomaschke, *Appl Phys Lett*, 44, 847-849 (1984). http://dx.doi.org/10.1063/1.94946

[5] Y. Xi, H. McKinstry, and L. E. Cross, *J. Am Ceram Soc*. 66, 637-644 (1983).
http://dx.doi.org/10.1111/j.1151-2916.1983.tb10612.x

[6] S. C. Bhatt and B. S. Semwall, *Solid State Ionics* 23, 77-80 (1987).
http://dx.doi.org/10.1016/0167-2738(87)90084-1

[7] A. El-Bachiri, F. Bennani and M. Bousselamti, *Spectrosc. Lett.*, 47, 374–380(2014).
http://dx.doi.org/10.1080/00387010.2013.857356

[8] N. Hirose and A. R. West, *J. Am. Ceram. Soc*, 79, 1633-1641 (1996).
http://dx.doi.org/10.1111/j.1151-2916.1996.tb08775.x

[9] F. Bennani and E. Husson, *J. Eur. Ceram. Soc*, 21, 847–854 (2001).
http://dx.doi.org/10.1016/S0955-2219(00)00285-5

[10] I. Bhaumik, S. Ganesamoorthy, R. Bhatt, A. K. Karnal, V. K. Wadhawan, P. K. Gupta, S.
Kumaragurubaran, K. Kitamura, S. Takekawa, and M. Nakamura, *J Appl Phy* **103**, 074106 (2008).
http://dx.doi.org/10.1063/1.2903907

[11] J.R. Macdonald, *Solid State Ionics* 13, 147 (1984).
http://dx.doi.org/10.1016/0167-2738(84)90049-3

[12] P. Kumari, R. Rai, A.L. Kholkin, *J Alloys and Cmopd* 637, 203–212 (2015).
http://dx.doi.org/10.1016/j.jallcom.2015.02.149

Dielectric Materials and Applications: ISyDMA'2016   Materials Research Forum LLC
Materials Research Proceedings 1 (2016) 288-290   doi: http://dx.doi.org/10.21741/2474-395X/1/71

A comparative study of electrical conductivity behaviour in polyamides/polyaniline composites

D. MEZDOUR

Laboratoire d'études des matériaux, Université Mohamed Seddik Ben Yahia, Jijel, Algeria

Corresponding author: E-mail:d_mezdour@univ-jijel.dz

Keywords: Conductivity, Polymer, Polyaniline, Polyamide

Abstract. In this study, the electrical properties of polyamide/polyaniline composites are investigated by ac measurements of the electrical conductivity of polyamide 6 and 12 matrices containing various concentrations of polyaniline. A percolative behavior was detected for both types of films. Low percolation threshold was recorded for PA12 matrices. The temperature effect was also investigated. It has been shown that there exist two thermally activated processes in PA12 matrices.

Introduction

A considerable attention was given to the fabrication and characterization of Schottky diodes, organic light emitting diodes (OLED), organic field effect transistors and solar cells, using conducting polymers and their derivatives as polyacetylene, polypyrrole, polythiophene and polyaniline. The electrical properties of these devices are in a major part controlled by the doping process, the nature of the substituants along the aromatic ring and by the preparation mode of the polymer. Polyaniline (PANI) is a versatile substance which has potential applications in corrosion prevention, as sensors, in electronics and electrochromic devices, and in batteries [1-4]. Not only tunable morphology, polyaniline can be synthesized with tunable properties like electrical conductivity and optical band gap. In present study, we report on the electrical properties of polyamide/polyaniline composite films. films were obtained by a simple in situ polymerization of aniline in the presence of polyamide in powder and film form.

Experimental Part

Formation of the layered PA6/PANI films

PA6/PANI films were prepared by oxidative polymerization process of PANI in the polyamide 6 matrix. The previously weighed film samples of PA6 were placed in the aniline solution until the desired aniline film content is reached. Swelled polymer films were then subjected to the chemical aniline polymerization in oxidant solution. Only one surface was concerned by the process. The obtained green transparent films were washed by distilled water and placed in soxlet apparatus for 24 h to extract with n-hexane by-products and unreacted aniline. Finally, films were dried under dynamic vacuum for 24 h.

Elaboration of polyamide12/PANI films

PA12/PANI films were synthesized by dissolving PA12/PANI powders of different concentrations in m-cresol. The solutions were deposited on glass slides and left drying. The synthesis of PA12/PANI powders is described elsewhere [5].

Results and Discussion

Electrical characterization was performed by AC measurements in the low frequency (10^{-1}-10^6 Hz) range, using a Novocontrol broad band dielectric spectrometer. Temperature was varied between -160 °C and 30 °C. Gold circular electrodes are positioned onto both flat sides of the samples. In the case of PA6 films, samples were sandwiched between two electrodes making the contact with the insulating side of the films.

Dielectric Materials and Applications: ISyDMA'2016   Materials Research Forum LLC
Materials Research Proceedings 1 (2016) 288-290   doi: http://dx.doi.org/10.21741/2474-395X/1/71

Effect of the polyaniline concentration on the electrical conductivity

The plot of the electrical conductivity versus the amount of aniline used to elaborate PA6/PANI films (Figure 1) shows a percolative behavior of the formed PANI clusters. The insertion of the PANI layer causes dipole layers established on the surface resulting in the reduction of the energy barrier.

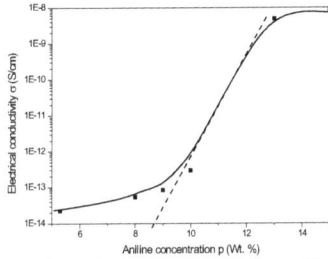

Fig. 1. Evolution of the electrical conductivity σ versus the aniline concentration p for PA6/PANI films at T=-10°C.

The PANI layer also affects the electron-hole transition probability giving rise to the conductivity up to 10^{-8} S/cm at 15 % of aniline, 6 orders of magnitude higher than the film elaborated with 5.3 % of aniline. In the other hand, PA12/PANI films exhibit a two step percolation behavior as shown in figure 2.The percolation threshold is estimated to be about 8.7 %. The first increase in conductivity is probably due to tunneling of carriers through the insulating matrix at low concentrations of PANI (< 1 wt. %). The second increase is observed arround 3 wt. % (the second percolation threshold).

Effect of the temperatureon the electrical conductivity

Figure 3 shows the evolution of conductivity versus frequency for PA6 films elaborated with 5.3 % of aniline. This concentration is chosen because it is below the percolation threshold. For comparative purposes, PA12 film containing the same concentration of PANI was studied versus temperature (Figure 4). One can observe a continuous increase in conductivity versus concentration for PA6 matrices indicating a thermally activated process of conduction. In contrast, the conductivity in the PA12 matrix decreases with increasing temperature until a temperature of -40 °C where it starts to increase slightly indicating two different conduction processes occurring at low (Fig.4.a) and high temperatures (Fig.4.b) respectively. Further investigations are needed to elucidate such a behavior.

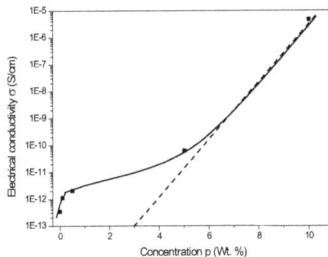

Fig. 2. Evolution of the electrical conductivity σ versus the PANI concentration p for PA12/PANI films at room temperature.

Dielectric Materials and Applications: ISyDMA'2016   Materials Research Forum LLC
Materials Research Proceedings **1** (2016) 288-290   doi: http://dx.doi.org/10.21741/2474-395X/1/71

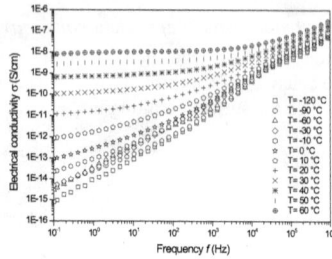

Fig. 3. Evolution of the   electrical conductivity
σ versus frequency for a PA6 film elaborated with 5.3 % of aniline at various temperatures T.

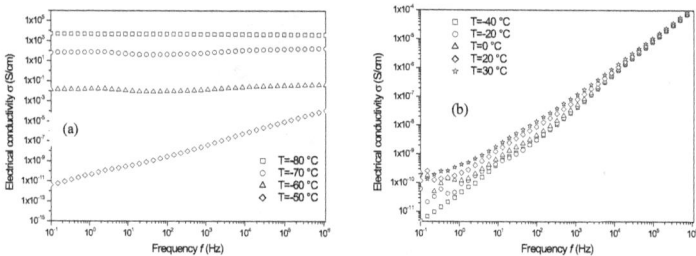

Fig. 4. Evolution of the electrical conductivity σ versus frequency for a PA12 film containing 5 %
of PANI at (a) low temperatures, (b) high temperatures.

Conclusions

The electrical conductivity of polyamide films was measured and found to depend strongly on the fraction of polyaniline in the composite films. The composite conductivity was found to increase as the percentage of PANI increased in the composite. Also the conductivity of composites films increases with the increase of temperature, except for PA12 matrices where it decrease below -40 °C.

References

[1] C.-Ho Chen, "Thermal and Morphological Studies of Chemically Prepared Emeraldine-Base-Form Polyaniline Powder," J. Appl. Polym. Sci., vol. 89, pp. 2142–2148, 2003. http://dx.doi.org/10.1002/app.12361

[2] S.K.Shukla, A. Bhradvaja, A.Tiwari, S. Pilla, G.K. Parashar, and G.C.Dubey, "Synthesis and Characterization of Highly Crystalline Polyaniline Film Promising for Humid Sensor," Advanced mater. Lett., vol.1(2), pp. 129-134, September 2010.

[3] R.H. Goncalves, W.H. Schreiner, and E.R. Leite, "Synthesis of TiO2 Nanocrystals with a High Affinity for Amine Organic Compounds," Langmuir, vol.26(14), pp. 11657-11662, June 2010. http://dx.doi.org/10.1021/la1007473

[4] J. Huang, and R.B. Kaner, Handbook of Conducting Polymers Third Edition Conjugated Polymers: Theory, Synthesis, Properties, and Characterization, Edited by T,A. Skotheim and J,R. Reynolds, Boca Raton: CRC Press   Taylor & Francis Group, 2006.

[5] D. Mezdour, M. Tabellout, S. Sahli, and K. Fatyeyeva, "Electrical Properties Investigation in PA12/PANI Composites," Macromol. Symp., vol. 290, pp. 175-184, April 2010. http://dx.doi.org/10.1002/masy.201050402

Dielectric Materials and Applications: ISyDMA'2016　　　　Materials Research Forum LLC
Materials Research Proceedings 1 (2016) 291-293　　　　doi: http://dx.doi.org/10.21741/2474-395X/1/72

UV-VIS spectroscopic studies of some skin of tomatoes and peppers

Cheyma NACEUR ABOULOULA*, Amane OUERIAGLI, Abdelkader OUTZOURHIT

Nanomaterials for Energy and Environment Laboratory (N2EL), Facultyof Sciences, Semlalia, P.O Box. 2390, 40000 Marrakesh, Morocco

*Corresponding author: E-mail: cheyma.naceur@ced.uca.ac.ma

Keywords: UV-Visible Spectroscopy, Skin, Tomatoes, Pepper, Bio-Polymers

Abstract. Vegetable skins have many properties that can be used to produce new materials. In this study, our work looks to exploit the bio-resources in the industry. We study the optical property of fresh and dried skin of tomato and pepper.

Introduction

Today, scientists are committed to increase the respect of the environment and exhaustible fossil resources. For that, an interesting alternative is the valorization our waste for the development of new bio-degradable materials for industry.

Skin of most of fruits and vegetables finished as waste in agro-industrial. These vegetable raw materials are rich in polymers having particularly interesting properties for the plastic industry, such as biodegradability, biocompatibility, and some editable physico-mechanical properties [1,2,3].

Our work focuses on the study of some physico-chemical properties of some fruits to use it in construction of new materials. In this context, it was led to make a first characterization of tomatoes skin, peppers skin: red, yellow and green. This fruit consumed highly in our daily lives and which have a tangible reject rate usable in the industry for the manufacture of product with altered characteristics.

Materials

Tomatoes and pepper were obtained from the local market and maintained at room temperature (25°C) until their use. Subsequently, the outer skin was separated from the rest of the sample with a blade, and then was placed on a glass substrate in a uniform manner. The samples are then used without any additional treatment. The optical transmittance measurements were performed using a Shimadzu UV-PCspectrophotometer in the 200–2800nm range. It was used to study the optical properties of our samples.

Results and discussion

At first, we use the UV spectrometry to define optical properties of the samples chose (Figure 1). We find that our samples absorb in the same frequency.

After that, we follow the behavior of our skins by takingthe transmission spectrum of the simple during 3 weeks: In the first week, spectrums are taken every day (D1, D2, D3, D4, D5, D6). In the second, they are taken every other day (D8, D10, D12). At the last week, we take a spectrum at the end of the week (D19, D26) (Figure 2).

Dielectric Materials and Applications: ISyDMA'2016
Materials Research Forum LLC
Materials Research Proceedings 1 (2016) 291-293
doi: http://dx.doi.org/10.21741/2474-395X/1/72

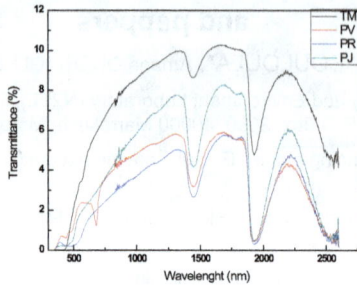

Fig.1. Transmission spectrum of fresh skins (TMT: tomato, PV: green pepper, PR: red pepper and PJ: yellow pepper)

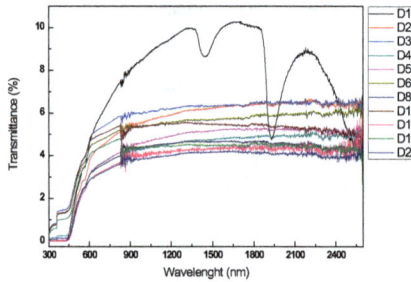

Fig.2. Evolution of transmission of tomato skin function of time

We note that the two peaks disappear at the 2nd day (D2), the disappearance of these peaks, after dryly membranes, is attributed to the water molecule. The same remark is noted for all species studied. Using liquid water spectrum, it absorbs into two frequencies bands close to those of the skins studied with a shift due to the chemical groups associated with it (Figure 3).

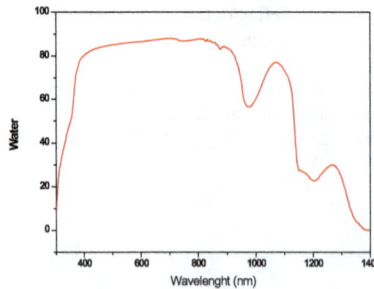

Fig.3. Transmission spectrum of liquid water

Dielectric Materials and Applications: ISyDMA'2016   Materials Research Forum LLC
Materials Research Proceedings **1** (2016) 291-293   doi: http://dx.doi.org/10.21741/2474-395X/1/72

To verify this hypothesis, we emerge dry skins in boiling water and taken up spectrum again (Figure 4). Comparing results of dry skin tip in water and fresh skin: we found that the peaks messing in drying state appear again; it shows that the peaks are related to the water molecule.

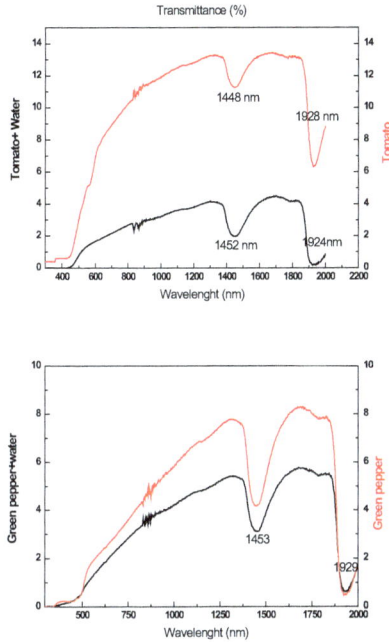

Fig.4. Absorption spectrum of dry skin immersed in boiling water

We can say that our skins absorb in the same wavelength of liquid water absorption.

Conclusion

As we have mentioned early, the studied skins have many useful properties in new manufacturing materials[4]. Studied skins absorb in the same absorption bands of the molecule of water, as they have the same chemical groups in infrared study.

Looking ahead, we will study mechanical and electrical/ dielectrical properties and the chemically composition of these membranes for their use in construction of new materials.

References

[1] Thielen, M.: Industrial Composting, "bioplastics MAGAZINE", Vol. 4., Issue 02/2009,

[2] Gandini, Alessandro, "polymers from renewable resources, polymer chemistry", 2012,

[3] Belgacem, Monomers, "Polymers and Composites from Renewable Resources", 2008,

[4] http://triblive.com/business/headlines/6259292-74/heinz-based-ford

Dielectric Materials and Applications: ISyDMA'2016   Materials Research Forum LLC
Materials Research Proceedings **1** (2016) 294-297   doi: http://dx.doi.org/10.21741/2474-395X/1/73

Characterization of microstructure of coriander seeds (coriandrum sativum)

L. KADIRI*, Y. ESSAADAOUI, E.H. RIFI, A. LEBKIRI

Laboratory of Organic Synthesis and Extraction Processes – Department of Chemistry, University Ibn Tofail, Kénitra, Morocco

*Corresponding author: E-mail: kadiri.lamya@gmail.com

Keywords: Coriander, Characterization, X-Ray Diffraction(XRD), Fourier Transform Infrared (FTIR), Extraction, Cu^{2+}, Atomic Absorption Spectroscopy (AAS)

Abstract. Coriander seeds have, over their culinary and medicinal benefits, a significant adsorbent power of heavy metals. They are characterized by a high reactivity due to: Their high adsorption capacity and their adaptation in aqueous medium. Our present work relates, on one hand, to the characterization of the material in order to determine its morphology by using techniques based on spectroscopy such as Fourier Transform Infrared (FTIR) and the X-ray Diffraction (XRD), on the other hand, to the study of acid-base behavior of seeds of coriander in contact with aqueous solutions in the presence and absence of metal ions. Also, an application of this material in the liquid-solid extraction of copper ions Cu^{2+} from aqueous solutions by coriander seeds has been realized using Atomic Absorption spectroscopy (AAS) in order to highlight the importance of coriander seeds as a potential tool in the treatment of wastewaters containing heavy metals.

Introduction

Coriander (Coriandrum sativum L. Umbelliferae) is a plant which originated from the Mediterranean area but is extensively cultivated in North Africa, Central Europe and Asia as a culinary and medicinal herb [1]. Furthermore, it is successfully grown in a wide range of conditions [1]. The seeds contain an essential oil (up to 0.7%) and the monoterpenoid, linalool, is the main component [2]. Coriander seed is a popular spice and finely ground seed is a major ingredient of curry powder. The seeds are mainly responsible for the medical use of coriander and have been used as a drug for indigestion [3, 4], against worms, rheumatism and pain in the joint [5].

This study will demonstrate that Coriander seeds can be used also to treat wastewaters in the treatment plants like an effective natural biosorbant of heavy metals. For that all, the characterization of the material has been studied so as to determine its morphology and acid – basic rearrangement by using distilled water, also, the extraction kinetic of copper ions Cu^{2+} by coriander seeds using Atomic absorption spectroscopy (AAS) and through following the evolution of pH and ion concentration in order to determine the material capacity to fix heavy metals.

Materiels and methods:

1. Materials:

Coriander seeds were washed to eliminate the impurities, and were dried to 40 °C in order to prevent a possible deterioration of the physical and chemical properties of material. Then, they were grinded, the particles between 112 and 250 μm were collected and used for analysis.

2. Methods

a) Characterization:

During this study, different technical analyses were employed, such as Fourier Transform Infrared (FTIR), X-Rays Diffraction (XRD) in order to define structure and texture of our material.

Dielectric Materials and Applications: ISyDMA'2016   Materials Research Forum LLC
Materials Research Proceedings 1 (2016) 294-297   doi: http://dx.doi.org/10.21741/2474-395X/1/73

b) Extraction:

The extraction of copper ions Cu^{2+} by coriander seeds was effectuated by using Atomic absorption Spectroscopy (AAS) in order to determine concentration of Cu^{2+} existing in the studied solutions.
A pH meter electrode was used for measuring the pH of solutions.

Results and discussions:
The curve represented in the **figure 1** which were a plot of measured infrared intensity versus wavenumbers of infrared light is a Fourier Transform Infrared (FTIR) spectrum of coriander seeds treated by distilled water and filtrated by a Buchner funnel.
The bond observed under 3302.5 cm^{-1} corresponding to O-H bond was observed due to the presence of carbolic acid (phenol). Under a wavenumber 2923.05 cm^{-1}, the C-H asymmetric stretches were appeared. However, from a wavenumber 2853.58 cm^{-1} the C-H symmetric stretches were appeared.
Around a wavenumber 1744.81 cm^{-1}, C=O stretches associated with the carboxylic acids and / or esters were appeared.
From a wavenumber 1458.8 cm^{-1}, a light deformation due to the presence of C-H was observed.
Around a wavenumber 1144.5 cm^{-1}, C-C stretches were appeared.
Under a wavenumber 1028.75 cm^{-1}, C-O stretches were observed due to the presence of cellulose [6] (**figure 1**).
The results observed have shown that the seeds of coriander contain chemical functions comprising oxygen due to the presence of carboxylic acid, esters and ethers. Moreover phenols and cellulose were dominated.

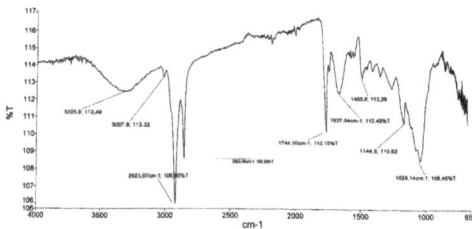

Figure.1.Infrared spectrum of coriander seeds

The curves of the X-ray Diffraction (XRD) of coriander seeds (rough and washed by distilled water) were represented in the **figure 2**. The results have shown a clear domination of the amorphous form which makes the extraction possible [7].

Dielectric Materials and Applications: ISyDMA'2016
Materials Research Proceedings 1 (2016) 294-297

Materials Research Forum LLC
doi: http://dx.doi.org/10.21741/2474-395X/1/73

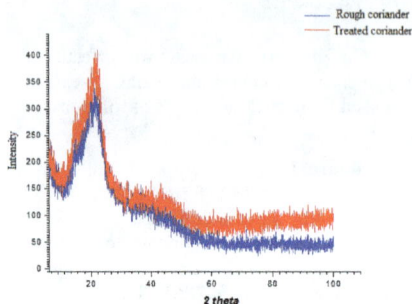

Figure.2. the X-ray spectrum of coriander seeds rough and treated by distilled water

When 0.5 g of the solid support was immersed in 100mL of the distilled water which pH= 5.8, an acid-base rearrangement of the solution was observed due to a rapid increase of the pH of the aqueous solution which results by using H^+ protons, after 3 hours the pH was stabilized on 6.9 (**figure 3**).

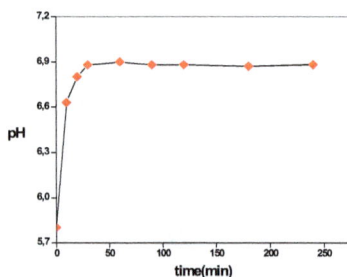

Figure.3. Change of pH versus time of the support immersed in distilled water

The study of the copper ions Cu^{2+} (V = 100 ml, $[Cu^{2+}]$ = 10ppm*) extraction by coriander seeds (m=0.5g) through following the evolution of the concentration and the pH versus time has shown on the **figure 4** that the fixation of Cu^{2+} on the area of our material has been detected on a short time (chemical balanced solution around 60 minutes). The result has been manifested by the decrease of the Cu^{2+} concentration. The pH measurements have increased with time; this is the acid – base balance of the material with the aqueous solution of the metal Cu^{2+}.

(* 1 ppm= 1 mg/L)

Dielectric Materials and Applications: ISyDMA'2016   Materials Research Forum LLC
Materials Research Proceedings **1** (2016) 294-297   doi: http://dx.doi.org/10.21741/2474-395X/1/73

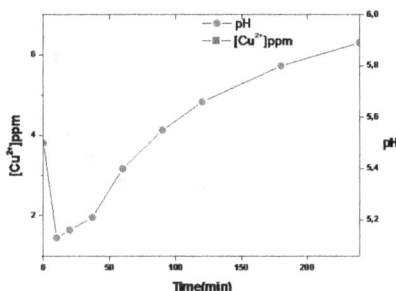

Figure.4. Evolution of pH and concentration of Cu^{2+}

Conclusion

This study has shown that coriander seeds contain cellulose which makes the adsorption of protons very easy. This result was obtained by the Fourier Transform Infrared (FTIR) analysis of the material.

Moreover, the X-ray Diffraction (XRD) analysis results have demonstrated that it is possible to extract total crystallinity from the rough and treated coriander seeds.

Also the results obtained by the acid-base rearrangement in the absence of metal coriander seeds have shown an acid-base rebalancing the material with the distilled water.

The study of the copper ions Cu^{2+} extraction by coriander seeds through following the evolution of the concentration and the pH versus time has shown that the fixation of Cu^{2+} on the area of our material has been detected on a short time (chemical balanced solution about 60 minutes). The pH measurements have increased with time; this is the rebalancing of the acid - base of the support with the solution of the metal Cu^{2+}.

To summarize, coriander seeds can be an effective adsorbent potential of heavy metals contaminating wastewaters.

References

[1] Seidemann J. (2005) World spice plants: economic, usage, botany, taxonomy. Berlin Heidelberg: Springer-Verlag; 591p.

[2] Wichtl, M. W. (1994) Herbal drugs and phytopharmaceuticals. Stuttgart: Medpharm GmbH Scientific Publishers.

[3] Dias MI, Barros L, Sousa MJ, Ferreira IC (2011) Comparative study of lipophilic and hydrophilic antioxidants from in vivo and in vitro grown Coriandrum sativum. 66(2):181–186p.

[4] Dhanapakiam P, Joseph JM, Ramaswamy VK, Moorthi M, Kumar AS (2008) The cholesterol lowering property of coriander seeds (Coriandrum sativum): mechanism of action. J Environ Biol 29(1):53–56p.

[5] Flora of the USSR [in Russian], Acad. Sci. USSR, Moscow (1964), Vol. 16, pp. 36 – 40.

[6] Elabed A., Doctorate of the University Mohammed V Agdal, Reduction, Morocco (2007).

[7] Marcovich N. E., Reboredo M. M., Aranguren M. I., J. Appl. Polym. Sci. 61(1) (1996) 119-124. http://dx.doi.org/10.1002/(SICI)1097-4628(19960705)61:1<119::AID-APP13>3.0.CO;2-2

Dielectric Materials and Applications: ISyDMA'2016   Materials Research Forum LLC
Materials Research Proceedings 1 (2016) 298-302   doi: http://dx.doi.org/10.21741/2474-395X/1/74

Characterization of the microstructure of bark of eucalyptus "eucalyptus camaldulensis"

Y. ESSAADAOUI*, L. KADIRI, E.H. RIFI, A. LEBKIRI

Laboratory of Synthesis Organic and Extraction processes, Department of Chemistry, Faculty of Science, University Ibn Tofail, Kenitra, Morocco

*Corresponding author: E-mail: essaadaouiy@gmail.com

Keywords: Lignocellulosic Material, Bark of Eucalyptus, Chemical Treatment, Capacity of Adsorption, Characterization

Abstract. The lignocellulosic residues can be developed in the treatment of wastewaters as clean and low costs natural adsorbents. In the present study, our approach consists to develop the preparation and physico-chemical characterization methods of an adsorbing material, starting from the barks of eucalyptus "eucalyptus camaldulensis". The preparation of the adsorbent material comprises the sieving of barking, extraction of extractable, treatment with formaldehyde as activating agent, and chemical treatment with acrylic acid by the grafting reaction. The chemical modification is checked using means of analysis such as: Fourier transform infrared (FTIR), X-rays Diffraction (DRX) (Index of crystallinity) and method of Boehm. The chemical modification of the lignocellulosic material induced by a change in the microstructure of the material, especially in the case of grafting of acrylic acid and the crosslinking with formaldehyde in favor of the rise in the potential of adsorption capacity .

Introduction

The bark of eucalyptus can be developed in the treatment of wastewaters as low costs biosorbants low [1-4]. However, the valorization of these barks like adsorbent supports for the purification of the effluents contaminated by micropolluants requires the knowledge of its structure and its texture. The capacity of the bark of eucalyptus to fix the pollutants can be largely improved in him to undergo a chemical treatment [5-9].

In this context, and because of the economic and environmental importance of the valorization of barks of eucalyptus in the treatment of wastewaters, we were interested, initially, on the chemical modification effect of the microstructure of eucalyptus barks "eucalyptus camaldulensis", within the framework of the formulation of adsorbent lignocellulosic materials.

Materiel

The barks of eucalyptus were collected from the area of Gharb-Kenitra-Morocco (close to the university Ibn Tofail). They were washed to eliminate the impurities such as sand and dust, and dried to 50 °C in order to prevent a possible deterioration of the physico-chemical properties of material. Then, they were crushed and sieved. The particles between 112 and 250 µm were collected and used for analysis.

Methods

Chemical treatment of the material

The extractable ones, which are likely to inhibit the préhydrolyse of the barks of eucalyptus, are eliminated by extraction in Soxhlet using sample of ethanol: toluene (rapport ½) in first time and acetone in second time as solvent, a process of backward flow uninterrupted. The solvents solubilize the extractable contents in the vegetable matter, which are finally concentrated in the balloon of recovery [10].

After that, the reaction of copolymerization with the acrylic acid is carried out in two times: preliminary oxidation of wood leading to the formation of macro-radicals followed by copolymerization with the acrylic acid [11].

The reaction pathway adopted for the grafting of acrylic acid on the bark of eucalyptus is the following [11]:

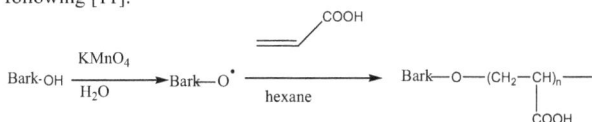

The oxygen of bark of eucalyptus found mainly in its components major with knowing cellulose, lignin and hemicellulose inform hydroxyle (- OH) is the seat of the reaction of copolymerization. Or oxygen oxidizes and become more accessible to react with overdrawn carbon in electrons of the double connection of the monomer of acrylic acid and consequently formed a covalent bond along the chain of the copolymer [12].

Technical analysis used

For analysis, different materials we are used different methods transform Fourier infrared (IRTF), Diffraction to X-Rays (DRX) and method of Boehm.

Results and discussions

The physicochemical of the rough bark of eucalyptus and treated results of the analysis are recapitulated in the table 1:

Table 1: Chemical composition and physicochemical analysis of barks of eucalyptus before and after chemical treatment

Support	BR	BS	BF	BAA
Moisture (%)	5,19	5,63	4,98	4,81
Mineral matter (%)	14,92	6,96	8,09	1,51
pH	4,25	4,56	3,18	3,28
Bulk density (g/cm3)	0.75	0,61	0,81	0,77
Cellulose (%)	37	-	-	-
Hemicellulose (%)	20	-	-	-
Lignin of klason (%)	26	-	-	-
Matter extractable (%)	4,03	-	-	-

BR: Rough bark; **BF**: Bark reticulated by formaldehyde
BS: Bark extracts by Solvent; **BAA**: Bark copolymerized by acrylic acid
[a] (%): g/100 g

Generally the bark of eucalyptus (rough and treated) are acid, which was explained by the contents of acidity found on the surface of all the materials on the one hand and on the other hand, by the pHeq which is acid (pH<5), as well as the absence of the basic functions.

The treatment of bark of eucalyptus by the acrylic acid caused an increase in the functions carboxyl of the surface (0.57 mmol/g). For other materials present close values, except for bark extracts by Soxhlet presents a low content of functions lactones (0.02 mmol/g), which is due to the loss of the extractable matters (resinous compounds and tannins) [13].

The results of proportioning of the functions of surface of materials by the method of Boehm are illustrated in the table 2.

Dielectric Materials and Applications: ISyDMA'2016   Materials Research Forum LLC
Materials Research Proceedings 1 (2016) 298-302   doi: http://dx.doi.org/10.21741/2474-395X/1/74

Table 2: Functions of surfaces of materials in various states

Material	Functions carboxyl (mmol/g)	Functions lactones (mmol/g)	Functions phenols (mmol/g)	Total of functions acids (mmol/g)	Functions basic (mmol/g)
BR	0,09	0,12	0,45	0,66	0
BS	0,08	0,02	0,6	0,7	0
BF	0,1	0,09	0,47	0,66	0
BAA	0,57	0,14	0,55	1,26	0

The spectra obtained by infrared analysis (IRTF) of the samples of rough and treated bark of eucalyptus (BR, BS, BF and BAA), are illustrated in the figure 1. The characteristic bands vibrations were allotted, mainly in agreement with the data of the literature, and by taking into account the principal differences between the spectra IR of material before and after treatment. At first sight, the spectra IR of rough and treated bark of eucalyptus have almost the same pace but with an increase in the intensity for the bark treated by formaldehyde (BF) and acrylic acid (BAA).

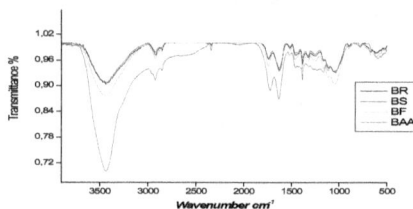

Figure 1: Spectra infrared of the bark of eucalyptus rough (BR) and treated (BS, BF and BAA).

The peak around 1720 cm^{-1} is characteristic of the stretching vibration of (C = O) carboxylic acids and / or esters of xylan, hemicellulose found in [14-16], with a net increase of intensity this peak for the compound treated with acrylic acid because of the occurrence of the C = O of carboxylic acids of acrylic acid grafted onto the bark.

In addition, the band observed around 1630 cm^{-1} corresponds to the deformation C=C aromatic rings of lignin. The bands observed at 1380 cm^{-1} and 1230 cm^{-1} assigned to the vibration v(CO) methoxy groups of lignin and the bond formed during the action of formaldehyde on cellulose (C-O-C). Moreover, the band appears at 1159 cm^{-1} in the bark treated with acrylic acid spectrum corresponds to the elongation of the C-O bond the acid group saw grafting acid functions on eucalyptus bark.

The peak at 1030 cm^{-1} corresponds to stretching vibrations of C-O and C-O-C bonds of cellulose [17], this peak becomes more intense in the case of bark treated with formaldehyde (BF) due to the action of formaldehyde on the OH groups of the cellulose and formation of the C-O-C bond.

The curves of diffraction of the X-rays of the samples of the rough and modified bark of eucalyptus "eucalyptus camaldulensis" are represented in the figure 2.

The analysis of these curves made it possible to extract total crystallinity from it from the rough and treated bark of eucalyptus.

Dielectric Materials and Applications: ISyDMA'2016   Materials Research Forum LLC
Materials Research Proceedings **1** (2016) 298-302   doi: http://dx.doi.org/10.21741/2474-395X/1/74

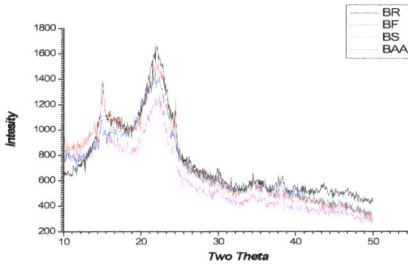

Figure 2: curves of diffraction to the X-rays of barks of eucalyptus rough and treated.

It appears that all the samples present the form characteristic of native cellulose of materials lignocellulosic, with a clear domination of the amorphous form [18].

The analysis of these curves made it possible to extract total crystallinity from it from the bark of eucalyptus butts and treated.

It appears that the bark was affected by the treatments used, where the index of crystallinity of the bark of eucalyptus treated by formaldehyde decreased under the effect of the reticulation of cellulose while passing from 46,57% for the rough bark (BR) with 45.50% for the bark treated with formaldehyde. As it passed to 41.44% for the bark treated by acrylic acid (BAA), this reduction because the cellulose was affected by the chemical treatment, where the groupings OH of cellulose were reacted with the acrylic acid. Therefore, their hydrogen is not available anymore to form intermolecular hydrogen bonds, and thereafter, the cellulose loses their crystalline form. But, the index of crystallinity for the bark extracts by solvent (BS) increased (48.44%), this later due to the loss of extractable (amorphous) of the bark during the extraction using organic solvents.

Table 3: Crystallinity Index of eucalyptus bark; rough and treated measured by X-rays diffraction.

Biomass	Index of crystallinity CrI (%)
BR	46,57
BF	45,50
BS	48,44
BAA	41,44

Conclusion

In this study, we have characterized the microstructure of eucalyptus barks before and after chemical treatment. The chemical modification, made it possible to influence the microstructure of the lignocellulosic material, which results in a change of its surfaces functions properties. We conclude the modified eucalyptus bark can be developed and find a potential application as adsorbent in the field of wastewater treatment.

References

[1] Natarajan Rajamohan, Manivasagan Rajasimman, Rajan Rajeshkannan, Viswanathan Saravanan. Equilibrium, kinetic and thermodynamic studies one the removal of Aluminum by modified Eucalyptus camaldulensis barks Original Research Article. Alexandria Engineering Journal, Volume 53, Issue 2, June 2014, Pages 409-415. http://dx.doi.org/10.1016/j.aej.2014.01.007

[2] Ilhem Ghodbane, Oualid Hamdaoui. Removal of Mercury (II) from aqueous media using eucalyptus bark: Kinetic and equilibrium studiesOriginal Research Article. Newspaper of Hazardous Materials, Volume 160, Issues 2-3, 30 December 2008, Pages 301-309. http://dx.doi.org/10.1016/j.jhazmat.2008.02.116

[3] Ilhem Ghodbane, Loubna Nouri, Oualid Hamdaoui, Mahdi Chiha. Kinetic and equilibrium study for the sorption of cadmium (II) ions from aqueous phase by eucalyptus barkOriginal Research Article. Newspaper of Hazardous Materials, Volume 152, Issue 1.21 March 2008, Pages 148-158

[4] Vikrant Sarin, K.K. Pant. Removal of chromium from industrial waste by using eucalyptus barkOriginal Research Article. Bioresource Technology, Volume 97, Issue 1, January 2006, Pages 15-20. http://dx.doi.org/10.1016/j.biortech.2005.02.010

[5] Randall JM, Hautala E, Waiss AC Jr, Tschernitz JL (1976) Modified barks as scavengers for heavy metal ions. For Prod J 27(11):51-56

[6] Deshkar AM, Dara SS (1988) Sorption of mercury of Techtona grandis bark. Asian Environ 10(4):3-11

[7] Deshkar AM, Bokada SS, Dara 88 (1990) Modified Hardwichia binata bark for adsorption of mercury (II) from water. Water Res 1011-1016. http://dx.doi.org/10.1016/0043-1354(90)90123-N

[8] Fujii M, Shioya S, Ito A (1988) Chemically modified coniferous wood barks as scavengers of uranium from sea water. Holzforschung 42:295-298. http://dx.doi.org/10.1515/hfsg.1988.42.5.295

[9] Freer J, Baeza J, Maturana H, Palma G (1989) Removal and recovery of uranium by modified Pinus radiata D. Don bark. J Chem Technol Biotechnol 46:4148. http://dx.doi.org/10.1002/jctb.280460105

[10] Mellouk H., Doctorate. University of the La Rochelle. France. (2007).

[11] Magali Geay, Véronique Marchetti, André Clément, Bernard Loubinoux, Philippe Gérardin. Decontamination of synthetic solutions containing heavy metals using chemically modified sawdusts bearing polyacrylic acid chains. J Wood Sci (2000) 46:331-333. http://dx.doi.org/10.1007/BF00766226

[12] J. Martin, PhD Thesis. University of Limoges. La France. 214 (2001) 44-2001.

[13] Koncsag et al., 2011C.I. Koncsag, D. Eastwood, A.E.C. Collis, S.R. Coles, A.J. Clark, K. Kirwan, K. Burton Extracting valuable compounds from straw degraded by Pleurotus ostreatus Resour. Conserv. Recy. (2011) doi:10.1016/j.resconrec.2011.04.007. http://dx.doi.org/10.1016/j.resconrec.2011.04.007

[14] Meyer K. H., Chem. Ber. 70 (1937) 266. http://dx.doi.org/10.1002/cber.19370700222

[15] O'Sullivan A. C., Cellulose. 4(3) (1997) 173–207. http://dx.doi.org/10.1023/A:1018431705579

[16] Sain M., Panthapulakkal S., Bioprocess 23(1) (2006) 1–8.

[17] Elabed A., Doctorate of the University Mohammed V Agdal, Reduction, Morocco (2007).

[18] Marcovich N. E., Reboredo M. M., Aranguren M. I., J. Appl. Polym. Sci. 61(1) (1996) 119-124. http://dx.doi.org/10.1002/(SICI)1097-4628(19960705)61:1<119::AID-APP13>3.0.CO;2-2

Dielectric Materials and Applications: ISyDMA'2016   Materials Research Forum LLC
Materials Research Proceedings **1** (2016) 303-305   doi: http://dx.doi.org/10.21741/2474-395X/1/75

Electrical proprieties of self-assembled polymers prepared by spin coating route

M. ADAR[1], N. AIARMOZI[1], M. MABROUKI*[1], A. OUERIAGLI[2],
A. OUTZOURHIT[2], R.M. LEBLANC[3]

[1]Laboratoire Génie Industriel Faculté des Sciences et Techniques, Université Sultan Moulay Slimane, Beni Mellal, Morocco

[2]Laboratoire physique des Solides et des Couches Minces, Faculté des sciences université Cadi Ayad, Marrakech, Morocco

[3]Supramolecular center, University of Miami, Corale Gables, USA

*Corresponding author: E-mail: mus_mabrouki@yahoo.com

Keywords: PPEEB, Schottky Diode, Polymer, I-V Charachteristic

Abstract. In the recent years, organic semiconductors are widely studied because they show a variety of electro-optical and electrical proprieties. Also it's easy and facile to fabricate and process such kind of materials. Which make them promising as photovoltaic and optoelectronic devices [1-3]. In this study, we characterize and investigate electrical proprieties of ITO/PPEEB/Al structure. This structure is prepared by spin-coating. First PPEEB polymer is dissolved in chloroform solvent then coated on the substrate. Coated films on glass are used for optical characterization like transmission spectra in order to extract gap energy value. The transition is observed around 420 nm which correspond to electron transition between the valence band and conduction band equal to the gap energy of polymer Eg equal to 4eV. Second ITO substrate is used for electrical characterization, the Al electrode is deposited by thermal evaporation in vacuum on the top of structure.

Introduction

A considerable and an important attention was given to the fabrication and characterization of schottky diodes, since the discovery of conductive polymers character in 1970[1] and π-conjugated polymer. This new electronics could allow to produce electronic [2] and optoelectronic devices [3] on flexible substrates, polymeric batteries [4], photovoltaic cells [5], energy storage [6] and display devices [7]. Furthermore, this technology should allow the manufacture of electronic circuits of large dimensions, with prices for low unit area [8].

Several authors[9-14] have already studied schottky diode based on conjugated polymers, They estimated various electronic parameters of the diodes such as the ideality factor, the rectifying ratio, the barrier height, the work function of the polymer, the Richardson constant and the saturation current density from current–voltage (I–V) and capacitance–voltage (C–V) measurements, this electrical properties are in a major part controlled by the doping process, the nature of the substituents along the aromatic ring and by the preparation mode of the polymer.

In this work electrical proprieties of ITO/PPEEB/Al structure have been investigated. The structure is prepared by spin-coating where PPEEB is dissolved in chloroform solvent then coated on the substrate, followed by Al evaporation in vacuum. From the specter transmission of PPEEB films deposited on glass..

Experimental procedure

Preparation of PPEEB thin film by spin coating and

An amount of PPEBB, dissolved in chloroform, is poured onto the ITO cleaned and rinsed with acetone and then dried by means of an absorbent paper, all rotating at a speed of 1400 rev / min.

Dielectric Materials and Applications: ISyDMA'2016   Materials Research Forum LLC
Materials Research Proceedings 1 (2016) 303-305   doi: http://dx.doi.org/10.21741/2474-395X/1/75

Elaboration of AL/PPEEB contact

Schottky-barrier type devices were fabricated using the PPEEB films deposited on ITO-coated glass substrates, which provided the back contact. The front (top) contact consisted of 2 mm diameter circular aluminum dots. The aluminum was deposited by vacuum thermal evaporation of 99.99% purity aluminum shots for 2 min at base pressure of 1.5×10^{-5} mbar.

Caracterisation of ito/ppeeb/al

The characterization of the elaborated diode will be based on the electronic scanning microscope and the current-voltage characteristic.

UV-visible spectroscopy

The optical transmission (absorption) of the PPEEB films was measured using the samples deposited on glass substrates. The measurements were performed using Shimadzu-3101PC double beam spectrophotometer in the wavelength range 220–3200 nm. The transmission spectrum of the PPEEB glass sample was normalized with the base substrate.

Electrical measurement

The electrical measurements were performed on the ITO/PPEEB/Al devices at room temperature.

The current–voltage characteristics, on other hand, were measured using a Keithly 410 programmable picoamperemeter, and 610C programmable micro-voltmeter.

All the instruments are controlled by a computer via a GPIB card

Results and discussion

In this part we will present and interpret the results found.

Optical properties of the spincoated PPEEB films

The films were uniform. The examination of their surface morphology by scanning electron microscopy did not reveal any presence of pinholes or porosity in the film as shown in Figure 1

Optical measurements are made in transmission mode on thin films, the energy of the bandgap is determined from the optical absorption spectrum Fig.2, through the law of Tauc.

The allowed direct optical gap of the PPEEB films is found to be on the order of 4.6 eV.

Electrical properties of ITO/PPEEB/Al device

We, positively, polarized the ITO relatively to the metal in order to obtain the direct polarization of the devices. The I–V characteristics of the prepared ITO/PPEEB/Al devices are shown in Fig. 3

The I–V curves is nonlinear, asymmetric and show a rectifying behavior of Al/PPEEB contact. The values of Js and η are: Js = 10.2 μAcm-2, η = 8.06 these values at room temperature are much higher than the theoretical values for a diode Schottky ($1 < \eta < 2$), The large value of ideality factor can be attributed to presence of barrier in homogeneities as well as the reactive nature of Al contact and recombination of electrons and holes in the depletion region and the presence of other conduction mechanisms such bulk limited currents in various voltage ranges and trap assisted of tunneling.

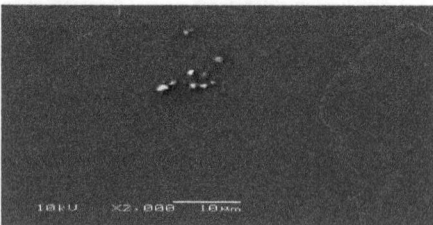

Fig. 1.The image obtained by SEM magnification 2000 X voltage 10 kV; shows the homogeneity of the thin layer.

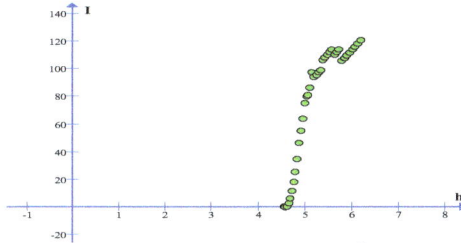

Fig. 2. Linear Dependence of $(\alpha h v)^2$ on hv.

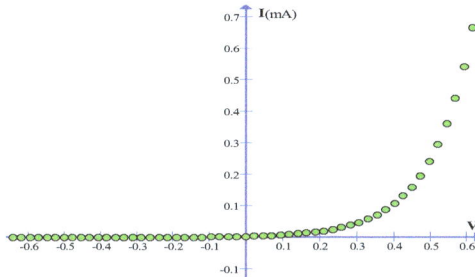

Fig .3. Current–voltage characteristic of ITO/PPEEB/Al device

References

[1] C. K. Chiang, C. R. Fincher, Y. W. Park, A. J. Heeger, H. Shirakawa, E. J. Louis, S. G. Gau and A. G. MacDiarmid, "Electrical Conductivity in Doped Polyacetylene", Physical Review Letters, vol. 39, pp. 1098 (1977). http://dx.doi.org/10.1103/PhysRevLett.39.1098

[2] G.B. Appetecchi, F. Alessandrini, S. Passerini, G. Caporiccio, B. Boutevin, F. Guida-Pietrasanta, Electrochem. Acta 50 (22) (2005) 4396–4404. http://dx.doi.org/10.1016/j.electacta.2005.02.003

[3] E.W. Paul, A.J. Ricco, M.S. Wrighton, J. Phys. Chem. 89 (1985) 1441. http://dx.doi.org/10.1021/j100254a028

[4] G.D. Sharma, S.K. Sharma, M.S. Roy, Thin Solid Films 468 (1–2) (2004) 208–215. http://dx.doi.org/10.1016/j.tsf.2004.04.059

[5] P. Camarulu, A. Cirpan, L. Toppare, J. Electroan. Chem. 572 (2004) 61–65. http://dx.doi.org/10.1016/j.jelechem.2004.06.002

[6] A.G. Mac Diarmid, S.L. Mu, N.L.D. Somasiri, W. Wu, Mol. Cryst. Liq. Cryst. 121 (1985) 187. http://dx.doi.org/10.1080/00268948508074859

[7] A. Kitani, J. Yano, K. Sasaki, J. Electroanal. Chem. 209 (1986) 227. http://dx.doi.org/10.1016/0022-0728(86)80200-5

[8] : "Organic and Printed Electronics", third edition of the OE-A brochure (2009)

Keyword Index

Author Index

313

About the Editors

ACHOUR Mohammed Essaid received the "Thèse de 3ème cycle" from The Bordeaux University (France) and "Thèse d'état" from the Moulay Ismail University of Meknes (Morocco) degrees in field of Physics in 1983 and 1991 respectively. From 1983 to 1992 he was an "Maitre assistant", at Sciences Faculty of Meknes (Morocco), "Maitre de Conférence" (1992-1996) and Professor. He joined the Sciences Faculty at Kenitra in 1999. From 1997 to 2011, he was also teacher with Royal Military Academy at Meknes. He is Expert evaluator, member of the scientific committee of the National Center for Scientific and Technical Research (CNRST), The Moroccan Center for Innovation (CIM) and he is Chair and foundater member of Moroccan Association of the Advanced Materials (A2MA). He is honoured as guest professor at Brest University in France and visiting scientist/researcher at different universities and research Centers in France, Canada, Portugal, Hungary, Italy and Tunisia. His research interests include electromagnetic and electrical properties, microwave characterization and dielectric responses of the composite materials : carbon dots, graphene, carbon nanotubes, carbon black, and natural fibers in the natural or synthetic polymers. Experience, modelling and numerical simulations. Pr. ACHOUR M.E. has co-authored peer-review more than 65 scientific papers published in leading refereed journals, about 100 congress communications, 2 Book Chapters and 6 Guest editorials. He participated in 12 cooperation projects. He was the chairs of the First International Symposium on Dielectric Materials and Applications "ISyDMA'2016" (Kenitra-Rabat, Morocco May 4-6, 2016), the Fourth Meeting On Dielectric Materials "IMDM'4" (Marrakech, Morocco May 29-31, 2013) and the Co-Chair of the International Symposium on the Advanced Materials for Optics Micro-Electronics and Nanoelectronics "AMOMEN'2011 (Kenitra, Morocco, October 27-29 2011).

TOUAHNI Raja received the Doctorat 3ème cycle in Microelectronics from the University Paul Sabatier , Toulouse, France, in September 1986. From 1987 to 2001, she was a Professor assistant at Sciences Faculty of Kénitra. In 2001, she received the "Thèse d'état" from the Ibn Tofail University of Kénitra (Morocco) degrees in field of Data Analysis and she is currently Professor in this University. Her current interests include Image Processing, Data Analysis, Antenna and Metamaterials. Professeur Touahni R. is the head of the research team in "Traitement de l'Information et Ingénierie de la Décision" and she has supervised PhD students who are currently pursuing research in academics institutions. She has co- authored peer-reviewed more than 40 scientifics papers publishing in refereed journals, 1Book Chapter and has been contributing to several projects in her field (ERASMUS, Volubilis, ...). Pr. Touahni R. was the co-chairs of the "First International Symposium on Dielectrics Materials and Applications May 2016" and "5th Doctoral Days for IT (JDTIC'2013)" and also founding member of the " Moroccan Association of the Advanced Materials (A2MA)" and "Moroccan Classification Society (SMC)".

MESSOUSSI Rochdi, received the Ph.D degree in solid state physics from the University of Nantes, France, in 1990. From 1993 to 1995 and from 1996 to 1997 he carried out a research stay in the field of semiconductors technology at the Tohoku University, Sendai, Japan. In 2001 he received the master degree in the use of IT for learning from the University of Strasbourg, France. He is currently professor at Ibn Tofail University, Kénitra, Morocco. For the last 15 years, he has been carrying out research, about Human Computer Interaction to promote learning and education. Pr. MESSOUSSI R. was the co-Chair of the 9 African Conference on Research in Computer Science (CARI'08) held at Kénitra – Rabat (October 2008) and co-chair of the 1th International Symposium on Dielectric Materials and Applications (Mai 20016). He also chaired the 5 Doctoral Days for IT (JDTIC-Kénitra 2013). He was Member of the Permanent Committee (20018-2012) of the African Symposium on Research in Computer Sciences (CARI and also founding member of the Moroccan Association of the Advanced Materials (A2MA). Pr. MESSOUSSI R. was the head of the physics Department from 2012 to 2014 and is currently Director of the IT Resources Center at Ibn Tofail University.

ELAATMANI Mohamed received the " These d'état " from The Bordeaux I University (France) of Solid State Chemistry and Materials Sciences in 1981. From 1981 to 1985 , he was an "Maitre de Conférences ", and Professor since 1985 at Semlalia Sciences Faculty of Cadi Ayyad University Marrakech (Morocco) . He joined the Semlalia Sciences Faculty in 1981 . Pr. ELAATMANI has co-authored of Patent. Ferroelectric oxyfluorinated materials and process for their production. Invention of Dossier No. 44411. Application No. 7911936, filed on May 10, 1979 in the name of the National Agency for the Promotion of Research (ANVAR) . He was editing a scientific work " Problems of solved exams with reminders of courses of general chemistry " February 1995, Africa, print edition 159 bis, Boulv. Yacoub Elmansour Casablanca (Morocco) . Since 1983 , He is foundater member of Moroccan Meeting on Solid State Chemistry (REMCES) . Member of the Organizing Committee of 24 th European Crystallographic Meeting (EMC 24) , Marrakech-Morocco 22-27 August 2007, of IMDM'4 Fourth International Meeting On Dielectric Materials Marrakech, Morocco, 29-31 May 2013 and several times as Professor invited to the University of South Toulon –Var, France. He participated in several cooperation projects, being coordinator of 15. Pr. ELAATMANI has co-authored peer-review more than 90 scientific papers published in leading refereed journals and about 150 congress communications. Pr. ELAATMANI is also referring to several articles in international newspapers , expertise of 12 PARS, projects member of the committee of reading, editing of the articles of the several Moroccan meetings on the REMCES, the supervision of several theses and several administrative responsibilities

AIT ALI Mustapha received the Doctorate "3^{ieme} cycle" and the Doctorate "d'Etat" degrees from Cadi Ayyad University Marrakech-Morocco in the field of chemistry (Organometallic and catalysis) in 1990 and 2001 respectively. From 1995 to 2001 he was a Professor Assistant, 2001 to 2005 as Professor and since 2005 as Professor titular of Organic Chemistry, Organometallic Chemistry and catalyze in the Department of Chemistry, Faculty of Sciences Semlalia Cadi Ayyad University Marrakech-Morocco. He is honoured as guest professor at Villeneuve d'Ascq University, France, ENS Chimie de Rennes France and University of Cergy Pontoise France. His research interests include Coordination Chemistry, asymmetric Catalysis, Green chemistry and nanoparticles and the chemistry of nanostructured materials: graphene; silicene and phosphorene. Pr. AIT ALI M. has co-authored peer-review more than 60 scientific papers published in leading refereed journals, about 70 congress communications and about 10 Thesis supervised. He participated in 10 international cooperation projects. He participated as active member in the organization of several international conferences: « Transmediterranean Meeting on Organometallic Chemistry and Catalysis » Morocco (RENACOM 2001, 2003, 2005, 2007, 2009); « International Conference on Nano-Materials and Renewable Energies, Safi, Morocco, 2010»; « Second Ecole Franco-Maghrebian of Nanosciences, Marrakech, Morocco, 2010»; « International Conference on Nano-Materials and Renewable Energies, Marrakech, Morocco, 2011»; the Fourth Meeting On Dielectric Materials "IMDM'4" (Marrakech, Morocco May 29-31, 2013); (Euro-Mediterranean Conference on Materials and Renewable Energies EMCMRE-3 Marrakech-Morocco 2015) and the First International Symposium on Dielectric Materials and Applications "ISyDMA'2016" (Kenitra-Rabat, Morocco May 4-6, 2016).

www.ingramcontent.com/pod-product-compliance
Lightning Source LLC
Chambersburg PA
CBHW071323210326
41597CB00015B/1332